Agri-food Waste Valorisation

Green Chemistry Series

Editor-in-chief:
James H. Clark, *Department of Chemistry, University of York, UK*

Series editors:
Catherine Birch, *AgriFood X Limited, UK*
Graham Bonwick, *AgriFood X Limited, UK*
George A. Kraus, *Iowa State University, USA*
Andrzej Stankiewicz, *Delft University of Technology, The Netherlands*

Titles in the series:
1: The Future of Glycerol: New Uses of a Versatile Raw Material
2: Alternative Solvents for Green Chemistry
3: Eco-Friendly Synthesis of Fine Chemicals
4: Sustainable Solutions for Modern Economies
5: Chemical Reactions and Processes under Flow Conditions
6: Radical Reactions in Aqueous Media
7: Aqueous Microwave Chemistry
8: The Future of Glycerol: 2nd Edition
9: Transportation Biofuels: Novel Pathways for the Production of Ethanol, Biogas and Biodiesel
10: Alternatives to Conventional Food Processing
11: Green Trends in Insect Control
12: A Handbook of Applied Biopolymer Technology: Synthesis, Degradation and Applications
13: Challenges in Green Analytical Chemistry
14: Advanced Oil Crop Biorefineries
15: Enantioselective Homogeneous Supported Catalysis
16: Natural Polymers Volume 1: Composites
17: Natural Polymers Volume 2: Nanocomposites
18: Integrated Forest Biorefineries
19: Sustainable Preparation of Metal Nanoparticles: Methods and Applications
20: Alternative Solvents for Green Chemistry: 2nd Edition
21: Natural Product Extraction: Principles and Applications
22: Element Recovery and Sustainability
23: Green Materials for Sustainable Water Remediation and Treatment
24: The Economic Utilisation of Food Co-Products
25: Biomass for Sustainable Applications: Pollution Remediation and Energy
26: From C–H to C–C Bonds: Cross-Dehydrogenative-Coupling
27: Renewable Resources for Biorefineries
28: Transition Metal Catalysis in Aerobic Alcohol Oxidation
29: Green Materials from Plant Oils

30: Polyhydroxyalkanoates (PHAs) Based Blends, Composites and Nanocomposites
31: Ball Milling Towards Green Synthesis: Applications, Projects, Challenges
32: Porous Carbon Materials from Sustainable Precursors
33: Heterogeneous Catalysis for Today's Challenges: Synthesis, Characterization and Applications
34: Chemical Biotechnology and Bioengineering
35: Microwave-Assisted Polymerization
36: Ionic Liquids in the Biorefinery Concept: Challenges and Perspectives
37: Starch-based Blends, Composites and Nanocomposites
38: Sustainable Catalysis: With Non-endangered Metals, Part 1
39: Sustainable Catalysis: With Non-endangered Metals, Part 2
40: Sustainable Catalysis: Without Metals or Other Endangered Elements, Part 1
41: Sustainable Catalysis: Without Metals or Other Endangered Elements, Part 2
42: Green Photo-active Nanomaterials
43: Commercializing Biobased Products: Opportunities, Challenges, Benefits, and Risks
44: Biomass Sugars for Non-Fuel Applications
45: White Biotechnology for Sustainable Chemistry
46: Green and Sustainable Medicinal Chemistry: Methods, Tools and Strategies for the 21st Century Pharmaceutical Industry
47: Alternative Energy Sources for Green Chemistry
48: High Pressure Technologies in Biomass Conversion
49: Sustainable Solvents: Perspectives from Research, Business and International Policy
50: Fast Pyrolysis of Biomass: Advances in Science and Technology
51: Catalyst-free Organic Synthesis
52: Hazardous Reagent Substitution: A Pharmaceutical Perspective
53: Alternatives to Conventional Food Processing: 2nd Edition
54: Sustainable Synthesis of Pharmaceuticals: Using Transition Metal Complexes as Catalysts
55: Intensification of Biobased Processes
56: Sustainable Catalysis for Biorefineries
57: Supercritical and Other High-pressure Solvent Systems: For Extraction, Reaction and Material Processing
58: Biobased Aerogels: Polysaccharide and Protein-based Materials
59: Rubber Recycling: Challenges and Developments
60: Green Chemistry for Surface Coatings, Inks and Adhesives: Sustainable Applications
61: Green Synthetic Processes and Procedures
62: Resource Recovery from Wastes: Towards a Circular Economy
63: Flow Chemistry: Integrated Approaches for Practical Applications
64: Transition Towards a Sustainable Biobased Economy

65: Transportation Biofuels: Pathways for Production: 2nd Edition
66: Challenges in Green Analytical Chemistry: 2nd Edition
67: CO_2-switchable Materials: Solvents, Surfactants, Solutes and Solids
68: Green Toxicology: Making Chemicals Benign by Design
69: Sustainable and Functional Redox Chemistry
70: Waste Valorisation Using Ionic Liquids
71: Natural Product Extraction: Principles and Applications: 2nd Edition
72: Renewable Resources for Surface Coatings, Inks and Adhesives
73: Green Corrosion Inhibition: Fundamentals, Design, Synthesis and Applications
74: Chemical Valorisation of Carbon Dioxide
75: Green Sample Preparation Techniques: Concepts, Novel Materials and Solvents, and Applications
76: Methane Conversion Routes: Status and Prospects
77: Green Gasoline: A Green Spark Transportation Fuel
78: Agri-food Waste Valorisation

How to obtain future titles on publication:
A standing order plan is available for this series. A standing order will bring delivery of each new volume immediately on publication.

For further information please contact:
Book Sales Department, Royal Society of Chemistry, Thomas Graham House, Science Park, Milton Road, Cambridge, CB4 0WF, UK
Telephone: +44 (0)1223 420066, Fax: +44 (0)1223 420247
Email: booksales@rsc.org
Visit our website at www.rsc.org/books

Agri-food Waste Valorisation

Edited by

Pankaj Chowdhary
CSIR Indian Institute of Toxicology Research, India
Email: pankaj161089@gmail.com

and

Abhay Raj
CSIR-Indian Institute of Toxicology Research, India
Email: araj@iitr.res.in

ROYAL SOCIETY
OF **CHEMISTRY**

Green Chemistry Series No. 78

Print ISBN: 978-1-83916-803-1
PDF ISBN: 978-1-83767-009-3
EPUB ISBN: 978-1-83767-010-9
Print ISSN: 1757-7039
Electronic ISSN: 1757-7047

A catalogue record for this book is available from the British Library

The Royal Society of Chemistry is a charity, registered in England and Wales, Number 207890, and a company incorporated in England by Royal Charter (Registered No. RC000524), registered office: Burlington House, Piccadilly, London W1J 0BA, UK, Telephone: +44 (0)20 7437 8656.

For further information see our website at www.rsc.org

Printed in the United Kingdom by CPI Group (UK) Ltd, Croydon, CR0 4YY, UK

Preface

The book Agri-food Waste Valorisation provides sustainable solutions for agriculture and food wastes for the betterment of our daily lifestyle. The wastes released from various food industries and agricultural waste cause severe threats to the ecosystem and are also a challenging task for proper management. Bio-waste (agriculture, forestry, and food industry) also contains toxic and beneficial chemical compounds, which need more attention to be identified for sustainable development. Governments around the globe are strictly advocating for the mitigation and valorisation of agri-food wastes to wealth. Hence, the management of agri-food wastes from different sources is of utmost importance to promote the sustainable development of our society with minimum environmental impacts. This book is an exercise to discuss the latest advancement in the area of waste reuse and valorization, which should meet the needs of undergraduate and graduate students, researchers, and policymakers working in this area. To cater the book to such a vast group of readers, the contents of the book broadly cover the two sections, *i.e.* waste reuse and valorisation. Furthermore, the last chapter of the book discusses the techno-economic analysis and life cycle assessment of value-added products from agri-food wastes.

Besides, this book also boosts up students, environmental scientists, environmental biotechnologists, microbiologists, and policymakers working in environmental microbiology, biotechnology, and environmental sciences with the basic and advanced facts about agri-food wastes and their management. Moreover, readers can also get state-of-the-art/background information in learning the existing environmental problems and their effects on human health and suggest ways to control or contain their effects by employing various effective management approaches.

The editors would like to express their sincere thanks to the contributors for submitting their work in a timely and proper manner. The editors are

Green Chemistry Series No. 78
Agri-food Waste Valorisation
Edited by Pankaj Chowdhary and Abhay Raj

Published by the Royal Society of Chemistry, www.rsc.org

also thankful to national and international reviewers for their evaluation and valuable suggestions and comments to heighten the book quality for readers. Dr. Chowdhary acknowledges the support received from his family, especially his Father (Mr. Ram Chandra) and Mother (Mrs. Malti Devi). Furthermore, the editors also acknowledge the cooperation received from the Royal Society of Chemistry team, for their guidance to finalize this book.

Pankaj Chowdhary
Environmental Microbiology Laboratory,
Environmental Toxicology Group
CSIR-Indian Institute of Toxicology Research (CSIR-IITR), Lucknow, India
Society for Green Environment, New Delhi-110067, India

Abhay Raj
Environmental Microbiology Laboratory,
Environmental Toxicology Group
CSIR-Indian Institute of Toxicology Research (CSIR-IITR), Lucknow, India
Society for Green Environment, New Delhi-110067, India

Editor Biographies

Pankaj Chowdhary is the Founder and President of the Society for Green Environment (SGE) (www.sgeindia.org), New Delhi, India. Currently, he is working as a Postdoctoral Fellow at the CSIR-Indian Institute of Toxicology Research. He received his post-graduation degree (2011) in Biotechnology from Deen Dayal Upadhyaya Gorakhpur University Uttar Pradesh (UP), India. Afterward, he obtained his PhD (2018) in the area of Microbiology from the Department of Microbiology at Babasaheb Bhimrao Ambedkar University (A Central University), Lucknow, Uttar Pradesh, India. During his PhD, his work mainly focused on the role of ligninolytic enzyme-producing bacterial strains in the decoloration and degradation of coloring compounds from distillery wastewater. His main research areas are Microbial Biotechnology, Biodegradation and Bioremediation of Environmental Contaminants in Industrial Wastewaters, Metagenomics, Biofuels, and Bioenergy production. He has edited nine (09) international books. Besides, he has also authored two (02) international books. He has published more than seventy (70) research publications including research/review papers and two book reviews in national and international peer reviewed journals of high impact factor published by Springer, Elsevier, Royal Society of Chemistry (RSC), Taylor & Francis Group, and Frontiers. He has also published many national and international book chapters and magazine articles on the biodegradation and bioremediation of industrial pollutants. He has presented many posters/papers relevant to his research areas at national and international conferences. He has also served

Green Chemistry Series No. 78
Agri-food Waste Valorisation
Edited by Pankaj Chowdhary and Abhay Raj

Published by the Royal Society of Chemistry, www.rsc.org

as a potential reviewer for various national and international journals in his respective areas of research. He is a life member of the "Association of Microbiologists of India (AMI)", the "Indian Science Congress Association (ISCA) Kolkata, India", and "The Biotech Research Society, India (BRSI)".

Abhay Raj was born in 1978 and he obtained his MSc (2001) and PhD (2009) in Environmental Science from Lucknow University, Uttar Pradesh, India. His major research work during his PhD focused on the bacterial degradation of lignin from paper mill effluents. Currently, he is a Senior Scientist in Environmental Microbiology Division at the CSIR-Indian Institute of Toxicology Research, Lucknow. His work involves the development of green technology through effluent treatment for sustainable environment and health. The major areas of his research work include Bioremediation, Toxicity and Genotoxicity Assessment, Metagenomics, Microbial Biotechnology, and Proteomics. He has published more than 60 research/review papers in peer reviewed SCI journals published by Springer, Elsevier, Wiley, and Frontiers. He has also published national and international book chapters and magazine articles. Also, he has presented his research work in India as well as abroad. He has also been serving as an active reviewer of various reputed national and international journals related to his field. He has successfully completed a major research project funded by the Department of Biotechnology (DBT), Department of Science & Technology (DST) & Council of Scientific and Industrial Research (CSIR). He has supervised many PhD and postgraduate students. He is a life member of scientific organizations like the American Society for Microbiology (ASM), the Association of Microbiologists of India (AMI), and the Biotech Research Society of India (BRSI).

Contents

Green Chemistry Series No. 78
Agri-food Waste Valorisation
Edited by Pankaj Chowdhary and Abhay Raj

Published by the Royal Society of Chemistry, www.rsc.org

CHAPTER 1

Valorisation of Agro-industrial Waste: Recent Advances in the Recovery of Bioactive Compounds and Environmental Perspectives

B. SANTIAGO,*[a] L. SILLERO,[a,b] M. T. MOREIRA,[a] G. FEIJOO[a] AND S. GONZÁLEZ-GARCÍA[a]

[a] CRETUS, Department of Chemical Engineering, School of Engineering, University of Santiago de Compostela, 15782 Santiago de Compostela, Spain; [b] Chemical and Environmental Engineering Department, University of the Basque Country, Plaza Europa, 1, 20018 Donostia-San Sebastian, Spain
*Email: beatriz.barahona@usc.es

1.1 Introduction

1.1.1 Connecting Biorefineries to Sustainable Development

Population growth has increased dramatically since the mid-20th century, tripling from 1950 to the present (from 2.5 billion to 7.9 billion).

According to United Nations' research, the world's population could grow to 11 billion by 2100. Figure 1.1 shows the size of the world's population and its growth over the years.[1]

As a result of widespread improvements in nutrition, personal hygiene and public health, among other factors, the world's population has grown

Green Chemistry Series No. 78
Agri-food Waste Valorisation
Edited by Pankaj Chowdhary and Abhay Raj

Published by the Royal Society of Chemistry, www.rsc.org

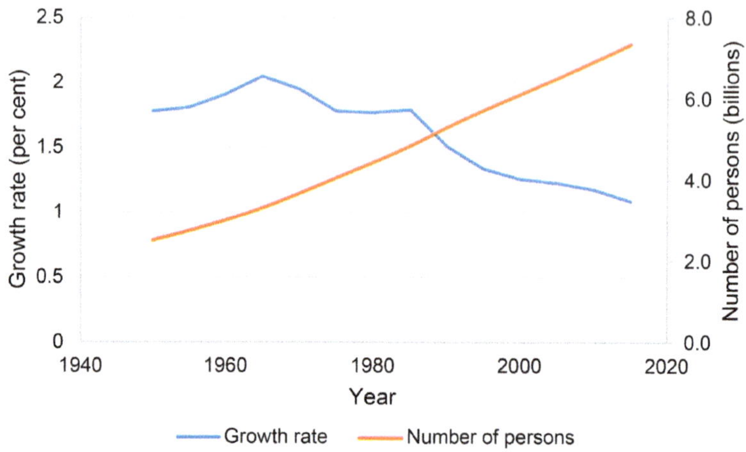

Figure 1.1 World population size and annual growth rate between 1950 and 2020.[1] Adapted from ref. 1 with permission from United Nations, Copyright 2022.

since 1950.[1] However, such rapid population growth has meant a rapid increase in the demand for food required to meet the needs of the population. Indeed, according to the Food and Agriculture Organization of the United Nations (FAO), it is estimated that by 2050, a 70% increase in food production will be needed to satisfy the food demand. Population growth is directly proportional to the increase in food demand and, consequently, to the increased amount of food production that is accompanied by the large-scale generation of food waste (edible and non-edible parts).[2] According to the FAO, one-third of all food produced globally is lost or wasted along the supply chain. The main stages at which this waste is produced are consumption and manufacturing, with 56% and 38% (approx.), respectively.[2] The total quantities of food waste generated at the European level are shown in Figure 1.2.

Of the total food waste generated globally, the beverage industry accounts for 26%, the dairy industry for 21%, the fruit industry for 15% and the cereal industry for 13% (see Figure 1.3).[3]

Food waste has economic (estimated losses of $1000 billion per year), environmental (water and biodiversity loss, land degradation and air and water pollution)[3] and social impacts.[4] These environmental, social and economic impacts, or problems, have arisen over time in both developed and developing countries. These problems have contributed to the deterioration of the environment and human well-being, as little effort has been made to mitigate them. Nevertheless, the most prominent effort made by the United Nations (UN) began in 2015, with the creation of the 2030 Agenda for Sustainable Development, which sets out a plan for improvement to be achieved in 15 years.[3,5] Sustainable Development Goals (SDGs) are guidelines for improving the global social, economic and environmental situations. In total, 17 SDGs have been proposed to end poverty, protect the planet and improve the quality of life of everyone everywhere.[5]

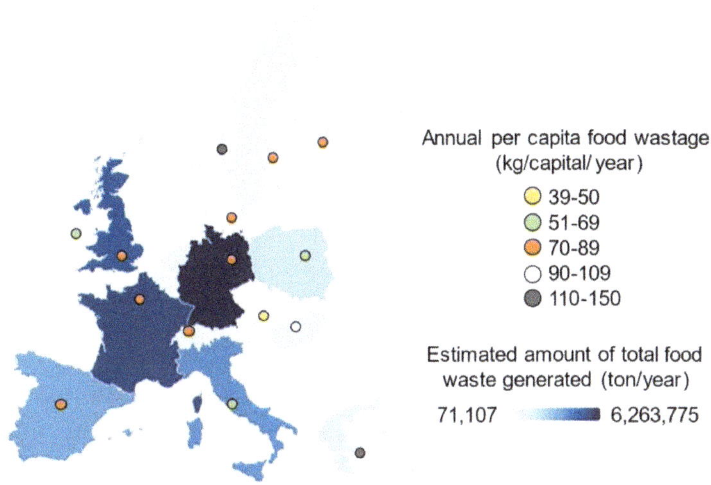

Figure 1.2 Estimates of food waste at the European level in 2019.[3] Data from ref. 3.

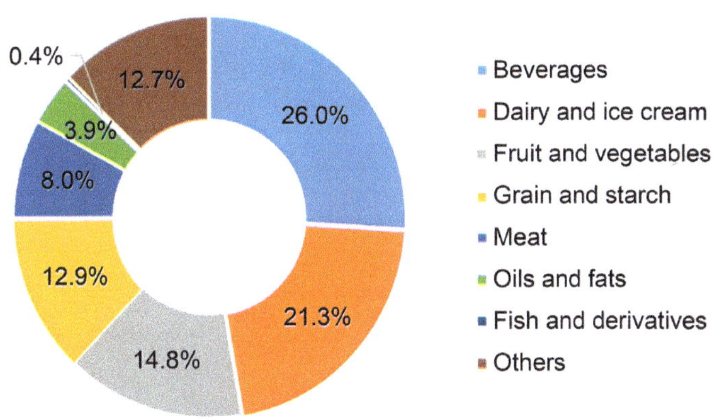

Figure 1.3 Percentage of different types of waste generated in different food industries.[17] Data from ref. 17.

The SDGs framework is a standardised way of organising the various strategies focused on specific issues, be they poverty, quality of life, environment and society. In addition, this framework leads to greater coherence in government policies in most UN member countries. Therefore, SDGs integrate concepts, principles and criteria, providing UN member countries with a structured analysis to enable the 2030 Agenda to be met.[3]

SDGs can be divided into three groups:[6]

- Natural resource use: the goals included in this group are considered the basis for the organisation of the SDGs, as they focus on the

protection, conservation, restoration and sustainable use of significant natural resources.
- Poverty and human well-being: SDGs related mainly to human well-being can be seen as an outcome of the fulfilment of other SDGs, because they are related to a more sustainable society.
- Sustainable production and consumption: these SDGs can be achieved by changing the current economic model from a linear to a circular one, for example, by shifting from fossil fuels to renewable energy sources.

As the SDGs focus on sustainable production and consumption, two new concepts are introduced: bioeconomy and biorefinery.

The bioeconomy concept is based on the need to overcome the damage caused to non-renewable resources. This concept can also be related to accomplishing the SDGs, since the transition towards a bioeconomy decreases the use of non-renewable resources. However, the concept of bioeconomy has been interpreted in different ways by different countries, organisations and/or institutions. For example, according to the FAO, bioeconomy is based on the generation of renewable biological resources and their conversion into high added-value products, such as bioproducts or bioenergy. Another concept, given by the Organisation for Economic Co-operation and Development, conceives of bioeconomy as "a world in which biotechnology contributes a significant share of economic input".[6] To really develop the concept of bioeconomy in any area, this concept and that of sustainability must be integrated to understand what is achieved when many bio-based products are obtained from biomass instead of non-renewable resources.

The biorefinery concept is an area of bioeconomy that is gaining much interest. Biorefineries use biomass as feedstock to produce one or several products.[7] The use of waste as a feedstock in biorefineries is seen as a potential strategy that contributes to the achievement of the SDGs and, simultaneously, to reduce the use of both fossil and bio-based resources, supporting the transition towards a circular bioeconomy, with a lower intensity of use of virgin materials.[8] The purposes of biorefineries are:

- To valorise biomass resources.
- To produce competitive and marketable products.
- To mitigate the environmental damage caused by the excessive use of fossil fuels.
- To encourage a sustainable future through the implementation of non-oil-based processes.
- To serve as the starting point to boost a bio-based economy in developing and developed countries.

Biorefineries can produce multiple high-value products such as proteins for food and feed,[9,10] biofuels,[11] biochemicals[12,13] and bioplastics.[14]

Table 1.1 Summary of the main SDGs that can be achieved through biorefineries.[6] Adapted from ref. 6 with permission from Elsevier, Copyright 2021.

SDG	Indicator	SDG	Indicator
1.1.1	Proportion of population below the international poverty line	7.1.1	Proportion of population with access to electricity
1.4.1	Proportion of population living in households with access to basic service	7.1.2	Proportion of population with primary reliance on clean fuels and technology
2.3.1	Volume of production per labour unit by classes of farming/pastoral/forestry enterprise size	8.4.1	Material footprint, material footprint per capita and material footprint per GDP
2.3.2	Average income of small-scale food producers	9.3.1	Proportion of small-scale industries in total industry value-added
2.4.1	Proportion of agricultural area under productive and sustainable agriculture	9.5.2	Researchers (in full-time equivalent) per million inhabitants
3.9.1	Mortality rate attributed to household and ambient air pollution	12.3.1	Global food loss index
3.9.3	Mortality rate attributed to unintentional poisoning	12.6.1	Number of companies publishing sustainability reports
6.4.1	Change in water-use efficiency over time		

In addition, the biorefinery concept can be fully integrated with established biotechnologies such as anaerobic digestion, which can produce bioenergy and biofertilisers.[9,15]

The implementation of biorefineries can directly or indirectly address the SDGs, as biomass conversion can increase the economic growth without sacrificing the environmental performance. Most of the SDGs address cleaner production, sustainable development, sustainable agriculture, affordable energy, business alternatives and responsible production and consumption. Biorefineries thus play a key role in driving the achievement of the SDGs, as renewable resources (*i.e.*, biomass) can be established as the new pillar of human development. Table 1.1 presents a summary of the main SDGs that can be achieved through biorefineries.

1.1.2 Biowaste Generation

Due to the incessant growth of the world's population and industrialisation, the amount of waste generated around the world is rising every year causing problems. The elimination and disposal of this untreated waste into the environment can lead to the deposition of pollutants in the ecosystem that will eventually affect humans and other living organisms.[16] Therefore, because of the impact that the lack of biowaste management generates at the social, economic and environmental levels,[17] it is necessary to propose a change of paradigm where residues are no longer waste and, instead, considered raw materials capable of providing high added-value products.

Solid waste management is a universal problem that affects everyone, making it necessary to look for more environmentally sustainable solutions from a circular economy standpoint. According to the data reported by Kaza *et al.* (2018),[18] 2 billion tonnes of municipal solid waste were already produced annually in 2016; however, the projection is that with the increase in population and social development, by 2030 it will exceed 2.5 billion tonnes of municipal solid waste, reaching 3.4 billion tonnes of waste by 2050. Figure 1.4 depicts the overall average composition of global municipal solid waste. This composition varies greatly depending on the development level of the country, with the amount of organic waste being higher in lower-income countries. However, higher-income countries are found to have a wider range of waste, as they include more materials such as paper, wood and plastic.[18]

Biowaste accounts for more than 60% of the waste generated annually (see Figure 1.4). Biowaste refers to biodegradable residues, such as fruit, vegetable, animal or forestry waste, as well as some household or industrial waste, among others.[19] Throughout this chapter, emphasis will be placed on the importance of the valorisation of vegetable waste, from fruits and vegetables to trees and their residues.

Agriculture, by definition, includes the set of techniques and procedures for cultivating land with the aim of producing food of plant origin, such as fruits, vegetables and cereals, among others.[20] The agriculture sector not only generates lignocellulosic waste such as dried crop stalks, pruning

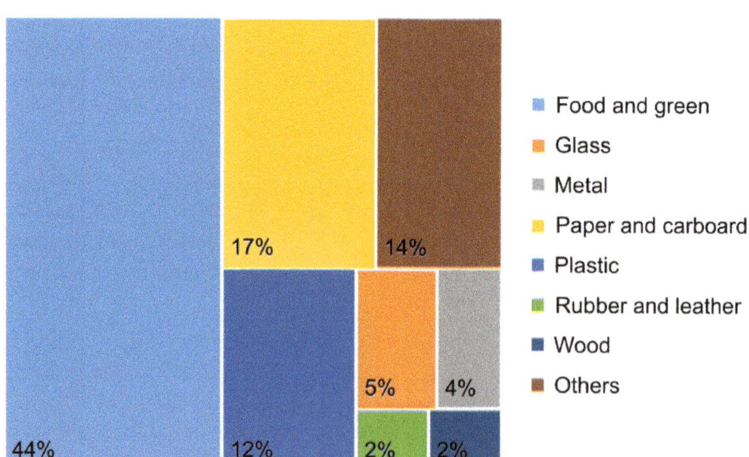

Figure 1.4 Global waste composition.[18] Adapted from ref. 18, https://openknowledge.worldbank.org/entities/publication/d3f9d45e-115f-559b-b14f-28552410e90a, under the terms of the CC BY 3.0 IGO license, https://creativecommons.org/licenses/by/3.0/igo/.

residues, molasses, husks (corn, wheat and rice) and shells (walnuts, almonds, peanuts, *etc.*),[19] but also produces a minor amount of food waste, such as that of fruits or vegetables due to their discarding when they do not fulfil product quality criteria. Food loss is regarded as the decrease in the quantity and quality of food that is produced due to the decisions and actions of the suppliers in the food chain (producers and the food industry).[21] However, food waste is the decrease in the quantity and quality of food due to the decisions and actions of retailers, food service providers and consumers.[21]

Food loss and waste occur during the five main stages of the food supply chain: production, handling and storage, processing, distribution and marketing, and consumption.[22] The sum of both, food loss and waste, corresponds to one-third of the total food produced (around 1.3 billion tonnes per year).[17] According to the Foodwaste Index Report, ∼931 million tonnes of food waste were generated in 2019 (61% household, 26% food service and 13% retail waste).[23] The main food groups contributing to nutrient and food waste or loss are cereals and pulses, fruits and vegetables (22% loss), meat and animal products, and roots/tubers/oil crops (26% loss).[21]

An analysis of the distribution of food waste in Europe shows that consumption generates 52% of waste, followed by production (23%), handing and storage (11%), processing (5%) and distribution and sale (9%), resulting in a waste generation rate of 180 kg per inhabitant per year.[22] Some of the most common wastes generated in the food industry are seeds, leaves, roots, tubers, fruits including their peel/skin and pomace, among others.[24] Food waste generally consists of carbohydrates (between 30 and 60%), proteins (between 5 and 20%), lipids (between 15 and 40%) and nutraceuticals.[17,22] Vegetable waste is richer in carbohydrates, while proteins and lipids are mainly obtained from meat and egg by-products.[25] This chapter focuses on the valorisation of the different vegetable wastes produced along the whole food chain, so animal waste and waste produced by animals will not be analysed.

The forestry sector involves forest resources and the production, trade and use of forest products and services.[26] This includes forestry and the different related industries, such as wood and furniture, paper and pulp, and bioenergy production. The timber industry, in addition to the uses derived from wood, also produces other non-wood forest products, such as cork and resin. The waste generated along the entire chain in this sector is mostly lignocellulosic waste, which consists of a three-dimensional polymeric network synthesised by plants through photosynthesis. Lignocellulosic biomass consists mainly of hemicellulose, cellulose and lignin (structural compounds), in combination with other minor compounds (pectins, inorganic compounds, proteins and extractives).[27]

Forestry residues (chips and dust from tree discards, such as branches, bark, foliage and stumps) can reach 50–68%.[28] Moreover, Europe harvests around 426 million m^3 of logs per year, which results in the generation of an annual large amount of waste that needs to be managed.[29] Pulp and paper production generates around 50% residues of wood pulp, since this process implies the removal of lignin, in addition to hemicelluloses and other extractives, carried out during the pulping and bleaching processes.[30] Timber production, on the other hand, represents up to 60% of waste from saw-logs due to the discard of different parts from the sawmill, such as branches, bark or other trimmings, in addition to the production of sawdust.[31]

1.1.3 Potential Products that Could be Manufactured

The diversity in the constituents of the studied biowaste allows a wide range of valorisation processes, leading to the production of a countless number of compounds. There are many high added-value compounds that can be recovered from each of the different residues (see Figure 1.5), and their applications can be versatile, ranging from the production of biomaterials to the pharmaceutical, cosmetics, food, textile and chemical industries, among others.[32] This fact consolidates the idea of biorefineries as the sustainable future pathway for moving towards a circular economy.

Lignocellulosic residues, with their high cellulose, hemicellulose and lignin contents, represent a natural source of these high added-value compounds with an unlimited number of applications, from the production of

Figure 1.5 Different high added-value compounds that can be recovered from different residues.[32] Adapted from ref. 32, https://doi.org/10.1016/j.biortech.2020.123575, under the terms of the CC BY 4.0 license https://creativecommons.org/licenses/by/4.0/.

biomaterials to the production of biofuels or food supplements.[33–35] This type of waste is also formed by other compounds, which means that polyphenolic compounds (like in bark), pigments (mainly in the leaves), as well as other compounds in smaller quantities can also be obtained. Oil crops produce a wide variety of residues (olive leaves, olive pomace, empty palm bunch, rapeseed meal, *etc.*), which in turn will facilitate the acquisition of various high added-value products such as polyphenolic compounds, monosaccharides and polysaccharides, dietary fibres, pectin and organic acids, among others.[32]

Vegetable and cereal manufacturing industries produce a lot of lignocellulosic by-products during the agricultural production stage; however, a large number of potential by-products are also generated in the vegetable and cereal preparation chain. Vegetable processing results in waste such as broccoli or asparagus leaves and stalks, and cabbage waste, among many others. By using different fractionation technologies, diverse compounds with high added-value such as phenolic compounds, pectin, glucosilates, carotenoids, saponins and sterols could be produced from these raw materials. The cereal processing industry could also be able to obtain sugar monomers, dietary fibres, phenolic compounds, xylo-oligosaccharides, organic acids, enzymes and proteins, among other possible high added-value products, derived from wheat bran, corncob, rice husks, brewers' spent grain, *etc.*[32]

Considering the wide range of products that can be obtained from this type of by-product, it is worthwhile to exploit this opportunity to achieve the goal of a circular economy promoting sustainable development. To this end, it is essential to carefully preselect the technologies to be employed for the best fractionation of each different biomass. The most critical stages in a biorefinery are commonly the separation processes. These stages are usually the most resource-intensive, and consequently, are the most expensive stages[36] with the highest environmental impact. Thus, there is a need to choose the right technologies to be used in biorefineries.

The aim of this chapter is to analyse and evaluate the different available techniques for the valorisation of biomass and its manufacturing industries, from the point of view of environmental sustainability. For this purpose, together with introducing the different available techniques, both conventional and alternative, an environmental impact assessment is presented considering the life cycle assessment (LCA) approach by assessing the environmental burdens associated with the processes.

1.2 Extraction Techniques

1.2.1 Conventional Techniques

Conventional techniques have been widely applied to isolate solid matrix compounds, both in the laboratory and in industry. These methods are based on solid/liquid extraction of compounds according to their polarity differences using suitable solvents (*e.g.*, methanol, ethanol or water).[37]

Table 1.2 Summary of the main advantages and disadvantages of the conventional extraction techniques.

Pretreatment	Advantage	Disadvantage
Pressing extraction	• Free from solvent contamination • Low initial capital cost • Minor consumable cost • Low-risk process	• Time-consuming process • Low capacity per batch • High power consumption • High dependent on moisture content
Maceration	• Operational simplicity	• Long extraction times • Low extraction yields
Soxhlet extraction	• Preservation of the sample • Operational simplicity • High yield	• Long extraction times • Large amount of solvent • No use of agitation • Need for subsequent evaporation • Thermal decomposition of the target compound
Steam distillation	• Good extraction yields • Low degradation of compounds	• Long extraction times • High equipment and operating cost • Difficulty separating water from the product
Hydrodistillation	• Non-toxic environmentally friendly solvent • Simple instrumentation	• Long extraction times • Low compound recovery • Thermal degradation of compounds • High consumption of energy

The most influential operating factors in these techniques are temperature, extraction times and the sample/solvent ratio.[38] The advantages and disadvantages of each of the methods explained below are listed in Table 1.2.

1.2.1.1 *Pressing Extraction*

It is a solvent-free method that has been traditionally used to extract edible oil. The two main stages involved in this method are the sample preparation and the extraction stage.[39] Sample preparation consists of applying various pre-treatments (*e.g.*, crushing, grinding or boiling) to the sample to destroy its cellular structure, thus reducing the moisture content in order to ultimately achieve higher extraction yields from the next stage.[40] Although it is a solvent-free method, solvents are sometimes used to continue the extraction process due to the amount of oil (20–30%) remaining in the press cake.[39] The pressing stage can produce between 70% and 80% oil.[41] Yields are affected by some factors, such as the type of seed sample, the extraction temperature, the pressure exerted during pressing and the operating time.[40]

Extraction by pressing is an inefficient process, but it is still the only extraction method in some rural areas. Furthermore, this type of extraction is only carried out in small industries due to limitations in scaling up, which has a negative effect on production costs.[42]

In these small industries, there are two types of extraction by pressing: cold pressing and hot pressing. As the name suggests, the main difference between the two methods is that cold pressing does not involve any heating during the pressing process, apart from the heat generated by friction.[41] Cold pressing is considered the first choice in the industries because this technique can produce higher quality oil products due to its thermal sensitivity. This means that, as oil is a heat-sensitive product, the heat applied in the hot-pressing process can promote the oxidation process of the oil, reducing its quality.[41] However, cold pressing has a lower extraction yield than the hot-pressing method.[39] Because of its good quality, cold-pressed oil usually commands a higher price on the market. However, given the wide use of this technique in the industry, the low extraction yield in cold pressing needs to be addressed. Therefore, various types of pressing equipment, such as hydraulic presses, have been applied to optimise the extraction process.[41]

1.2.1.2 Maceration

It is considered the simplest extraction process used for the extraction of thermolabile compounds. The process to extract bioactive components consists of pre-treatment of milling, increasing the contact surface of the feed to improve the extraction efficiency, and then immersing the sample in a solvent for a long period of time. Finally, the liquid is filtered and the remaining solid is pressed to recover as much of the dispersed extracts as possible.[43] Among the most used solvents are hexane, benzene and toluene. The range of maceration times is very wide, from one hour to 15 days.[44,45] In addition to the long extraction times, one of the main disadvantages of maceration is the low extraction yields. However, adding agitation to the process results in homogenisation of the system, which provides an increase in the extraction yields. It is also well known that one of the other parameters that has great influence on the extraction yield is the sample/solvent ratio.[46] In general, the extraction yield increases with time until the extract concentration equilibrium between the raw material and the solvent is reached.

Therefore, the extraction process can be optimised by collecting at set times to determine the time to equilibrium.[46] Ćujić *et al.* (2016) achieved high yields of total phenols and total anthocyanins from chokeberry fruits under optimised conditions (50% ethanol, a solid–solvent ratio of 1 : 20, and 0.75 mm particle size), which suggested that maceration was a simple and effective method for the extraction of phenolic compounds from chokeberry fruits.[47] Albuquerque *et al.* (2017) conducted a study on the extraction of catechin from *Arbutus unedo* L. fruits using maceration, microwave-assisted extraction and ultrasound techniques.[48] Their results showed that microwave-assisted extraction was the most effective. However, if maceration was performed at low temperatures, similar extraction yields were obtained, which could translate into economic benefits.

Jovanović *et al.* (2017) evaluated the extraction efficiency of polyphenols from *Serpylli herba* using different extraction techniques (maceration, heat-assisted extraction and ultrasonic-assisted extraction).[49] Based on the content of total polyphenols, ultrasound-assisted extraction resulted in the highest total yield of flavonoids, with no statistical difference between maceration and heat-assisted extraction. In all cases, the method with the lowest extraction results was the maceration method.

1.2.1.3 Soxhlet Extraction

It is widely used in a variety of industry sectors, such as waste treatment, food engineering and pharmaceuticals. Solvent extraction is a separation process that uses solvents (either polar or non-polar) to separate liquids from solid–liquid samples. The most commonly employed solvents are hexane, ethanol, methanol, chloroform, diethyl ether, petroleum ether and acetone.[39] This extraction method is based on refluxing and siphoning by combining the processes of percolation and refluxing, using heat. In Soxhlet extraction, the solvent is evaporated by a heat source and then condensed in a reflux condenser, dripping onto the sample compartment. When the solvent charged with the extracted product reaches the top of the sample chamber, the solvent is discharged back to the bottom through a siphon.[50]

Solvent extraction efficiency could be affected by diverse independent factors, such as the residence time, moisture content of the sample, extraction temperature, sample size and selected solvent type. The sample should be crushed to a smaller size during the sample preparation stage to enhance its surface area. This helps to increase the contact area between the solvent and the sample, increasing the extraction yield. Some of the advantages of Soxhlet extraction are (1) conservation of the sample in contact with a fresh solvent, (2) the relatively high extraction temperature being maintained due to the heat of the distillation flask, (3) no requirement of filtration after lixiviation and (4) the simple and cheap operation of the system.

Nevertheless, the main disadvantages of Soxhlet extraction are the long extraction time, the use of a large amount of solvent, the inability to speed up the process by agitation, the need for a concentration step by evaporation, and the possible thermal decomposition of the target compounds.[51]

1.2.1.4 Steam Distillation and Hydrodistillation

In these methods, the applied heat is the main cause of vegetable cellular disruption, allowing the release of essential oils. Therefore, the heating temperature must be enough to break down the vegetable cellular structure and to release the desired compounds.[52] In steam distillation, water is boiled in a chamber, and the resulting steam flows through plant matter in another compartment. Alternatively, steam can be introduced directly into the sample for extraction. On the other hand, in hydrodistillation, water and the raw material are placed in the same vessel and boiled together. Both techniques obtain the extracts dissolved in the steam for extraction. Then, the steam is condensed by indirect contact with cold water, and the mixture is separated into two fractions: extracted compounds and water. This separation is usually performed by decantation based on the difference in density between the compounds.[43]

Despite the similarities between these methods, steam distillation could improve the yield of essential oils and the capture of volatile compounds. The small amount of volatile compounds recovered in hydrodistillation can be due to the hydrolysis reactions caused by the combined action of water and high temperature, which can lead to the degradation of some compounds. Nonetheless, steam distillation can avoid this degradation and allow the production of better-quality essential oils because the vegetal material is not in direct contact with water. Another important factor affecting the extraction yield is the extraction time, which varies on the scale of hours. While Santos *et al.* (2019)[53] needed only 6 h, Meullemiestre *et al.* (2017)[54] needed 8 h for the extraction. Both techniques, hydrodistillation and steam distillation, allow the recovery of compounds with different properties, mainly depending on the employed raw material.

1.2.1.5 Anaerobic Digestion

This method is a biological process in which organic matter is metabolized and converted into biogas through complex reactions in the absence of oxygen.[55] Anaerobic digestion is common in nature, for example in animal digestive systems and wetlands. Anaerobic digestion is most-commonly performed worldwide in several ways, such as the digestion of primary and secondary sewage sludge, in anaerobic sludge bed reactors and upstream from activated sludge plants. The process consists of four steps: hydrolysis, acidogenesis, acetogenesis, and methanogenesis, which can occur sequentially or simultaneously in one step.[3] Anaerobic digestion of food waste generates products, such as methane (CH_4), volatile fatty acids (VFAs), acetic acid, iso-butyric acid, valeric acid and hydrogen (H_2), among others.

Anaerobic digestion can be performed in single or two-stage operations. For single-stage configurations, all reactions are performed in one reactor, helping to reduce the operating costs and reactor complexity. However, the formation of intermediate products in the reactor could accelerate the inhibition of subsequent processes. Therefore, with this reactor configuration, a lower product yield and conversion efficiency are achieved.

In general, process instability, reactor acidifications and co-production of hydrogen and methane are common problems in single-stage reactor configurations.[56] The two-stage process in which the acidogenic and methanogenic processes are physically separated appears to be efficient and can overcome the problems associated with primary digestion.[57]

Anaerobic digestion of food waste mixtures, poultry litter and sewage sludge increased the biogas production to 640 $L\,kg^{-1}$ VS with a mixture ratio of 2 : 1 : 1 (sewage sludge : food waste : poultry litter).[58] Anaerobic digestion of food waste produced a methane yield of 276.5 mL $CH_4\,g^{-1}$ VS at mesophilic temperature (34 °C), while 307.5 mL $CH_4\,g^{-1}$ VS was obtained at the thermophilic temperature of 55 °C.[59] Lukitawesa *et al.* (2020) observed that VFA production can be increased by controlling the pH of the acidogenesis process (at pH 6) during the anaerobic digestion of food waste.[60] The highest VFA production (0.8 g VFA g^{-1} VS) was obtained at 1 : 3 ratio (inoculum : substrate).[60] The combination of dark fermentation (acidogenesis) and methanogenesis of food waste resulted in a 1.22-fold increase in biohethane ($H_2 + CH_4$) production.[61] The amount of methane produced by anaerobic digestion of one ton of food waste can reach 90.6 m^3.[62]

The anaerobic digestion reaction configuration is strongly influenced and controlled by the process parameters (pH, acidity, temperature, substrate composition, C/N ratio, reaction time and inoculum) and the desired final product. Therefore, by optimising the process parameters, the yield of the desired final product can be improved.

1.2.2 Non-conventional Alternatives

Conventional methods have several disadvantages such as long reaction times, high volumes of solvent, additional processes for purification and concentration of the compounds,[63] and the generation of degradation products due to the reaction conditions,[64] among others. These drawbacks make the processes expensive and have a high impact on the environment.[65] Therefore, in order to overcome these problems, other so-called non-conventional methods are currently being studied. The main objective of these technologies is to eliminate the drawbacks of conventional methods in order to achieve competitive and environmentally sustainable processes. These non-conventional methods need to fulfil certain requirements, such as a lower energy and chemical solvent requirement and low target compound degradation.[65] The advantages and disadvantages of each of the non-conventional methods explained below are listed in Table 1.3.

The pre-treatments necessary for the fractionation of biomass are the most expensive processes in a biorefinery, so it is essential to make a good process selection. The purpose of the pre-treatments is to facilitate the release of the desired compounds in a selective way, avoiding contamination by insoluble macromolecules or micro-organisms.[66] Thus, it is necessary to determine what end products can be obtained from the different biomass in order to choose the most suitable pre-treatment depending on the type of biomass and the desired product. The selection of the ideal pre-treatment should satisfy the following requirements according to Jesus *et al.* (2022):[67] simple and economical operations, low energy, water and chemical consumption, low corrosion, ability to modify the biomass structure, high

Table 1.3 Summary of the main advantages and disadvantages of the non-conventional extraction techniques.

Pretreatment		Advantage	Disadvantage
Physical pre-treatments	Extrusion	• Negligible inhibitor formation • Low degradation products • Cellulose decrystallisation • Moderate temperature	• Costly process • High energy consuming • Low solubilisation • Bad cooling capacity • Limited residence time
	Ultrasound-assisted extraction	• Short reaction time • High extraction yield • Low severity • Less chemicals • Different solvents can be used • Low process temperature	• Energy-intensive process • Small particle size • Long sonication period may cause adverse effects • Target compounds can be degraded • Filtration stage needed
	Microwave-assisted extraction	• High extraction yield • Short reaction time • Mild pre-treatment condition • Low operating cost • Low energy consumption • Minimum inhibitors formation • Homogenous heating • Minimum solvent use • Simple in operation	• High capital investment • Not suitable for thermosensitive compounds • Solvent-related damage to equipment
	Pulse electrical field	• Fast method • Mild pre-treatment conditions • No fermentation inhibitors • Low energy requirement • Non-toxic • High extraction yield • High intracellular compounds selectivity	• Energy intensive process • Not suitable for all kinds of substrates • High capital investment • High voltage pulse • Corrosion can occur • Medium-dependent method

Table 1.3 (*Continued*)

Pretreatment		Advantage	Disadvantage
Physico-chemical pre-treatments	Supercritical fluid extraction	• Low cost • Low chemical consumption • Faster biomass penetration • CO_2 as solvent: non-toxic, cheap, inert, low polarity, clean solvent, *etc.* • High extraction yield	• Complex equipment • Water presence is a problem • High capital cost investment • High pressure • CO_2 as solvent: non-polar solvent, modifier requirement • High energy requirement
	Subcritical fluid extraction	• Short time • High extraction yield • Low-polar and non-polar compounds can be extracted • Water as solvent: green, cheap, available	• Corrosion problems • Temperature degradation may occur • Difficult to clean the equipment • High energy requirement • High solvent consumption • Inhibitors compounds can be generated
	Pressurised liquid extraction	• Fast • No additional step of filtration is required • High extraction yield	• Equipment limitations
	Steam explosion	• Short extraction time • Low energy consumption • Low solvent consumption • High energy efficiency • No recycling cost	• Inhibitors formation • High reaction time • Process affected by a lot of parameters
Chemical pre-treatments	Acidic	• High reaction yield • Simple process • Quick and direct	• Corrosion problems • High cost • Sugar degradation • Toxic process • Neutralisation required • Filtration stage needed

	Alkaline	• Mild pre-treatment conditions • Less sugar degradation • Increase accessible surface area • High surface area increase	• Long reaction time • High cost • Filtration stage needed • Neutralisation required • High amount of water required
Chemical pre-treatments	Organosolv	• High-quality lignin • Simple solvent recovery stage • High surface area increase • Low inhibitor formation	• High solvent requirement and cost • To be conducted under controlled conditions • Solvents can act as inhibitors in subsequent stages
	Ionic liquids	• Mild pre-treatment conditions • No inhibitor generation • Less energy requirements • High extraction yield • High selectivity	• Solvent high cost • High solvent viscosity • Toxicity of ILs needs to be studied • Recycling stage required
	Deep eutectic solvents	• High extraction yield • High selectivity • Mild pre-treatment conditions • Biodegradable solvents • Biocompatible solvents	• Recycling stage required • High solvent viscosity

selectivity, high extraction yield, low production of degradation products and limited waste generation.

The different pre-treatments can be grouped into four large categories defined by the way in which they act on the biomass to modify its structure: physical, chemical, physicochemical, and biological pre-treatments. Physical treatments are those that only modify the form of presentation/structure of the biomass. Some physical pre-treatments have the unique objective of reducing the particle size of the biomass in order to increase the contact surface and decrease the polymerisation degree (*e.g.*, grinding), while others have a direct effect on the biomass structure (*e.g.*, ultrasound, extrusion, *etc.*).[68] Particle size reduction pre-treatments are often considered mandatory since they support subsequent fractionation steps. Chemical pre-treatments are those that use catalysts to separate the compounds of interest, such as acids, bases or oxidising agents. Physico-chemical pre-treatment, on the other hand, is a combination of chemical and physical methods applied simultaneously to favour the extraction of the desired compounds. On the other hand, biological pre-treatments are based on the use of biocatalysts (biological agents) to separate the target compounds.[67]

1.2.2.1 *Physical Pre-treatments*

Physical treatments include not only mechanical operations, but also different types of radiation, such as ultrasound or microwave radiation.

1.2.2.1.1 Extrusion. The extrusion process is based on two unitary operations, mixing and heating. First, the raw material injected under pressure is vigorously agitated using up to two rigid screws, mixing everything perfectly. Subsequently, this mixture is heated during further intensive mixing, thereby allowing the mechanical action to create a change in the structure of the starting biomass when it passes through the grate.[63] This technique permits continuous processing of biomass, which saves both space and energy. The most important advantages of this pre-treatment are shorter time, less water consumption and low corrosion, among others.[69]

Delvar *et al.* (2019) reported an extraction process using a twin-screw extruder for the extraction of polyphenolic compounds from passion fruit residues.[70] Using this technique, the group was able to extract up to 67% of the total polyphenol content of the different studied biomass.[70] A recent study evaluated the combined effect of the freeze-drying process and extrusion to improve the quality of flour from banana waste.[71] The improvement in flour composition resulted from the retention of phenolic compounds in the flour due to the process combination.

Regarding lignocellulosic waste, the work carried out by Wang *et al.* (2020) with corn stover confirms that extrusion pre-treatment not only favours the subsequent stages for biomass recovery, but also achieves the extraction of both proteins and cellulose.[72] Duque *et al.* (2020) evaluated the benefits of combining extrusion with an alkaline pre-treatment followed by an enzymatic pre-treatment of barley straw.[73] The bioextrusion favours the mixing of the media with the biomass, increasing the efficiency of the separation processes.

1.2.2.1.2 Ultrasound-assisted Extraction. Ultrasound-assisted extraction (UAE) pre-treatments are based on the principle of acoustic cavitation generated by ultrasound radiation, which intensifies the transfer and transport phenomena between the solvent and the biomass. This effect occurs due to the propagation of the ultrasound waves in the medium, which generates bubbles that increase in size until they reach a maximum, when these bubbles collapse. The collapse of the bubbles generates different physical effects, such as micro-jets, shock waves and turbulence.[74] When these phenomena occur close to the biomass fractions, it promotes cell disruption, which favours the transfer of matter between the solvent and the biomass, facilitating the separation of the compounds. The main advantages of this pre-treatment are the improvement in extraction yield, time and energy savings, and lower working temperatures, among others.[50] This technique is usually used in combination with solvents, which can be conventional solvents, in general, different organic solvents, or more recent solvents, such as ionic liquids (ILs) or deep eutectic solvents (DES).

Banožić *et al.* (2019) conducted a research study to optimise the extraction of both phenolic compounds and solanesol from different tobacco plant residues using UAE.[75] Leaves were the residues from which most phenolic compounds and solanesol were extracted using pre-treatment at temperatures below 45 °C and with a duration below 43 min. This confirmed UAE as a potential technique for the extraction of these compounds. Medina-Torres *et al.* (2019) used UAE to maximise the extraction of phenolic compounds from Persian lemon residues, requiring only 10 minutes.[76] This technique is also being widely studied on residues generated from the forestry industry. An example of this is the results published by Cetera *et al.* (2019) on their work related to the extraction of polyphenolic compounds with antioxidant activity from Turkey oak.[77] Using this pre-treatment, a considerable amount of polyphenolic compound was extracted in just one hour, reducing considerably the extraction time of conventional methods.

1.2.2.1.3 Microwave-assisted Extraction. Microwave-assisted extraction (MAE) pre-treatments are based on electromagnetic irradiation (frequencies between 0.3 and 300 GHz), usually using 2.45 GHz for these processes to take advantage of the dielectric heating principle leading to a more efficient material heating. This heating principle is based on the dipole moment rotation mechanisms and ionic conduction,[78] so the heating will depend on the ability of the material and the solvent to absorb microwave energy. Microwaves interact with polar molecules by making them vibrate, so that the dipoles of the molecules are constantly being aligned with the magnetic field produced by the microwaves, thus resulting in the heating of the solvent or mixture.[79] In addition, they can penetrate into the cell matrix of the biomass, since water and other polar compounds present in the cells are also affected by microwaves causing cell disruption.[50] In this way, the use of microwaves, in addition to being a more efficient heating mechanism, enhances the penetration of the solvent into the matrix, making it easier to separate the desired compounds. The most outstanding advantages of this technology are the high reaction yield, the reduction of reaction time and the reduction of the required solvent volume.[80]

In recent years, MAE has been one of the most studied non-conventional methods, mainly applied to intensify different processes. Jesus *et al.* (2019), for example, used MAE to extract polyphenolic compounds more efficiently from pruning residues.[81] Recently, Dávila *et al.* (2021) carried out optimisation to maximise the extraction of oligosaccharides from vine shoots by intensifying the autohydrolysis process with MAE.[82] They concluded that the use of MAE required less reaction time, which translates into less energy consumption, making the process more environmentally sustainable. Among the latest work published on the valorisation of pumpkin peel and pulp waste, Sharma and Bhat (2021) investigated the extraction of carotenoids using MAE.[83] In this study, after only 30 min, the carotenoid content extracted was almost double compared to conventional methods, in addition to exhibiting higher oxidative stability, thus confirming the potential of this technique.

In recent years, the simultaneous use of UAE and MAE techniques is also being evaluated in order to improve the limitations of these techniques separately. The simultaneous use of microwaves and ultrasound generates a synergistic effect that induces the improvement of both heat (microwaves) and mass (ultrasound) transfer.[84] The synergistic effect generated by the simultaneous use of both techniques results in higher extraction yields in shorter reaction times.[85]

In a recent study conducted by Estrada-Gil *et al.* (2022), they confirmed that the simultaneous use of microwaves and ultrasound has a better extraction yield of phenolic compounds from Mexican Rambutan peel residues than the separate use of those techniques.[86] On the other hand, Sillero *et al.* (2020), by optimising the extraction of pine bark by simultaneous use of microwave and ultrasound, concluded that this pre-treatment not only improves extraction yields, but also reduces reaction times by up to 47 times compared to the conventional method.[85] Recently, the use of this pre-treatment has also been studied to favour depolymerisation by acid treatment of lignin.[87] The results of this work show that the hybrid use of these techniques favours the yield of bio-oil, and monomer production, which converts it into a technique with potential application in this field.

1.2.2.1.4 Pulse Electrical Field (PEF). This non-thermal pre-treatment is based on the phenomenon of electroporation, which can be defined as the increase of cell membrane permeability due to the application of a high-voltage pulse. It facilitates the recovery of intracellular compounds by diffusion. During the treatment, the biomass to be processed is suspended in a solvent, and then the mixture is placed between two electrodes where the pulses are generated, using an electric current of 0.1–100 kV cm^{-1} for a short period (microseconds or milliseconds).[64] The electric potential applied on biomass, whose cell membranes have a dipolar nature, results in the separation of the membrane molecules forming pores and increasing the permeability of the membrane, thus enhancing the movement of intracellular compounds.[24,63] The resulting membrane changes, depending on the adjustment of the reaction parameters, could be reversible or irreversible depending on the objective. The main advantages of this pre-treatment are related to the ability to conduct this process at room temperature and the short reaction times, which reduce energy consumption.[68]

Lončarić *et al.* (2020) reported research on the extraction of polyphenolic compounds from blueberry pomace using PEF.[88] In this study, the results confirmed that PEF achieved a higher yield of phenolic compounds, compared to those obtained by UAE, with an energy input of only 41.03 kJ kg^{-1}. In the study carried out by Visockis *et al.* (2021) to obtain betalain from red beet, it was shown that only 3 pulses of 1 Hz and a pulse duration of 100 µs for each resulted in extraction of up to 70% of the total initial betalain.[89]

1.2.2.2 Physico-chemical Pre-treatments

Physico-chemical methods are carried out due to the combination of effects, which tends to make them more efficient in biomass.

1.2.2.2.1 Supercritical Fluid Extraction. Supercritical fluid extraction (SFE) pre-treatment is based on the thermodynamic properties of the fluids, meaning that through the application of pressures and temperatures over the critical point, a fluid acquires the properties of both gas and liquid.[90] In this way, the fluids maintain the properties of surface tension, high diffusivity and low viscosity characteristic of gases, and the solvating power of liquids.[91] Therefore, the separation and extraction of the target compounds are enhanced. The solvents that can be used include ammonia ethylene, toluene and carbon dioxide, which are used at a pressure of 10–30 MPa and at temperatures of 40–50 °C.[92]

This pre-treatment has been extensively studied for the extraction of oil from different agricultural residues. In the study performed by Ferrentino *et al.* (2020), it was confirmed that the use of SFE with CO_2 achieved similar results to Soxhlet extraction but in less than half time, saving both energy and the required amount of solvent.[93] Thus, using SFE, they achieved an oil yield from apple seeds of more than 20% in only 140 min at 40 °C. In the work carried out by Marić *et al.* (2020) for the treatment of raspberry seeds, they also managed to reduce the reaction time by half compared to Soxhlet extraction, and the reaction temperature was only 50 °C.[94]

In a recent work developed by Bukhanko *et al.* (2020) for the valorisation of different Norway spruce residual fractions (branches, needles and bark),

SFE with CO_2 has been proven to slightly improve the extraction yield compared to Soxhlet for similar reaction times, but with the advantage of solvent savings, as well as the use of green solvents.[95] In addition, this technique proved to be more efficient for the extraction of both resins and sterols. In another recent study, up to 50% of the total extractives have been removed from different forest residues by CO_2 SFE, which favoured the subsequent pyrolysis treatment to obtain charcoal.[96]

1.2.2.2.2 Subcritical Fluid Extraction. Subcritical fluid extraction (SubFE) pre-treatment, also considered an environmentally friendly and non-toxic technique, is based on the variation of the working temperature as well as the pressure, which directly influences the dielectric constant of the solvent, but always works below the critical point.[80] Different polar solvents can be used; however, water is the most common of all due not only to its properties, but also to its green solvent character. Thus, when water is heated to temperatures between 100 and 320 °C, at pressures between 20 and 150 bar, the dielectric constant of the solvent is modified and can be equal to that of methanol or ethanol. Thanks to this decrease in the dielectric constant, other properties such as surface tension, viscosity and polarity are also reduced, which favours the dissolution of the target compounds.[74] The major advantages of using this pre-treatment are the reduction of the extraction time, the selectivity of the procedure and the use of a green solvent.[80]

Subcritical water pre-treatment has been used for the extraction of different compounds, from polysaccharides to peptins and polyphenolic compounds.[97] In the recent work carried out by Muñoz-Almagro *et al.* (2019) for the valorisation of cocoa pod husk using subcritical water conditions, they reached an extraction yield of almost 11%, compared to the 8% obtained by conventional extraction.[98] In addition to improving the yield, the extraction time was reduced by up to 3 times, obtaining a pectin-richer product. In another study carried out for the extraction of pectin from jackfruit peel residue, it was concluded that the use of subcritical water is an efficient method, which reduces the extraction time and has a lower environmental impact.[99] Gonçalves Rodrigues *et al.* (2019) have also used this technique for the extraction of polyphenolic compounds from papaya seeds as can be seen in their recent work.[100]

1.2.2.2.3 Pressurised Liquid Extraction. Pressurised liquid extraction (PLE) pre-treatment, also known as accelerated solvent extraction, as well as the previous two, is based on the relationship between the temperature and pressure variables. In this case, different solvents are used at high temperatures and pressures in order to enhance the extraction processes, thanks to the improvement of the kinetics and distribution coefficient.[101] The increase in pressure allows working at temperatures higher than the boiling temperature at atmospheric pressure. The increase in temperature (between 50 and 200 °C), which improves the diffusivity of the solvent, together with the high pressure (10–15 MPa) facilitates both the solubility and the mass transfer rate, thereby promoting the extraction of the target compounds.[50] The main advantages of this technique are the reduction of reaction times, the reduction in solvent use, and the lack of subsequent filtration steps.[66,101]

It has recently been demonstrated that the use of PLE for the valorisation of chestnut wood leads to an almost doubling of the extraction yield of polyphenolic compounds using only 15 min of extraction, compared to the 7 h required in conventional methods.[102] The use of this pre-treatment on other agro-industrial wastes also allows an increase in the extraction yield of polyphenolic compounds. One example of this is the work published by Alexandre *et al.* (2019), in which researchers improved the amount of extracted phenolic compounds from pomegranate peel with a treatment at 300 MPa for only 15 min.[103] In the work carried out by Pereira *et al.* (2019), PLE was implemented for the valorisation of grape marc in order to maximise the phenolic compound content.[104] This only required the use of ethanol–water (50/50 w/w) as the solvent at 100 °C, without using any catalyst and with only 20 min of reaction time.

1.2.2.2.4 Steam Explosion. Steam explosion pre-treatment involves exposure to high-pressure saturated water vapour for short periods, followed by immediate depressurisation while the biomass is forced through a small hole. This process applies high thermo-mechanical energy to the biomass, allowing the alteration of the cell wall,[65] thus enhancing the solubilisation of the target compounds. The ranges of the working parameters in this process are temperatures between 160 and 270 °C and pressures of 0.7–5 MPa.[105] The main advantages of this pre-treatment are the reduction of reaction time and energy consumption and the fact that the process is free of chemicals.[68] This process is widely used to obtain hemicelluloses from lignocellulosic biomass.

In the work developed by Álvarez *et al.* (2020) to obtain xylooligosaccharides from barley straw, the good performance of steam explosion pre-treatment was confirmed.[106] Using this technology, high-purity oligosaccharides were obtained using only 130 min of reaction time at 180 °C. The result of this technique for oligosaccharide extraction is so promising that, in a recent study, Swart *et al.* (2022) performed the scale-up of this pre-treatment in order to process brewery wastes.[107] Under the optimised scale-up conditions, they were able to extract more than 70% of the total oligosaccharides from the waste in only 10 min, working at 180 °C. This result is promising for the industrialisation of this technique, since the low time required will permit the processing of large amounts of waste. Bhatia *et al.* (2020) also carried out a pilot plant study of steam explosion treatment of herbaceous plants, where they managed to extract more than 50% of the total xylooligosaccharides from the biomass under the following reaction conditions: 10 min, 15 bar, and 200 °C.[108]

1.2.2.3 Chemical Pre-treatments

Chemical methods tend to be more aggressive in biomass fractionation, causing a permanent alteration of the biomass structure. Within the different pre-treatments, it is possible to separate, on the one hand, those carried out with acids, bases or organic solvents and, on the other hand, the more recent ones that are carried out with ionic liquids (ILs) or deep eutectic solvents (DES). Moreover, it has been confirmed that the use of co-solvents at low concentrations can promote the extractions. The most commonly used co-solvent is carbon dioxide as it reaches its supercritical state at 7.38 MPa and 31 °C.[63] In addition to being easy to operate with, carbon dioxide is inert in nature and is non-toxic and non-inflammable, making it a potential green solvent. Apart from the advantage derived from the use of green solvents, this pre-treatment also reduces the amount of generated waste in the process, as well as decreasing reaction times and energy consumption.[74,105]

1.2.2.3.1 Acid Pre-treatments. Acid pre-treatments are carried out by mixing an acid solution with the biomass to be treated, at a given temperature for a controlled time. This treatment always requires a subsequent filtering stage to separate the solid and liquid fractions. Acid pre-treatments lead to hydrolysis of the biomass allowing the separation of the desired compounds. For this purpose, acid solutions are usually used at low concentrations, between 0.2 and 2.5 w/w%, mixing at temperatures between 130 and 210 °C.[68] The most used acids are sulphuric acid and nitric acid, although in recent years organic acids are also being used. These types of pre-treatments, although yielding good results for the separation of compounds, have the disadvantage of generating toxic products, as well as corrosion in the equipment.[105] This technique is widely used as a delignification method for lignocellulosic biomass.

In a recent work conducted by Morais *et al.* (2020), the use of a phosphoric acid solution to perform acid hydrolysis of sugarcane biomass was investigated.[109] The results showed an improvement in the saccharification of the biomass using a 4.95% acid solution at 80 °C for more than 6 h. Although a long period of time was required, the use of the pre-treatment resulted in a hydrolysate with 98% of glucose. Previously, Chiranjeevi *et al.* (2018) developed an acid pre-treatment of rice straw in order to facilitate the delignification of this lignocellulosic biomass.[110] In this work, where they combined the use of sulphuric acid and boric acid, at 150 °C for 20 min, they achieved the total elimination of the hemicellulosic fraction, in addition to removing 44% of the lignin. This process, besides saving time and energy, facilitated the subsequent bioethanol production stage.

1.2.2.3.2 Alkaline Pre-treatments. Alkaline pre-treatment, as in the case of acid treatment, consists of mixing the biomass with an alkaline solution at a given temperature for a given period. The most commonly used solutions are those prepared with potassium hydroxide, sodium hydroxide, calcium hydroxide, and ammonium salt, among others.[65] The process causes the biomass to expand, which affects the physical properties of the biomass, including porosity, crystallinity and the surface area.[105] Thus, it facilitates the separation of the biomass components, leading to their extraction. Although this pre-treatment results in good yields, it has the disadvantage of requiring long reaction times and a subsequent filtering and neutralisation stage, which increases both the economic and energetic cost of the process.[68] This process is being widely used, mainly for the delignification of lignocellulosic biomass.

In a recent study conducted by Morales *et al.* (2020) to choose the best biorefinery sequence to obtain the highest yield from almond shells, alkaline hydrolysis with NaOH was selected to perform the delignification.[111] Using this technique, they were able to remove 40% of hemicelluloses and up to 35% of lignin from the biomass that had been previously submitted to an autohydrolysis step. Moreover, this technique made subsequent stages of the proposed biorefinery easier. Cavali *et al.* (2020) used alkaline hydrolysis (NaOH) for the recovery of sawdust, so that with a 170 °C treatment almost all the lignin in the biomass was removed.[112] This work confirmed that the sequential use of acid and alkaline treatments promotes the total elimination of hemicelluloses and lignin from the lignocellulosic biomass, achieving the removal of 87% of hemicelluloses and 94% of lignin.

1.2.2.3.3 Organosolv Pre-treatments. In organosolv pre-treatments, different organic solvents (*e.g.*, ethanol, acetone, methanol, *etc.*) are employed, usually in combination with water, for biomass defragmentation. This procedure sometimes additionally requires the use of different catalysts (*e.g.*, oxalic acid, salicylic acid, *etc.*) to increase the yield and purity of the separated compounds.[113] Organosolv pre-treatments are usually carried out at temperatures above 150 °C, with a duration of 1–3 h.[114] Once the temperature treatment has been completed, a subsequent filtration process is required for phase separation. This pre-treatment is mainly used with lignocellulosic biomass in the delignification process, achieves a high yield and high purity of the separated compounds, and facilitates the subsequent processes.[113] This approach, in addition to high efficiency, has other advantages such as the use of mild-operation conditions, the ease to recycle solvents and the simplicity of the processes.[105]

The recent work published by Morales *et al.* (2022) validates organosolv pre-treatment with an ethanol/water mixture as a very good technique for the delignification of walnut shells.[115] Here, up to 94% of the lignin was removed by the consecutive application of three organosolv delignifications, with the first one having the highest yield, removing up to 60% of the lignin. In another study performed for the revalorisation of olive pomace, they used an ethanol/water solution (50%) with 1% H_2SO_4 as a catalyst.[116] The use of a catalyst led to a pre-treatment at lower temperatures, 140 °C being the best temperature in this case, where a 53% delignification was achieved and high-purity lignin was produced. Rivas *et al.* (2021) used a 1-butanol/water mixture for the delignification of vine shoot.[117] By intensifying the organosolv pre-treatment using microwaves in addition to a catalyst, solubilisation of both part hemicelluloses and lignin (almost 68%) was achieved in a single step. The optimum reaction conditions of the microwave-assisted organosolv pre-treatment were 2% catalyst, 52% 1-butanol and 190 °C. Other investigations have been performed with different alcohols with good results, such as pentanol, propanol, butanol, and glycerol.[118,119]

1.2.2.3.4 Ionic Liquid Pre-treatment. One of the novelties of the last decades is the use of ionic liquids (ILs) for biomass fractionation. ILs consist of organic cations combined with organic or inorganic anions in a wide range of possible combinations. This new class of non-molecular solvents has unique properties that convert them into potential green solvents for biomass fractionation. Their most important properties include negligible vapour pressure, low combustibility, good thermal and chemical stability and adaptable solubility, among others.[120,121] In general, ionic liquids tend to have a high viscosity, so it has been confirmed that the use of ILs in combination with a not very high percentage of water not only improves viscosity, but also makes the pre-treatment more effective.[113] Pre-treatments with ILs consist of mixing the solvent (IL) with the biomass to be treated at a given temperature for a defined period. In this case, the working temperatures, due to the properties of the ILs, range between 90 and 130 °C at ambient pressure.[68] The advantages of this pre-treatment are good efficiency, the use of mild operation conditions, easy recovery and recycling of the solvent and energy savings, among others.[105] This chemical pre-treatment has been investigated not only for delignification processes, but also for the separation of other compounds from the biomass, such as polyphenolic compounds, polysaccharides, *etc.*

Sillero *et al.* (2021) investigated the use of ionic liquids in combination with water for the extraction of flavonoid compounds from pine bark.[122] The use of the ionic liquid (1-butyl-3-methylimidazolium bromide) gave an extraction yield of more than 15% with a total flavonoid content of 779 mg CE g^{-1} DBE, but this pre-treatment also solubilised part of lignin and hemicelluloses. Li *et al.* (2020) used 1-dodecyl-3-methylimidazolium chloride IL for protein extraction from tobacco leaves.[123] This pre-treatment improved protein extraction by almost 2-fold compared to a conventional technique. In a work where the use of ILs is combined with UAE, it was demonstrated that acetate-based ILs were able to dissolve bamboo biomass in only 40 minutes.[124] This reduces energy and solvent costs, since the IL was found to be recyclable. Recently, Rauf *et al.* (2022) studied the use of IL triethylammonium hydrogen sulphate for the pre-treatment of corn straw, achieving removal of 80% of the lignin from the initial biomass at 78 °C only.[125] Many ILs can be used for delignification, and proof of this is the work done by Neubert *et al.* (2020), where they examined the use of eight different ILs for the valorisation of sugar cane bagasse.[126] The delignification yields range from 23% to 86%, when 1-butyl-3-methylimidazolium H_2SO_4 was used. Another promising IL is HTEA H_2SO_4, with an efficiency close to 80%, thereby extending the IL possibilities.

1.2.2.3.5 Deep Eutectic Solvent Pre-treatment. Deep eutectic solvents (DES) are formed by the complexation of a hydrogen acceptor (HBA) and a hydrogen donor (HBD), usually formed by non-ionic species. DES are formed by mixing two or more non-toxic compounds under mild-conditions for a certain period of time (minutes or hours)[127] forming a eutectic mixture that allows these solvents to be liquid at temperatures below 70 °C. The compounds that form DES can be natural compounds, such as choline derivatives, organic acids and amino acids, among others, forming the so-called natural DES (NADES). Similar to ILs, DES have good chemical stability, adjustable solubility and negligible vapour pressure,[91] as well as being biodegradable and recyclable.[113] Pre-treatments with DES are carried out by mixing them with the biomass to be treated at fairly low temperatures, between 80 and 150 °C, at room pressure.[91] Due to the high viscosity of many DES, the use of co-solvents, especially water, has been confirmed in order to improve the efficiency of the extraction processes.[128] One of the most important advantages of DES pre-treatment is based on the inherent characteristics of DES (non-toxic and biodegradable products), which makes this technology more environmentally friendly.[114]

A recent study on the valorisation of pine bark using green solvents concluded that the use of choline chloride-based DES (ChCl:1,4-butanediol) in combination with a low percentage of water (25%) promoted the extraction of flavonoid compounds.[129] Recently, Elgaharbawy *et al.* (2019) optimised the pectin extraction process from grapefruit peel using NADES in combination with UAE.[130] The pre-treatment carried out at 80 °C for 60 min of sonication was efficient for pectin extraction, especially by using ChCl:malonic acid and ChCl:glucose–water as solvents, extracting up to 94% of the pectin.

DESs are also being widely investigated as green solvents for use in delignification processes. Kohli *et al.* (2020) demonstrated that DESs are effective for the delignification of different lignocellulosic biomass, extracting more than 85% of the starting lignin.[131] In this case, the DES providing the best results were ChCl:oxalic acid and ChCl:formic acid. Recently, the delignification process based on DES assisted with phosphotungstic acid catalysts has also been evaluated.[132] This pre-treatment resulted in the extraction of 86% of the lignin and 90% of the hemicelluloses at medium temperatures (100–150 °C), for periods of 3 h.

In addition to the pre-treatments mentioned here, other less-studied treatments could also be interesting from a biorefinery point of view. Examples of this are ozonolysis, high-pressure homogenisation, different biological treatments, hydrodynamic cavitation, plasma treatment, and gamma irradiation, among others.[64]

1.3 Environmental Assessment of Biorefineries

The biorefinery approach processes and supply chains are multi-sectoral, starting with raw material procurement, transport, and industrial processes up to the reuse or disposal of the sub-streams generated in each process. Therefore, in order to understand the biorefinery products' sustainability, it is important to consider the impacts of each sector involved in their production.[7] LCA methodology is a systemic evaluation of the environmental aspects of a product, process or activity during its life cycle stages, analysing its associated environmental impacts.[133] With regard to non-conventional techniques, it is interesting to know which ones are the most recommendable from an environmental point of view, as they are called "green" techniques. Nevertheless, the use of this term does not guarantee their sustainability, thus the environmental burdens arising from their operation should be considered.

The LCA methodology is driven by four phases: goals and scope definition, life cycle inventory (LCI), life cycle impact assessment (LCIA) and interpretation of results.[133,134] An attributional LCA approach was considered since averaged data were used in all analysis scenarios.[135]

Regarding the conventional and non-conventional techniques outlined above, it is interesting to know which are the most environmentally friendly. Therefore, this study aims to consider the different extraction sequences of phenolic compounds considering the LCA methodology from an environmental perspective. This methodology takes into account the entire life cycle of the extraction sequence (cradle-to-gate approach) and allows the identification of the main processes or key steps responsible for environmental burdens.

1.3.1 Definition of the Scope and System Boundaries

The objective of this study is to assess the environmental performance of various scenarios of valorisation routes to obtain phenolic compounds as a target product. The first step in performing an LCA is to define the functional unit, as it supplies the reference to which the inputs and outputs of the system under study are related.[133,134] The selection of the functional unit in a biorefinery study can be on the basis of the feedstock or the quantity of the target product.[136] In this case, due to the variety of biomass used as feedstock and that the main objective is the comparative environmental study between conventional and non-conventional extraction techniques, one kilogram of the target product (*e.g.*, phenolic compounds) has been chosen as the functional unit. This functional unit would allow effective comparison of extraction methods in terms of their environmental performance.

In order to ensure that the input and output of the valorisation scenarios are as similar as possible, the impacts associated with the construction and installation of the biorefinery plant during its lifetime stayed out of the study. It has not been considered to include biomass storage or any chemical compounds within the system boundaries. These assumptions have been previously considered in other similar LCA studies.[13,137–139] The system boundaries of this environmental study include the process units of each of

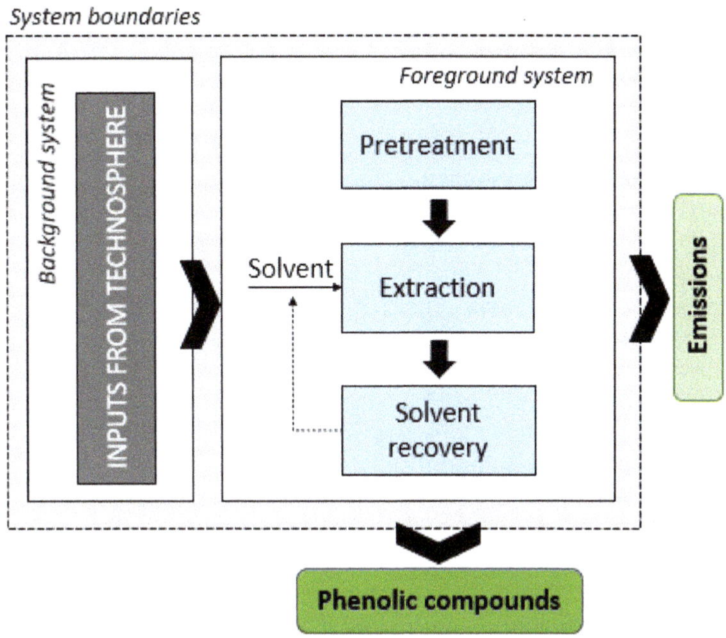

Figure 1.6 Simplified description of the phenolic extraction system boundary.

the studied scenarios (see Figure 1.6). The general composition of all scenarios is three subsystems: pretreatment, extraction and solvent recovery.

1.3.1.1 Description of Production Facilities

The use of non-conventional extraction processes is widely discussed as the "green" future. However, it is necessary to conduct an environmental impact assessment of the different processes before confirming this. Conventional methods are assumed to have high environmental impacts, mainly due to high energy consumption and the use of high amounts of solvents. Nevertheless, environmental impact studies need to be completed to confirm this and to be able to compare with more modern processes.

Therefore, in this study previously published different extraction processes have been selected to obtain phenolic compounds. In the work developed by Barjoveanu *et al.* (2020), three extraction methods are mainly described for obtaining phenolic compounds from spruce bark (*Picea abies*).[140] The authors use three extraction methods, Soxhlet extraction (SE-1), solid–liquid extraction with NaOH (NaOH-SLE) and ultrasound-assisted extraction (UAE). SE-1 was carried out at 78 °C for 6 h and 34 min. The solvent selected in this case was ethanol (70%), and the solid–liquid ratio (SLR) used was 1 : 20. The extraction with NaOH (1%) was conducted at 90 °C for only 1 h, with an SLR of 1 : 10. Finally, the UAE was performed using an ultrasonic bath (35 kHz and 320 W) at 50 °C for 1 h. The solvent used in this case was ethanol (70%), with an SLR of 1 : 10. NaOH extraction was the best in terms of yield, with 21.66 mg GAE g^{-1} bark. Soxhlet extraction was the poorest in phenolic compounds (12.39 mg GAE g^{-1} bark), while UAE yielded 19.1 mg GAE g^{-1} bark.

In the study carried out by Solana *et al.* (2015), a comparative study of the use of three techniques for the extraction of polyphenolic compounds from asparagus stalks was carried out.[141] The first extraction technique used was Soxhlet extraction (SE-2) at 100 °C, for 4 h. The solvent used was ethanol, according to the modification proposed by Santiago *et al.* (2021).[13] This method was compared with the optimised pressurised liquid extraction (PLE). The reaction parameters were 65 °C, 10 MPa, 30 min and a 50% ethanol/water mixture, with a flow rate of 2 ml min^{-1}. Finally, supercritical fluid extraction (SFE) was tested, using an ethanol/water mixture (50%), but with 8% CO_2 as an additive. This reaction was conducted for 1 h, at 65 °C and 15 mPa, with a CO_2 flow rate of 0.25 kg h^{-1}. The highest concentration of phenolic compounds was obtained by Soxhlet extraction (4 mg g^{-1}); while looking at the antioxidant capacity, the best result was achieved with SFE (2.4 mg rutin equivalent g^{-1}).

The last non-conventional extraction process studied is that developed by Murugan *et al.* (2021).[142] In this work, a green solvent (choline chloride:ethylene glycol) was used for the extraction of phenolic compounds from the seeds of *Dacryodes rostrata*. To this end, the ground seeds were mixed with the solvent keeping the SLR 1 : 10 and heated at 40 °C, for 100 min. To enhance the extraction and reduce the viscosity of the solvent, 10% water was added to the solvent. The extraction achieved excellent results, 105 mg GAE g^{-1} sample. The use of this type of solvent allows working at lower temperatures with good final results, which suggests that its environmental impact should be lower than that of conventional methods. However, this needs to be confirmed by calculating the environmental impact of the process.

1.3.2 Life Cycle Inventory

The life cycle inventory stage involves collecting primary and secondary quantitative input and output data for the system under consideration for the environmental assessment. In some cases, data are obtained from publications, research or official databases. In this study, the inventory data, such as electrical demand for all equipment, chemical consumption, process water and thermal energy, have been obtained from environmental studies previously developed by Barjoveanu *et al.* (2020),[140] Santiago *et al.* (2021)[13] and Murugan *et al.* (2021).[142] Table 1.4, which is a summary of the inventory tables of the works described above, is employed to build the life cycle inventory. The input–output structure of the inventory entries, as well as their associated impacts, were sourced from the Ecoinvent® database.[143] SimaPro v9.3.0.3 software was used for the computational implementation of the life cycle inventory data.[144] Due to the differences between the alternative valorisation scenarios (*e.g.*, extraction technique, raw material, location and

Table 1.4 Life cycle inventory of the different valorisation scenarios proposed to obtain phenolic compounds using traditional and non-traditional techniques. Functional unit: one kg of phenolic compound.

	Barjoveanu *et al.*, 2020[140]			Santiago *et al.*, 2021[13]			Murugan *et al.*, 2021[142]
	SE-1	NaOH-SLE	UAE	SE-2	PLE	SFE	DES
Inputs from technosphere							
Pretreatment stage							
Electricity	30.2 W h	51.8 W h	19.6 W h	26 kW h	26 kW h	26 kW h	38.4 kW h
Extraction stage							
Tap water	0.40 kg	—	—	9.6 kg	15.8 kg	18.4 kg	14.5 g
Ethanol	0.18 g	—	0.14 mg	303.0 kg	22.5 kg	31.3 kg	—
Carbon dioxide	—	—	—	—	—	300.0 kg	—
Sodium hydroxide	—	0.014 g	—	—	—	—	—
Electricity	0.03 kW h	0.06 kW h	0.3 kW h	0.08 kW h	2.9 kW h	305.0 kW h	1.26 kW h
Heat energy	—	—	2.2 kW h	1251.2 kW h	29.2 kW h	618.9 kW h	—
Cooling energy	—	—	—	715.7 kW h	—	663.8 kW h	—
Solvent recovery							
Electricity	1.1 W h	1.5 W h	0.6 kW h	14.9 kW h	14.9 kW h	14.9 kW h	3.4 kW h
Heat energy	—	—	18.3 kW h	351.7 kW h	749.5 kW h	509.9 kW h	—
Cooling energy	—	—	—	350.6 kW h	788.2 kW h	512.4 kW h	—
Outputs to technosphere							
Emissions to air	0.18 g EtOH 0.29 g H_2O	0.49 g H_2O	0.14 mg EtOH 0.19 g H_2O	—	—	—	18.8 g H_2O
Biowaste treatment	88.3 mg	92.3 mg	20.9 mg	30.0 kg	30.0 kg	30.0 kg	14.5 g
Phenolic compounds	1 mg	1 mg	1 mg	70.3 g	65.4 g	68.6 g	104.6 mg

Table 1.5 Listing of the main Ecoinvent® database processes considered in the comparative environmental study.

Database and processes	
Tap water	Tap water {RER} | market group for | cut-off, U
Ethanol	Ethanol, without water, in 99.7% solution state, from ethylene {RER} | market for ethanol, without water, in 99.7% solution state, from ethylene | cut-off, U
Sodium hydroxide	Sodium hydroxide, without water, in 50% solution state {GLO} | market for | cut-off, U
Electricity	Electricity, medium voltage {RER} | market group for | cut-off, U
Heat energy	Heat, district or industrial, natural gas {RER} | market group for | cut-off, U
Cooling energy	Cooling energy {RoW} | from natural gas, at cogen unit with chiller 100 kW | cut-off, U
Biowaste treatment	Biowaste {RoW} | treatment of biowaste, open dump | cut-off, U

consequent energy and chemical production), two assumptions have been considered: (1) all processes are located in Europe, including background processes such as the production of chemicals and energy (electrical and thermal), as Saavedra *et al.* (2021) considered in their environmental study,[14] and (2) the heat source is obtained from steam from the chemical industry, while the cooling energy consists of the recovery of the cooling utility in a CHP unit. This information is accessible in the Ecoinvent® database.[143] The main processes of the Ecoinvent® database considered in the comparative environmental study are listed in Table 1.5.

1.3.3 Life Cycle Impact Assessment Methodology

For the environmental analysis, the hierarchical midpoint method of ReCiPe Europe v1.13 (2016) was used to perform the environmental profiling.[145] The research focused only on the global warming (GW) category, without including other categories such as eutrophication or land use, as would be contemplated in a holistic LCA. Thus, the characterisation factors reported by this method were considered to estimate the environmental loads in terms of a single impact category: global warming (GW).

1.3.4 Environmental Results

In order to assess the sustainability of different extraction processes employed for the extraction of phenolic compounds from different biomass, in this work the GW of previously published processes has been calculated (see Figure 1.7). This figure provides a straightforward interpretation of the GW burden of each of the processes. The conventional extraction methods that have been studied are shown in yellow. This group includes two Soxhlet extractions (SE-1 and SE-2) and one extraction with NaOH. This last process

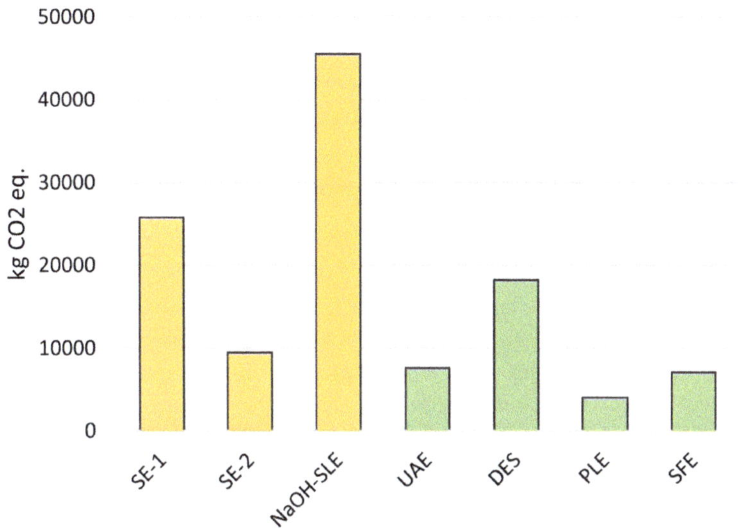

Figure 1.7 Results of the environmental characterisation of GW per kg CO_2 eq. and functional unit (1 kg of phenolic compounds) for the extraction techniques; SE-1, SE-2, NaOH-SLE, UAE, DES, PLE and SFE. (■) Conventional technique, (■) non-conventional technique.

is the one with the highest impact on the GW category, with a total of 45.5 tonnes of CO_2 equivalent per kg of product. According to the LCA calculations performed by the authors of the research, the impact of this process is entirely due to the energy production demanded in the extraction stage, as temperatures of up to 90 °C must be reached, and with the processes involved in the solvent recovery stage.[140]

Proceeding with the results of the conventional methods, it could be noted that there is a considerable difference between the impacts associated with the two Soxhlet extractions. Furthermore, compared to the NaOH-SLE scenario, the GW environmental burden of the SE-1 and SE-2 scenarios decreases by 43% and 79% respectively. As in the previous case, energy consumption has the greatest impact on the GW of the Soxhlet extraction processes, with an impact of more than 90%.[140] The extraction process described for the extraction of polyphenolic compounds from spruce bark (in Figure 1.7 presented as SE-1) was found to have a higher impact than the extraction process described for asparagus stems, being 63% higher. Despite the difference in the temperature at which each of the processes is performed, with 78 °C in SE-1 and 100 °C in SE-2, the main reason for this large difference is due to the scale at which the environmental analysis is performed, *i.e.*, SE-1 was studied at the laboratory scale while SE-2 was performed at the pilot scale. When data are provided at the laboratory scale, it results in a higher impact compared to data at larger scales.[146]

Figure 1.7 illustrates in green the results of the GW impact of the non-conventional extraction processes investigated in this study. It is evident at first view that both UAE and PLE are technologies with lower environmental loads. The UAE conducted at 60 °C for only 1 h, in addition to achieving a good extraction yield, as mentioned before, is also a process with a low environmental impact (7.5 tonnes of CO_2 eq. per kg of product). It was found to have a burden of less than 83% of that generated in the extraction with NaOH of spruce bark and up to 71% less than the Soxhlet extraction (SE-1) of the same residue. For the GW impact, electricity production is once again the protagonist, in particular due to the processes involved in solvent recovery. The ethanol recovery process requires energy, but water recovery is more energy demanding. On the other hand, PLE (4.0 tonnes of CO_2 eq. per kg of product), although it has a lower impact than UAE, also improves the results achieved by conventional methods. This decrease in the environmental load is due to the lower energy required for the process compared to the rest, in addition to the lower solvent demand required in the extraction stage.[13]

The SFE scenario has an environmental burden similar to the UAE scenario (7.1 tonnes of CO_2 eq. per kg of product). The main cause of this environmental burden is the production of electrical and thermal energy demand for the process of converting CO_2 into a supercritical fluid.[13]

Finally, the DES scenario has been found to have a higher environmental burden than Soxhlet extraction (SE-2), being approximately 48% higher. However, as it is a non-conventional extraction technique, a lower environmental impact was expected in this scenario. Since this extraction process was carried out at a low temperature (40 °C), the environmental load was expected to be lower. However, in this case, apart from energy consumption, chemical reagents also appear as the main problems. Therefore, the GW result rises to 18.2 tonnes kg CO_2 eq. per kg of product. The calculation of the GW in this case was based on existing processes in the Ecoinvent® database, which limited some of the calculations. The chemicals were blamed for the unexpected increase in the environmental load of this process. This may be due to the fact that since the DES production process does not exist in the database, and the data required to create a process were not available, it was decided to assume a process of a conventional organic chemical. However, choline chloride:ethylene glycol is a compound obtained by mixing pure compounds of choline chloride and ethylene glycol. The eutectic mixing of these two compounds, unlike the processes required for the synthesis of many organic compounds, is carried out at temperatures not exceeding 80 °C for 1–2 h.[129] Therefore, the loads involved in the production process of the compound are assumed to be lower (this needs to be confirmed), and the properties of organic conventional compounds are different from those of DES, which means that the impact will also be different. One of the most important properties of these compounds, as already mentioned, is that they are not volatile, with the benefits that the use of these compounds in diverse processes can have for the reduction of pollutant emissions. In conclusion, it is also necessary to carry out an environmental analysis of the DES preparation processes, so that the results of the extraction processes with this type of compounds fit more adequately with the reality.

1.4 Conclusions

According to the FAO statistics, around 33% of the food produced globally is lost or wasted along the supply chain. The world's population is increasing, so is food consumption as well as waste generation due to its production, so it is necessary to develop techniques to minimise, eliminate and treat this food waste. SDGs were created, among other things, as an opportunity to implement recycling and reuse strategies in the production and supply chains to reduce food loss. Therefore, in addition to the interest in reducing the amount of food waste, the biorefinery concept aims to convert biomass into high added-value products, which can even be commercially exploitable such as biofuels, biofertilisers or proteins, among others. Depending on the desired final product, as well as the quality, quantity and composition of the biomass, several parameters determine the choice of the production process in a biorefinery. So far, maceration, pressing, hydrodistillation or Soxhlet extraction, known as traditional or conventional techniques, have been the most common techniques for obtaining high added-value products. However, to overcome the limitations of conventional methods for the extraction of these high added-value compounds, advanced techniques and strategies are being continuously developed. The advantages of these techniques, also known as "green" techniques, are the production of extracts with high quality and yield, low solvent and energy consumption and short extraction time. Recent research has shown that "green" extraction methods offer excellent alternatives to traditional methods. Many studies are still ongoing in this field, to further improve these new green extraction techniques, to reduce the extraction cost and extraction time and to improve the extract quality, environmental safety and health. However, the wealth of information available on the extraction of high added-value compounds from different types of biomass makes it difficult to draw general conclusions. Fortunately, the understanding of the process and how variables can be used to control is rapidly increasing, leading to innovative approaches to maximise yields and reduce degradation while minimising the environmental impact.

In addition to the study of the different techniques available for the extraction of high added-value compounds, this study has also focused on the comparative environmental analysis of different biomass valorisation scenarios, based on the LCA methodology, to obtain phenolic compounds as the main high added-value product, under a biorefinery approach. The global warming impact caused by the use of different techniques, both traditional and non-conventional, in the production of one kilogram of phenolic compound has been calculated. The overall results show that the PLE ("green" technique) has the lowest environmental burdens, while the extraction using soda (conventional technique) has the highest value, with a difference between them of 91%. When comparing the two scenarios focusing on the Soxhlet extraction technique, a difference of 67% could be seen in the climate change category. The main reason for this is the difference in the scaling of the environmental analysis. When comparing the non-conventional techniques with each other, a similarity between the UAE and SFE scenarios could be seen, reporting an approximate value of 7 tonnes of CO_2 eq. with respect to the GW category. The DES scenario reports the highest value in the GW category, being approximately 60–80% higher depending on the scenario with which it is compared. Furthermore, this scenario reports a higher environmental load value (48%) when compared to the conventional Soxhlet technique, namely the one using asparagus residue as the raw material. This is because the DES production process does not exist in the database used and traditional organic chemistry processes have to be used. The LCA methodology makes it possible to establish a comparative framework that can represent a useful tool for identifying the advantages and disadvantages of each selected technique, as well as the possibility to determine the operating parameters on which to act in the short and medium term. However, it is important to carry out the comparative environmental analysis on the same scale, trying to avoid the lab-scale, as there will be a noticeable difference between the environmental impacts for the different scenarios analysed. For this reason, plant and larger scale studies to ensure the environmental viability of the processes are needed. Considering the environmental results, it could be confirmed for the researched studies that the non-conventional extraction methods have lower environmental burdens, except for the DES scenario, as well as better extraction yields. This confirms the promising future of these methods. Finally, the production of energy demand is the main hotspot in many processes. Therefore, special attention should be paid when using renewable energy sources (instead of taking it directly from the national grid) to satisfy and optimise the energy demand, as well as when reusing some of the internal steams to reduce the environmental impact of all scenarios. In addition, once the environmental analysis of these techniques has been carried out, it is suggested that a cost analysis be carried out as a future project to determine their economic viability.

Abbreviations

DES	Deep eutectic solvents
FAO	Food and Agriculture Organization of the United Nations
HBA	Hydrogen acceptor
HBD	Hydrogen donor
IL	Ionic liquids
LCA	Life cycle assessment
LCI	Life cycle inventory
LCIA	Life cycle impact assessment
MAE	Microwave-assisted extraction

PEF	Pulse electrical field
SDG	Sustainable development goal
SFE	Supercritical fluid extraction
SLR	Solid–liquid ratio
SubFE	Subcritical fluid extraction
PLE	Pressurised liquid extraction
UAE	Ultrasound-assisted extraction
UN	United Nations
VFA	Volatile fatty acid
VS	Volatile solid

Acknowledgements

This research has been partially supported by the SENSE project granted by the FEDER/Spanish Ministry of Science, Innovation and Universities, Spanish National Research Agency (CTQ2016-75136-P) and by the project Enhancing Diversity in Mediterranean Cereal Farming Systems (CerealMed) funded by the PRIMA Programme and FEDER/Ministry of Science and Innovation – Spanish National Research Agency (PCI2020-111978). We also like to thank the Spanish Ministry of Science, Innovation and Universities for the project RED2018-102623-T. B. Santiago would like to thank the Spanish Ministry of Science, Innovation and Universities for the financial support (grant reference BES-2017-081715). Dr Sillero would like to thank the Spanish Ministry of Universities for the Margarita Salas grant for the re-qualification of the Spanish University System funded by the European Union-Next Generation EU. The authors belong to the Galician Competitive Research Group GRC 2013-032 and to CRETUS Strategic Partnership (AGRUP2015/02). All these programmes are co-funded by Xunta de Galicia and FEDER (EU).

References

1. J. Wilmoth, C. Menozzi and L. Bassarsky, in *Why population growth matters for sustainable development*, United Nations Department of Economic and Social Affairs, 2022.
2. J. Gustavsson, C. Cederberg, U. Sonesson, R. van Otterdijk and A. Meybeck, *Extent, Causes Prev.*, 2011, p. 1.
3. B. Tsegaye, S. Jaiswal and A. K. Jaiswal, *Foods*, 2021, **10**(6), 1174.
4. A. Seberini, *SHS Web Conf.*, 2020, **74**, 03010.
5. A. Asadikia, A. Rajabifard and M. Kalantari, *World Dev.*, 2021, **140**, 105269.
6. J. C. Solarte-Toro and C. A. Cardona Alzate, *Bioresour. Technol.*, 2021, **340**, 125626.
7. C. Vance, J. Sweeney and F. Murphy, *Renewable Sustainable Energy Rev.*, 2022, **159**, 112259.

8. S. Dahiya, A. N. Kumar, J. Shanthi Sravan, S. Chatterjee, O. Sarkar and S. V. Mohan, *Bioresour. Technol.*, 2018, **248**, 2.
9. J. Cristóbal, C. Caldeira, S. Corrado and S. Sala, *Bioresour. Technol.*, 2018, **259**, 244.
10. A. Corona, M. Ambye-Jensen, G. C. Vega, M. Z. Hauschild and M. Birkved, *Sci. Total Environ.*, 2018, **635**, 100.
11. H. Cai, J. Han, M. Wang, R. Davis, M. Biddy and E. Tan, *Biofuels, Bioprod. Biorefin.*, 2018, **12**, 815.
12. B. Santiago, A. Arias Calvo, B. Gullón, G. Feijoo, M. T. Moreira and S. González-García, *Chem. Eng. J.*, 2020, **392**, 123772.
13. B. Santiago, G. Feijoo, M. T. Moreira and S. González-García, *Food Bioprod. Process.*, 2021, **129**, 176.
14. M. Saavedra del Oso, M. Mauricio-Iglesias and A. Hospido, *Chem. Eng. J.*, 2021, **412**, 127687.
15. C. Drosou, K. Kyriakopoulou, A. Bimpilas, D. Tsimogiannis and M. Krokida, *Ind. Crops Prod.*, 2015, **75**, 141.
16. P. R. Yaashikaa, P. Senthil Kumar and S. Varjani, *Bioresour. Technol.*, 2022, **343**, 126126.
17. E. Capanoglu, E. Nemli and F. Tomas-Barberan, *J. Agric. Food Chem.*, 2022, **70**, 6787.
18. S. Kaza, L. C. Yao, P. Bhada-Tata and F. Van Woerden, in *What a Waste 2.0: A Global Snapshot of Solid Waste Management to 2050*, World Bank Group, Washington, 2018.
19. Inamuddin and A. Khan, *Sustainable Bioconversion of Waste to Value Added Products*, Springer, Cham, 2021.
20. Real Academia Española, Diccionario de la lengua española, 23.a ed., [Version 23.5 online] (last accessed July 2022).
21. FAO, in *The State of Food and Agriculture 2019: Moving Forward on Food Loss and Waste Reduction*, Rome, 2019.
22. R. Yukesh Kannah, J. Merrylin, T. Poornima Devi, S. Kavitha, P. Sivashanmugam, G. Kumar and J. Rajesh Banu, *Bioresour. Technol. Rep.*, 2020, **11**, 100524.
23. United Nations Environment Programme, in *Food Waste Index Report 2021*, UNEP, Nairobi, 2021.
24. A. Patra, S. Abdullah and R. C. Pradhan, *Bioresour. Bioprocess.*, 2022, **9**, 14.
25. K. Paritosh, S. K. Kushwaha, M. Yadav, N. Pareek, A. Chawade and V. Vivekanand, *Biomed Res. Int.*, 2017, **2017**, 1.
26. F. Thoma, H. Husson, N. Krajnc, N. N. Fhlatharta, T. Nordfejll, J. Picos and M. Prior, in *Sustainable Mobilisation of Forest Biomass. Minipaper 8: European map of the regional forest-based sector*, European Commission, 2016.
27. A. Devi, S. Bajar, H. Kour, R. Kothari, D. Pant and A. Singh, *Bioenergy Res.*, 2022, **15**, 1820–1841.
28. P. Baruya, *World forest and agricultural crop residue resources for cofiring*, IEA Clean Coal Centre, London, 2015.

29. UNECE/FAO, in *Forest Products: Annual Market Review 2020–2021*, United Nations, Genova, 2021.
30. W. Stafford, W. De Lange, A. Nahman, V. Chunilall, P. Lekha, J. Andrew, J. Johakimu, B. Sithole and D. Trotter, *Renewable Energy*, 2020, **154**, 461.
31. M. Moshkelani, M. Marinova, M. Perrier and J. Paris, *Appl. Therm. Eng.*, 2013, **50**, 1427.
32. C. Caldeira, A. Vlysidis, G. Fiore, V. De Laurentiis, G. Vignali and S. Sala, *Bioresour. Technol.*, 2020, **312**, 123575.
33. I. Dávila, B. Gullón, J. L. Alonso, J. Labidi and P. Gullón, *Carbohydr. Polym.*, 2019, **207**, 34.
34. S. Sugiarto, R. R. Pong, Y. C. Tan, Y. Leow, T. Sathasivam, Q. Zhu, X. J. Loh and D. Kai, *Mater. Today Chem.*, 2022, **26**, 101022.
35. T. Wang, J. Zhao, Z. Yang, L. Xiong, L. Li, Z. Gu and Y. Li, *Green Chem.*, 2022, **24**, 3605.
36. V. G. Zuin and L. Z. Ramin, *Top. Curr. Chem.*, 2018, **376**, 1.
37. P. Wen, T. G. Hu, R. J. Linhardt, S. T. Liao, H. Wu and Y. X. Zou, *Trends Food Sci. Technol.*, 2019, **83**, 138.
38. A. Sridhar, M. Ponnuchamy, P. S. Kumar, A. Kapoor, D. N. Vo and S. Prabhakar, *Environ. Chem. Lett.*, 2021, **19**(4), 3409.
39. C. H. Geow, M. C. Tan, S. P. Yeap and N. L. Chin, *Eur. J. Lipid Sci. Technol.*, 2021, **123**(4), 1.
40. O. Ezeh, M. H. Gordon and K. Niranjan, *Food Chem.*, 2016, **194**, 354.
41. N. A. Febrianto and T. A. Yang, *Adv. J. Food Sci. Technol.*, 2011, **3**, 317.
42. M. I. Jahirul, J. R. Brown, W. Senadeera, N. Ashwath, C. Laing, J. Leski-Taylor and M. G. Rasul, *Procedia Eng.*, 2013, **56**, 619.
43. M. Selvamuthukumaran and J. Shi, *Food Qual. Saf.*, 2017, **1**, 61.
44. A. M. Martínez-Gil, M. del Alamo-Sanza, G. Gutiérrez-Gamboa, Y. Moreno-Simunovic and I. Nevares, *Food Chem.*, 2018, **266**, 90.
45. L. Todaro, D. Russo, P. Cetera and L. Milella, *Ind. Crops Prod.*, 2017, **109**, 384.
46. S. Bostyn, E. Destandau, J. P. Charpentier, V. Serrano, J. M. Seigneuret and C. Breton, *Ind. Crops Prod.*, 2018, **126**, 22.
47. N. Ćujić, K. Šavikin, T. Janković, D. Pljevljakušić, G. Zdunić and S. Ibrić, *Food Chem.*, 2016, **194**, 135.
48. B. R. Albuquerque, M. A. Prieto, M. F. Barreiro, A. E. Rodrigues, T. P. Curran, L. Barros and I. C. F. R. Ferreira, *Ind. Crops Prod.*, 2017, **95**, 404.
49. A. A. Jovanović, V. B. Đorđević, G. M. Zdunić, D. S. Pljevljakušić, K. P. Šavikin, D. M. Gođevac and B. M. Bugarski, *Sep. Purif. Technol.*, 2017, **179**, 369.
50. M. B. Santos, L. Sillero, D. A. Gatto and J. Labidi, *Ind. Crops Prod.*, 2022, **186**, 115231.
51. L. Wang and C. L. Weller, *Trends Food Sci. Technol.*, 2006, **17**, 300.
52. P. Tongnuanchan and S. Benjakul, *J. Food Sci.*, 2014, **79**, 1231.
53. K. A. Santos, J. E. Gonçalves, L. Cardozo-Filho and E. A. da Silva, *Ind. Crops Prod.*, 2019, **130**, 428.

54. A. Meullemiestre, Z. Maache-Rezzoug, F. Chemat and S. A. Rezzoug, *J. Mater. Environ. Sci.*, 2017, **8**, 2608.
55. A. Kumar and S. R. Samadder, *Energy*, 2020, **197**, 117253.
56. A. G. Mari, C. L. Andreani, T. U. Tonello, L. C. C. Leite, J. R. Fernandes, D. D. Lopes, J. A. D. Rodrigues and S. D. Gomes, *Int. J. Hydrogen Energy*, 2020, **45**, 5165.
57. G. Srisowmeya, M. Chakravarthy and G. Nandhini Devi, *Renewable Sustainable Energy Rev.*, 2020, **119**, 109587.
58. S. P. Lohani, S. Shakya, P. Gurung, B. Dhungana, D. Paudel and B. Mainali, *Energy Sources, Part A*, 2021, 1.
59. A. Ghimire, V. Luongo, L. Frunzo, P. N. L. Lens, F. Pirozzi and G. Esposito, *Environ. Technol.*, 2022, **43**, 2190.
60. Lukitawesa, R. J. Patinvoh, R. Millati, I. Sárvári-Horváth and M. J. Taherzadeh, *Bioengineered*, 2020, **11**, 39.
61. O. Sarkar, J. Santhosh, A. Dhar and S. Venkata Mohan, *Int. J. Hydrogen Energy*, 2021, **46**, 18832.
62. J. Kuo and J. Dow, *J. Air Waste Manag. Assoc.*, 2017, **67**, 1000.
63. F. Chemat, M. Abert Vian, A. S. Fabiano-Tixier, M. Nutrizio, A. Režek Jambrak, P. E. S. Munekata, J. M. Lorenzo, F. J. Barba, A. Binello and G. Cravotto, *Green Chem.*, 2020, **22**, 2325.
64. A. Sidana and S. K. Yadav, *J. Cleaner Prod.*, 2022, **335**, 130286.
65. H. Chandel, P. Kumar, A. K. Chandel and M. L. Verma, *Biomass Convers. Biorefin.*, 2022, 1.
66. R. Castro-Muñoz, E. Díaz-Montes, E. Gontarek-Castro, G. Boczkaj and C. M. Galanakis, *Compr. Rev. Food Sci. Food Saf.*, 2022, **21**(1), 46.
67. M. Jesus, A. Romaní, F. Mata and L. Domingues, *Polymers*, 2022, **14**(9), 1640.
68. B. R. Prasad, R. K. Padhi and G. Ghosh, *Int. J. Environ. Sci. Technol.*, 2023, **20**, 6929–6944.
69. D. Konan, E. Koffi, A. Ndao, E. C. Peterson, D. Rodrigue and K. Adjallé, *Energies*, 2022, **15**(9), 3002.
70. A. Delvar, P. de Caro, L. Candy, Y. Caro, A. S. Cheong Sing and C. Raynaud, *J. Food Eng.*, 2019, **263**, 388.
71. J. Pico, K. Xu, M. Guo, Z. Mohamedshah, M. G. Ferruzzi and M. M. Martinez, *Food Chem.*, 2019, **297**, 124990.
72. Z. Wang, X. He, L. Yan, J. Wang, X. Hu, Q. Sun and H. Zhang, *Ind. Crops Prod.*, 2020, **143**, 111960.
73. A. Duque, P. Doménech, C. Álvarez, M. Ballesteros and P. Manzanares, *Renewable Energy*, 2020, **158**, 263.
74. M. Pattnaik, P. Pandey, G. J. O. Martin, H. N. Mishra and M. Ashokkumar, *Foods*, 2021, **10**, 1.
75. M. Banožić, I. Banjari, M. Jakovljević, D. Šubarić, S. Tomas, J. Babić and S. Jokić, *Molecules*, 2019, **24**, 1611.
76. N. Medina-Torres, H. Espinosa-Andrews, S. Trombotto, T. Ayora-Talavera, J. Patrón-Vázquez, T. González-Flores, Á. Sánchez-Contreras, J. C. Cuevas-Bernardino and N. Pacheco, *Molecules*, 2019, **24**, 3541.

77. P. Cetera, D. Russo, L. Milella and L. Todaro, *Ind. Crops Prod.*, 2019, **130**, 380–388.
78. P. Panja, *Curr. Opin. Food Sci.*, 2018, **23**, 173.
79. A. Ali, B. L. Chua and Y. H. Chow, *TrAC – Trends Anal. Chem.*, 2019, **118**, 338.
80. S. A. Gokulakrishnan, G. Arthanareeswaran, G. Gnanasekaran, Z. László, G. Veréb, S. Kertész and W. Taweepreda, *Biomass Convers. Biorefin.*, 2022, 1.
81. M. S. Jesus, Z. Genisheva, A. Romaní, R. N. Pereira, J. A. Teixeira and L. Domingues, *Ind. Crops Prod.*, 2019, **132**, 99.
82. I. Dávila, P. Gullón and J. Labidi, *Waste Manag.*, 2021, **120**, 146.
83. M. Sharma and R. Bhat, *Foods*, 2021, **10**, 787.
84. C. M. Cova, L. Boffa, M. Pistocchi, S. Giorgini, R. Luque and G. Cravotto, *Molecules*, 2019, **24**, 2681.
85. L. Sillero, R. Prado and J. Labidi, *Chem. Eng. Process. – Process Intensif.*, 2020, **156**, 108100.
86. L. Estrada-Gil, J. C. Contreras-Esquivel, C. Flores-Gallegos, A. Zugasti-Cruz, M. Govea-Salas, M. A. Mata-Gómez, R. Rodríguez-Herrera and J. A. Ascacio-Valdés, *Molecules*, 2022, **27**, 1592.
87. M. Rana, T. Nshizirungu and J.-H. Park, *Sustainable Energy Fuels*, 2022, **6**, 861.
88. A. Lončarić, M. Celeiro, A. Jozinović, J. Jelinić, T. Kovač, S. Jokić, J. Babić, T. Moslavac, S. Zavadlav and M. Lores, *Foods*, 2020, **9**, 1521.
89. M. Visockis, R. Bobinaitė, P. Ruzgys, J. Barakauskas, V. Markevičius, P. Viškelis and S. Šatkauskas, *Innovative Food Sci. Emerging Technol.*, 2021, **73**, 102761.
90. C. Jimenez-Lopez, M. Fraga-Corral, M. Carpena, P. García-Oliveira, J. Echave, A. G. Pereira, C. Lourenço-Lopes, M. A. Prieto and J. Simal-Gandara, *Food Funct.*, 2020, **11**, 4853.
91. M. Cvjetko Bubalo, S. Vidović, I. Radojčić Redovniković and S. Jokić, *Food Bioprod. Process.*, 2018, **109**, 52.
92. B. Gullón, M. Gagaoua, F. J. Barba, P. Gullón, W. Zhang and J. M. Lorenzo, *Trends Food Sci. Technol.*, 2020, **100**, 1.
93. G. Ferrentino, S. Giampiccolo, K. Morozova, N. Haman, S. Spilimbergo and M. Scampicchio, *Innovative Food Sci. Emerging Technol.*, 2020, **64**, 102428.
94. B. Marić, B. Pavlić, D. Čolović, B. Abramović, Z. Zeković, M. Bodroža-Solarov, N. Ilić and N. Teslić, *LWT*, 2020, **130**, 109627.
95. N. Bukhanko, T. Attard, M. Arshadi, D. Eriksson, V. Budarin, A. J. Hunt, P. Geladi, U. Bergsten and J. Clark, *Ind. Crops Prod.*, 2020, **145**, 112096.
96. G. R. Surup, A. J. Hunt, T. Attard, V. L. Budarin, F. Forsberg, M. Arshadi, V. Abdelsayed, D. Shekhawat and A. Trubetskaya, *Energy*, 2020, **193**, 116696.
97. J. Zhang, C. Wen, H. Zhang, Y. Duan and H. Ma, *Trends Food Sci. Technol.*, 2020, **95**, 183.
98. N. Muñoz-Almagro, L. Valadez-Carmona, J. A. Mendiola, E. Ibáñez and M. Villamiel, *Carbohydr. Polym.*, 2019, **217**, 69.

99. W. Li, Z. Fan, Y. Wu, Z. Jiang and R. Shi, *J. Sci. Food Agric.*, 2019, **99**, 5283.
100. L. G. Gonçalves Rodrigues, S. Mazzutti, L. Vitali, G. A. Micke and S. R. S. Ferreira, *Biocatal. Agric. Biotechnol.*, 2019, **22**, 101367.
101. M. T. Tena, in *Encyclopedia of Analytical Science (Third Edition)*, ed. P. Worsfold, C. Poole, A. Townshend and M. Miró, Elsevier, 3rd edn, 2019, p. 78.
102. M. D'auria, M. Mecca, M. R. Bruno and L. Todaro, *Forests*, 2021, **12**, 1.
103. E. M. C. Alexandre, S. Silva, S. A. O. Santos, A. J. D. Silvestre, M. F. Duarte, J. A. Saraiva and M. Pintado, *Food Res. Int.*, 2019, **115**, 167.
104. D. T. V. Pereira, A. G. Tarone, C. B. B. Cazarin, G. F. Barbero and J. Martínez, *J. Food Eng.*, 2019, **240**, 105.
105. S. Periyasamy, V. Karthik, P. Senthil Kumar, J. B. Isabel, T. Temesgen, B. M. Hunegnaw, B. B. Melese, B. A. Mohamed and D.-V. N. Vo, *Environ. Chem. Lett.*, 2022, **20**, 1129.
106. C. Álvarez, A. González, J. L. Alonso, F. Sáez, M. J. Negro and B. Gullón, *Food Bioprod. Process.*, 2020, **124**, 131.
107. L. J. Swart, O. K. K. Bedzo, E. van Rensburg and J. F. Görgens, *Biomass Convers. Biorefin.*, 2022, **12**, 1295.
108. R. Bhatia, A. Winters, D. N. Bryant, M. Bosch, J. Clifton-Brown, D. Leak and J. Gallagher, *Bioresour. Technol.*, 2020, **296**, 122285.
109. W. G. Morais, T. F. Pacheco, P. S. Corrêa, A. A. Martins, T. M. Mata and N. S. Caetano, *Energy Rep.*, 2020, **6**, 18.
110. T. Chiranjeevi, A. J. Mattam, K. K. Vishwakarma, A. Uma, V. C. R. Peddy, S. Gandham and H. Ravindra Velankar, *ACS Sustainable Chem. Eng.*, 2018, **6**, 8762.
111. A. Morales, F. Hernández-Ramos, L. Sillero, R. Fernández-Marín, I. Dávila, P. Gullón, X. Erdocia and J. Labidi, *Bioresour. Technol.*, 2020, **315**, 123896.
112. M. Cavali, C. R. Soccol, D. Tavares, L. A. Zevallos Torres, V. Oliveira de Andrade Tanobe, A. Zandoná Filho and A. L. Woiciechowski, *Bioresour. Technol.*, 2020, **316**, 123884.
113. Y. Zhang, S. Ni, R. Wu, Y. Fu, M. Qin, S. Willför and C. Xu, *Ind. Crops Prod.*, 2022, **177**, 114451.
114. S. Banu Jamaldheen, M. B. Kurade, B. Basak, C. G. Yoo, K. K. Oh, B.-H. Jeon and T. H. Kim, *Bioresour. Technol.*, 2022, **346**, 126591.
115. A. Morales, J. Labidi and P. Gullón, *J. Environ. Manage.*, 2022, **310**, 114730.
116. I. Gómez-Cruz, I. Romero, M. del, M. Contreras, J. Labidi, F. Hernández-Ramos, L. B. Roseiro, L. C. Duarte, E. Castro and F. Carvalheiro, *Adv. Sustainable Syst.*, 2022, **6**, 1.
117. S. Rivas, L. López, C. Vila and J. C. Parajó, *Bioresour. Technol.*, 2021, **342**, 125967.
118. D. S. Zijlstra, C. W. Lahive, C. A. Analbers, M. B. Figueirêdo, Z. Wang, C. S. Lancefield and P. J. Deuss, *ACS Sustainable Chem. Eng.*, 2020, **8**, 5119.

119. C. Sun, H. Ren, F. Sun, Y. Hu, Q. Liu, G. Song, A. Abdulkhani and P. Loke Show, *Bioresour. Technol.*, 2022, **344**, 126264.
120. S. K. Singh and A. W. Savoy, *J. Mol. Liq.*, 2020, **297**, 112038.
121. S. Roy and S. P. S. Chundawat, *BioEnergy Res.*, 2023, **16**, 263–278.
122. L. Sillero, R. Prado, T. Welton and J. Labidi, *J. Cleaner Prod.*, 2021, **308**, 127286.
123. Y. Li, F. Fang, M. Sun, Q. Zhao, Y. Hu, Z. Sui, Z. Liang, L. Zhang and Y. Zhang, *Talanta*, 2020, **213**, 120848.
124. A. H. A. Rahim, Z. Man, A. Sarwono, N. Muhammad, A. S. Khan, W. S. W. Hamzah, N. M. Yunus and Y. A. Elsheikh, *Cellulose*, 2020, **27**, 2135.
125. A. Rauf, A. Shafeeq and K. Shahzad, *Chem. Eng. Technol.*, 2022, **45**, 1106.
126. L. Neubert, J. Sunthornvarabhas, M. Sakulsombat and K. Sriroth, *Cellul. Chem. Technol.*, 2020, **54**, 301.
127. M. Zhou, O. A. Fakayode, A. E. G. Ahmed Yagoub, Q. Ji and C. Zhou, *Renewable Sustainable Energy Rev.*, 2022, **156**, 111986.
128. E. K. New, S. K. Tnah, K. S. Voon, K. J. Yong, A. Procentese, K. P. Yee Shak, W. Subramonian, C. K. Cheng and T. Y. Wu, *J. Environ. Manage.*, 2022, **307**, 114385.
129. L. Sillero, R. Prado, T. Welton and J. Labidi, *Sustainable Chem. Pharm.*, 2021, **24**, 100544.
130. A. Elgharbawy, A. Hayyan, M. Hayyan, M. Mirghani, H. Salleh, S. Rashid, G. Ngoh, S. Liew, M. Nor, M. Y. Z. bin Mohd Yusoff and Y. Alias, *Processes*, 2019, **7**, 416.
131. K. Kohli, S. Katuwal, A. Biswas and B. K. Sharma, *Bioresour. Technol.*, 2020, **303**, 122897.
132. J. Xie, J. Xu, Z. Cheng, S. Zhu and B. Wang, *Ind. Crops Prod.*, 2021, **172**, 114058.
133. ISO 14040, *Environmental Management-Life Cycle Assessment – Principles and Framework*, Geneve, Switzerland, 2006.
134. ISO 14044, *Environmental management/Life cycle assessment/Requirements and guidelines*, Geneve, Switzerland, 2006.
135. M. J. Bastante-Ceca, J. L. Fuentes-Bargues, L. Hufnagel, F. Mihai and C. Iatu, in *Sustainability Assessment at the 21st century*, IntechOpen, London, 2020, p. 198.
136. S. Ahlgren, A. Björklund, A. Ekman, H. Karlsson, J. Berlin, P. Börjesson, T. Ekvall, G. Finnveden, M. Janssen and I. Strid, *LCA of biorefineries identification of key issues and methodological recommendations*, The Swedish Knowledge Centre for Renewable Transportation Fuels, Sweden, 2013.
137. A. Cortés, M. T. Moreira and G. Feijoo, *Waste Manag.*, 2019, **95**, 70.
138. H. K. Jeswani, T. Falano and A. Azapagic, *Biofuels, Bioprod. Biorefin.*, 2015, **9**, 661.
139. H. Karlsson, P. Börjesson, P. A. Hansson and S. Ahlgren, *J. Cleaner Prod.*, 2014, **83**, 420.

140. G. Barjoveanu, O. Alexandra, C. Teodosiu and I. Volf, *Sci. Rep.*, 2020, **10**, 13632.
141. M. Solana, I. Boschiero, S. Dall'Acqua and A. Bertucco, *J. Supercrit. Fluids*, 2015, **100**, 201.
142. M. D. Murugan, L. H. Tee and K. S. Oh, *J. Phys. Conf. Ser.*, 2021, **2120**, 012005.
143. G. Wernet, C. Bauer, B. Steubing, J. Reinhard, E. Moreno-Ruiz and B. Weidema, *Int. J. Life Cycle Assess.*, 2016, **21**, 1218.
144. PRé Sustainability, in *SimaPro database manual – Methods library*, SimaPro, 2020.
145. M. Huijbregts, Z. J. N. Steinmann, P. M. F. M. Elshout, G. Stam, F. Verones, M. D. M. Vieira, M. Zijp and R. van Zelm, Natl. Inst. Public Heal. Environ., 2016.
146. F. Piccino, R. Hischier, S. Seeger and C. Som, *J. Cleaner Prod.*, 2016, **135**, 1085–1097.

CHAPTER 2

Agri-food Waste: An Adjuvant for the Management of Oxidative Stress-related Disorders?

INÉS MÁRMOL*[a] AND DANIEL PELLICER[b]

[a] Institute for Health Research Aragón (IIS Aragón), Zaragoza, Spain; [b] Research group on Rare Respiratory Diseases (ERR), Department of Physiology, School of Medicine, University of Valencia, Valencia, Spain
*Email: ines.marmol9@gmail.com

2.1 Oxidative Stress in Disease

2.1.1 Physiology of Reactive Oxygen Species

The intracellular redox balance is constantly changing, since most metabolic reactions in living cells are of the oxidation–reduction type. Given that cellular physiological functions strongly depend on reducing power, proper maintenance of the redox balance is crucial. Redox homeostasis, also called homeodynamics, requires constant monitoring of redox parameters to ensure that the redox balance is maintained even in the presence of both intracellular and extracellular changes. Oxygen peroxide, H_2O_2, is considered to play a key role in the redox signalling metabolism.[1,2]

H_2O_2 is generated upon the metabolization of molecular oxygen, O_2. Other products derived from this reaction include hydroxyl radical ($OH^•$) and superoxide radical ($O_2^{•-}$), and this group of molecules is called reactive oxygen species (ROS). These intermediate products of O_2 reduction are highly

Green Chemistry Series No. 78
Agri-food Waste Valorisation
Edited by Pankaj Chowdhary and Abhay Raj

Published by the Royal Society of Chemistry, www.rsc.org

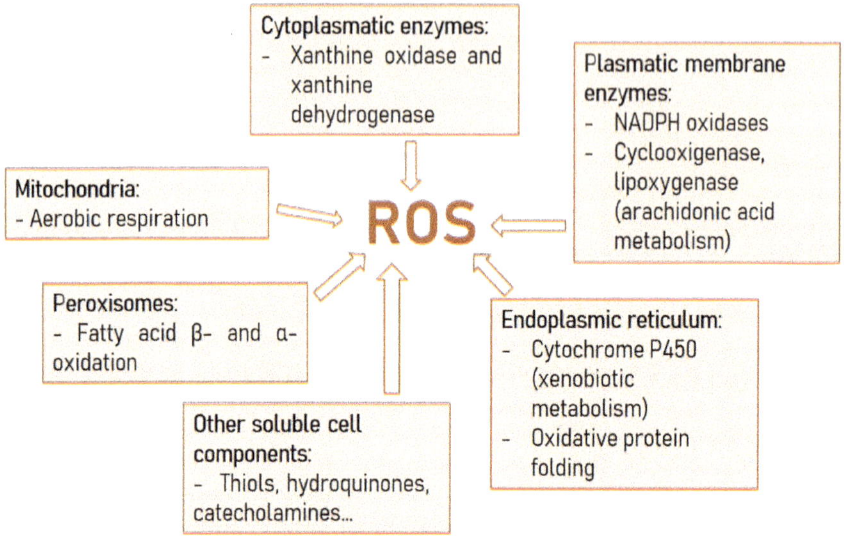

Figure 2.1 Main cellular sources of ROS.

reactive molecules with a reduced lifespan, although H_2O_2, as a non-radical ROS, has the longest among them.[3] Mitochondria are the main source of prooxidant molecules; however, they are not the only ones.[4] Oxidation–reduction reactions at peroxisomes and the endoplasmic reticulum might also result in ROS production[3,5] (Figure 2.1).

At low levels, ROS act as signalling molecules and play key roles in cellular and system physiology. Regarding their role in cell processes, ROS are involved in cell proliferation and death, gene expression and protein post-translational modifications among others. Furthermore, inflammatory response and blood pressure regulation depend on ROS levels as well.[6–8] Interestingly, some authors have noticed a cross-talk between ROS molecules generated from different sources, *e.g.*, cytoplasmic enzymes and mitochondria, to coordinate certain processes such as angiogenesis.[9] Nevertheless, the mechanism behind this communication remains unclear.

2.1.2 ROS-induced Damage

The finding of the physiological role of ROS is, however, a relatively recent discovery. Although the existence of these molecules has been known for decades, research has historically been mainly focused only on their negative effect on cell integrity. Excessive amounts of ROS lead to macromolecular damage and/or disruption of normal redox signalling and control.[10] Due to their high-reactive nature, ROS tend to oxidize lipids, proteins and DNA, leading to lesions and cellular dysfunction (Table 2.1).

Lipids, especially those on the cell membranes such as cholesterol or phospholipids containing polyunsaturated fatty acids, are highly sensitive to

Table 2.1 Consequences of ROS-induced macromolecular damage.

Target	Consequences
Lipids	- Cell membrane dysfunction - Indirect DNA damage - Indirect protein damage
Proteins	- Severe cell stress due to protein destabilization, fragmentation, aggregation or even enhanced degradation - Changes in the composition and function of the extracellular matrix
DNA	- Mutations, carcinogenesis - Impeded DNA-associated processes (replication, transcription, translation)

oxidation. Therefore, lipid oxidation leads to damage of cell membranes, which lose fluidity and selective permeability.[11] Lipid oxidation is mainly driven by hydroxyl and hydroperoxyl ($HOO^{\bullet}$) radicals and results in the generation of reactive carbonyl species (RCS). These molecules entail a greater risk than free radicals, since, as being uncharged, they display a lower mobility restriction. As a consequence, RCS are able to migrate through the cell and even enter the nucleus and/or the mitochondria, where they can react with nucleophilic groups of DNA and form adducts that trigger lesions. In addition, RCS might form protein adducts that lead to misfunction. The accumulation of lipid peroxidation by-products might be involved with age-related diseases and other pathologies, including cardiovascular disease, cancer or neurological disorders.[12,13]

Although it has previously been mentioned that ROS might be involved in post-translational modifications of proteins,[6] these are usually reversible and highly controlled. In contrast, excessive ROS amounts might lead to irreversible post-translational modifications that induce protein destabilization, fragmentation, aggregation or even enhanced degradation. Given that proteins are the most abundant macromolecules in living cells and the major molecules responsible for cell functioning, ROS-induced protein damage triggers severe cell stress. As an example, protein dysfunction leads to indirect DNA damage, since the DNA integrity maintenance machinery is impaired. Therefore, protein damage caused by ROS is closely related to aging and age-related diseases, as occurs with lipid peroxidation.[14,15]

The consequences of lipid and protein oxidation are not, however, limited to the intracellular environment. Atherosclerosis is caused by the oxidation of low-density lipoproteins, which involves lipid peroxidation as well as protein oxidation, on blood vessels. The products of lipid peroxidation induce damage to the surrounding endothelial cells which leads to chronic inflammation, resulting in advancement of the disease.[16] In line with this, the extracellular matrix (ECM) might also suffer from a systemic excess of ROS. This damage might be caused by direct oxidation of ECM components, mainly proteins, or by indirect damage to those enzymes involved in ECM turnover. Both types of damage result in changes in the ECM composition and function, thus leading or contributing to disorders such as rheumatoid arthritis.[17]

DNA damage triggered by an abnormal, high amount of ROS might be divided into direct and indirect lesions. As the name suggests, direct lesions are caused by the reaction of ROS and DNA. This kind of damage might be induced, in turn, in several ways. The so-called single-nucleobase lesions are derived by the formation of adduct radicals upon the interaction among the hydroxyl radical and a nucleobase, *e.g.*, guanine. Moreover, ROS are capable of inducing tandem DNA lesions through the formation of covalent bindings between adjacent nucleobases on the same strand.[18] On the other hand, as mentioned above, the products of lipid peroxidation[13] and protein damage[15] might trigger indirect DNA lesions. ROS-induced DNA damage is strongly correlated with the onset of several diseases, especially cancer, given the acquisition of mutations. Furthermore, this kind of lesion interferes with common DNA processes such as replication, transcription and translation, thus impeding correct cell functioning.[18]

2.1.3 The Antioxidant Defence System

In order to avoid severe damage, living cells undergo ROS scavenging mechanisms that efficiently transform free radicals into non-radical oxidants, the so-called antioxidant defence system (Figure 2.2). An antioxidant has the ability to avoid or delay the potential oxidation of a molecule, thus reducing the previously mentioned risks of macromolecular oxidative damage. This goal might be achieved by preventing ROS formation or by detoxifying the already generated ROS.[11]

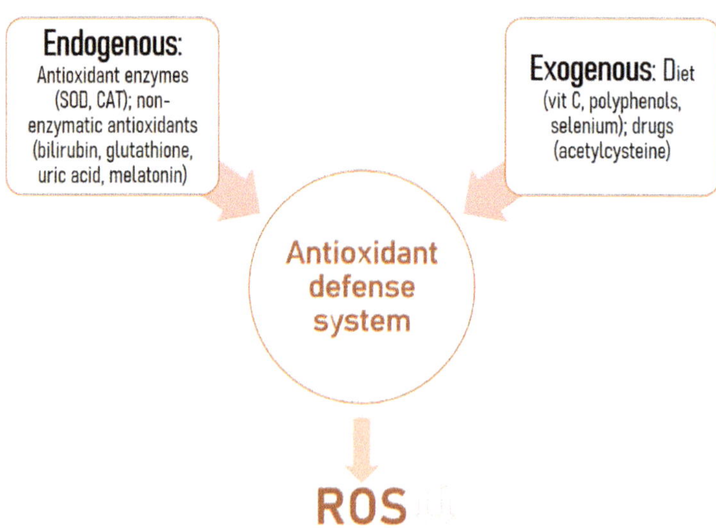

Figure 2.2 Main components of the antioxidant defence system.

2.1.3.1 Endogenous Antioxidant Defence System

2.1.3.1.1 Enzymatic Endogenous Antioxidant Defence System. Endogenous antioxidant defence system might be divided into enzymatic and non-enzymatic. Superoxide dismutase (SOD), catalase (CAT) and glutathione peroxidase (GPx) are the main enzymes involved in redox homeostasis maintenance. Both CAT and GPx are able to convert H_2O_2 into water, despite CAT displaying a higher affinity for this molecule. CAT is distributed in several intracellular locations, although the main concentration of this enzyme is found in peroxisomes. This enzyme contains a heme group to decompose hydrogen peroxide and it is of great importance for the correct function of erythrocytes. Although CAT is well-known for its role as an antioxidant defence enzyme, it is actually a multifunctional enzyme that can also catalyse further reactions that do not involve H_2O_2.[19,20] Certain polymorphisms on the CAT gene that lead to a deficiency or malfunction of this enzyme are strongly related to diseases such as diabetes mellitus, hypertension or anaemia, given the systemic abnormally increased levels of hydrogen peroxide.[21]

On the other hand, the most abundant isoform of the GPx family is GPx-1, a selenocysteine-containing enzyme that can be found in both cytosol and mitochondria. GPx-1 requires glutathione as a cofactor to reduce H_2O_2, which will be oxidized as a consequence of the detoxifying reaction. Therefore, oxidized glutathione must be further recycled by another enzyme, glutathione reductase, which requires NADPH as a cofactor. The glutathione antioxidant system is completed with glucose-6-phosphate dehydrogenase, responsible for converting $NADP^+$ into NADPH. GPx-1 is also able to reduce lipid hydroperoxides and further soluble hydroperoxides into alcohols.[19,20,22] GPx-1 expression is strongly correlated to selenium intake. Therefore, diets low in this mineral might lead to reduced levels of this enzyme, which might be the main cause of certain diseases, including cancer. Low levels of GPx-1 might also be a consequence of certain polymorphisms that might increase the risk of suffering certain types of cancer, diabetes mellitus or cardiovascular diseases among other disorders.[22]

The SOD family is considered the most relevant in terms of enzymatic antioxidant defence, since SOD catalyses the conversion of the highly reactive $O_2^{\bullet-}$ into H_2O_2, which will be further converted into water by CAT and/or GPx. Three isoforms of SOD have been identified. SOD1 and SOD3 require copper and zinc as cofactors. SOD1 is a cytosolic enzyme, whereas SOD3 is extracellular. Dysfunction of SOD1 is related to certain disorders, such as neurodegenerative disorders. On the other hand, SOD2 requires manganese as a cofactor and is located in mitochondria.[19] The lack of function of this isoform might be incompatible with aerobic life, according to experiments on mouse models of SOD2 knockout that resulted in severe neurodegeneration and early death.[23,24]

2.1.3.1.2 Non-enzymatic Endogenous Antioxidant Defence. Among the non-enzymatic endogenous antioxidants is bilirubin, a lipophilic, blood-circulating molecule able to prevent peroxidation of membrane lipids. Bilirubin is produced as a consequence of heme group degradation by heme oxygenase. The first product of this process is biliverdin, which is converted into bilirubin by biliverdin reductase (BLVR). Bilirubin might become biliverdin again by the interaction with lipophilic ROS that interact with membrane lipids, especially H_2O_2.[11,25] Bilirubin levels, as well as the activity of BLVR, are correlated to ROS levels and, as a consequence, to oxidative damage. Low levels of bilirubin might be related to diabetes mellitus, metabolic syndrome and neurodegenerative disorders among others.[25,26] Furthermore, defects in heme oxygenases, which might also result in reduced levels of bilirubin, might be related to atherosclerosis and other cardiovascular diseases.[27,28]

Whereas bilirubin mainly protects lipids from oxidative damage, soluble low molecular weight thiols reduce ROS-induced damage to proteins. Glutathione, previously mentioned as the cofactor of GPx, is the most relevant of these molecules. This tripeptide is able to protect proteins against irreversible oxidation by forming mixed disulphide bonds with cysteine residues. This post-translational modification, named *S*-glutathionylation, is reversible and enhanced in response to an increase in ROS levels. *S*-Glutathionylation might occur spontaneously or be driven by certain enzymes such as glutathione transferase isoform P1-1.[29,30] Low levels of glutathione, a possible consequence of certain polymorphisms of glutathione transferases, have been correlated to both male and female infertility.[31–33]

Uric acid is a water-soluble molecule produced by purine nucleotide metabolism. It is considered the main blood antioxidant due to its capacity to react with ROS. As a consequence of this interaction, uric acid is converted into allantoin. High levels of plasmatic uric acid might be related to a lower risk of cancer, glaucoma or neurodegenerative diseases, despite that an excessive concentration of this molecule is toxic. On the other hand, patients with rheumatism or Parkinson's disease among other disorders show a higher concentration of allantoin as a consequence of ROS-mediated transformation of uric acid.[11,34]

Melatonin, a tryptophan-derived molecule, is mainly produced in mitochondria. This molecule acts as a potent ROS scavenger through a cascade reaction. Through its interaction with ROS, melatonin is converted into further secondary and tertiary metabolites that are also capable of detoxifying ROS, thus amplifying the potential of melatonin as a ROS scavenger. Furthermore, melatonin exhibits an indirect antioxidant potential through the interaction with melatonin receptors on cell and organelle membranes that results in the stimulation of antioxidant enzymes such as GPx. Low levels of melatonin have been found in cardiovascular disease and in patients suffering from Alzheimer's disease.[35,36]

2.1.3.2 Exogenous Antioxidant Defence System

Diet is the main source of exogenous antioxidants. One of the likely reasons behind the benefits of dietary patterns rich in fruits and vegetables may be the high intake of plant metabolites with antioxidant potential.[37–39] Among these compounds are vitamins C and E, carotenoids and polyphenols (Table 2.2). Furthermore, certain studies suggest that enrichment of dietary antioxidants might be related to an improvement of disorders strongly related to ROS-induced damage.[38,40,41]

2.1.3.3 Repair Systems

Complementary to the antioxidant defence systems, cells also have mechanisms to repair ROS-induced damage to macromolecules. Upon identifying damaged macromolecules, repair systems are able to directly repair the lesions or to remove and replace the cellular structure.

Two enzymes are able to reverse certain types of oxidative damage in proteins: disulphide reductase, which reverses both intramolecular and intermolecular disulphide bonds as a consequence of cysteine oxidation, and methionine sulfoxide reductase, which regenerates oxidized methionine residues that might trigger protein malfunction.[42] However, the most common cellular strategy to overcome oxidative damage to proteins consists of selective proteolytic degradation and further *de novo* protein synthesis. Proteolysis is mainly mediated by the ubiquitin/proteasome system and, to a lesser extent, chaperone-mediated autophagy.[42–44] With an excessive amount of ROS production, the proteolytic systems might not be enough to

Table 2.2 Main endogenous antioxidants and their effects.

Endogenous antioxidant	Sources	Properties
Vitamin C (ascorbic acid)	Citrus	- Effective scavenger of $OH^•$ and $O_2^{•-}$ - Prevents lipid peroxidation (in combination with vitamin E)
Vitamin E (α-tocopherol)	Nuts, vegetable oils	- Protects the cell membrane from lipid peroxidation by scavenging lipid peroxyl radicals - Prevents LDL lipid peroxidation
β-Carotene	Carrots, apricots, mangoes, spinach	- Prevents DNA damage caused by UV-induced ROS (photoprotective)
Resveratrol	Grapes, plums, berries	- Potent ROS scavenger - Prevents lipid peroxidation - Downregulation of NADPH oxidase 1 (NOX1), involved in excessive ROS production

remove oxidized proteins, thus leading to protein aggregates that interfere with the regular cell function.[42]

Severe cell membrane damage due to lipid peroxidation, mainly polyunsaturated fatty acids, might result in ferroptosis, an iron-dependent type of cell death. GPx4 plays a great role in ferroptosis prevention due to its capacity to detoxify peroxidised phospholipids using glutathione as a co-factor. This is the only member of the glutathione peroxidase family with this capacity (Liu 2021 FEBS J; Chen 2021 Cell Death Differ). Similarly, peroxiredoxin 6 also plays a key role in membrane protection from lipid peroxidation. This enzyme is able to reduce both hydroxyperoxides and phospholipid hydroperoxides and might be the main responsible factor for reducing lipid peroxidation in organs with low expression of GPx4, *e.g.*, lungs.[45,46] Moreover, the endosomal sorting complex required for transport-III (ESCRT-III), composed of 12 subunits, also contributes to cell membrane repair by shedding damaged sections.[47–49]

Given the great relevance of DNA integrity for cell functioning, living cells have developed different mechanisms to repair oxidative damage towards this macromolecule. The main one is the base excision repair (BER) system, composed of four enzymes. Among other functions, this pathway removes and replaces oxidized bases through the sequential intervention of the following enzymes: DNA glycosylase, AP-endonuclease I, DNA polymerase β and DNA ligase. The BER system is also able to repair oxidative damage to the sugar residues of the DNA strand, which interfere with DNA replication and further processes as well.[50] Furthermore, the nucleotide excision repair (NER) pathway might also display a key role in oxidative damage protection. This complex cell machinery, which involves *circa* 30 proteins, repairs bulky DNA lesions.[51] The protective effects of NER proteins against DNA oxidative damage might be mediated through a close interplay with the BER system.[52,53]

Mitochondrial DNA (mtDNA) shows a higher sensitivity to ROS-induced damage than nuclear DNA. This might be mediated by, at least, three factors. mtDNA is close to the major production site of mitochondrial ROS and lacks the protection for the chromatin structure, and the nucleotide excision repair (NER) system is not found in this organelle. Given that mtDNA contains the genes involved in the expression of the components of the oxidative phosphorylation systems, the maintenance of its integrity is crucial. Furthermore, mutations in mtDNA might lead to uncontrolled ROS production and a variety of disorders, including cancer. The mitochondrial BER system is the main player in reversing ROS-induced damage to mtDNA, which is able to remove and replace oxidized bases among other functions.[54–56]

2.1.4 Oxidative Stress Theory of Disease

The term 'oxidative stress' was used for the first time by Paniker *et al.*[57] in a study published in 1970 that discussed the roles of glutathione reductase, oxygen peroxide levels and erythrocyte correct function. The classical definition of oxidative stress refers to an imbalance of the redox levels, which results in an excessive amount of ROS that trigger the previously discussed macromolecular damage. Furthermore, oxidative stress interferes with correct redox signalling.[58] This situation might be a consequence of (a) abnormally high levels of ROS due to an increased amount of exogenous ROS,[59] (b) defects in the antioxidant defence systems/repair systems or (c) both. Prolonged oxidative stress might contribute to the onset and/or worsening of the prognosis of different diseases, some of which have been already mentioned in previous sections: cancer, cardiovascular diseases, neurodegenerative disorders, *etc.*

Despite the great popularity of the oxidative stress theory of disease, to date few antioxidant drugs have been approved and used in daily clinical practice for the treatment of those diseases, and some clinical trials have provided disappointing results.[58,60] However, this might be the consequence of the misunderstanding of the role of oxidative stress in a disease. We will use cancer to explain this concept.

Chronic inflammation and oxidative stress might trigger carcinogenesis due to lesions in DNA that result in the activation of pro-oncogenes and reduced expression of tumour suppressor genes. However, once malignant transformation has been completed, tumour cells produce higher ROS amounts than non-tumour cells. This might be related to their physiological role in cell proliferation, thus needed to maintain the abnormally high growth rate of cancer cells. Moreover, some authors suggest that elevated ROS might contribute to tumour aggressiveness due to the continuous damage to DNA, which leads to further mutations. To avoid excessive ROS-induced damage and cell death, cancer cells display overexpressed antioxidant defence systems.[61–63] Therefore, in this case, the role of ROS changes as the disease progresses from an initiator to a secondary contributor to disease progression, and the designing of therapies must change in consequence. Therefore, while diets rich in antioxidant compounds might be related to a reduced risk of cancer,[64,65] the use of antioxidants once tumour mass has been already developed is controversial. Furthermore, the use of prooxidant therapies might display greater benefits than the antioxidant ones.[66,67]

In conclusion, despite cancer being an interesting example of the role of oxidative stress in disease, we must be extremely careful regarding the potential role of antioxidants in therapy. Multifactorial diseases such as cancer must be approached through multiple pathways to ensure clinical success. Reducing a disease to the contribution of oxidative stress would be a mistake. In line with this, other disorders in which oxidative stress might have a role would also benefit from this strategy. The use of antioxidants should be focused on prevention and as adjuvants rather than as a sole and definitely therapeutic treatment. Dietary interventions using antioxidant supplementation might result in symptom relief of certain disorders, thus enhancing the pharmacological effect of drug-based therapies.[40] In Section 2.2, the potential benefits of these interventions will be discussed in detail.

2.2 Potential of Dietary Intervention in Disease Management

2.2.1 Impact on Dietary Patterns Towards Health: the Case of the Mediterranean Diet

Lifestyle has a key role in redox balance. As previously mentioned, high levels of ROS from exogenous sources might result in redox imbalance, chronic oxidative stress and disease. Smoking,[68,69] high alcohol intake[70–72] and sedentary lifestyle[73,74] are closely related to high ROS levels. In contrast, a healthier lifestyle might result in reduced risk of disorders and 'better' aging.[75–77] In this chapter, we will focus on the role of diet in oxidative stress markers and related disorders.

The so-called Western diet (WD) (Figure 2.3) is characterized by high amounts of fat, salt and sugar, commonly known as ultra-processed foods that, in addition, contain low levels of fibre. Along with the low overall quality of this kind of food, this dietary pattern is associated with frequent eating.[78] As a consequence, WD is related to low, chronic systemic inflammation, oxidative stress and gut microbiota dysbiosis, thus resulting in a higher risk of suffering certain disorders.[79–82] Moreover, WD is often associated with other undesirable lifestyle factors previously mentioned as sedentarism, which increases the risk of disease onset.[78]

On the other hand, traditional dietary patterns, which are previous to the age of the ultra-processed food era, are characterized by the intake of

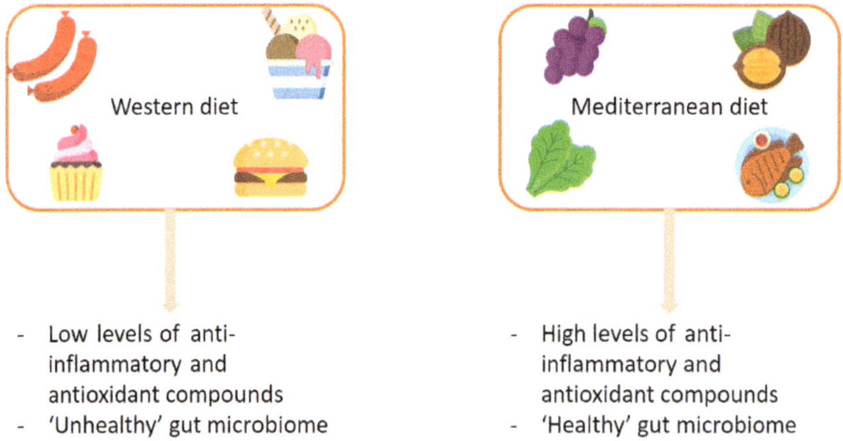

Figure 2.3 Schematic comparison of the main components and effects of two common dietary patterns, the Western and Mediterranean diets. Icons obtained from Flaticon. Cheeseburgers, cupcakes, sausages, ice cream and grapes designed by Freepik. Lettuce designed by Dinosoft-Labs. Walnuts designed by Monkik. Asian food designed by BZZEINCANTATION.

proximity products: seasonal fruits and vegetables, whole grains, legumes, fish, nuts, *etc.* This is related to a high intake of fibre and phytocompounds with antioxidant potential, with the consequent benefits for human health.[83] Mediterranean diet (MD) (Figure 2.3) is one of those traditional dietary patterns that have been widely studied for its potential capacity to reduce disease risk. MD, a term that became popular due to the findings of the Seven Countries Study,[84] is characteristic of Southern European countries such as Spain, Italy, Greece and Northern African countries. Although some variations in the type of food consumed might be found among countries, the main ingredients are mostly maintained and quite similar to the ones mentioned above. Therefore, MD is associated with a high intake of fruits, vegetables, legumes, nuts and whole grains, along with moderate amounts of fish, meat and dairy products. Moreover, extra virgin olive oil is one of the main sources of lipids, which is in particular contrast to the Western diet.[85]

Nowadays, it is widely accepted that MD might be related to a decreased risk of suffering inflammation-related disorders, thus being likely responsible for enhanced longevity in comparison with adherence to WD-like dietary patterns.[84] Regarding the likely health-promoting properties of MD, most studies have focused on the role of this dietary pattern in cardiovascular health. A meta-analysis by Papadaki *et al.*[86] that included eighty-four previous studies concluded that adherence to MD might result in a lower risk of cardiovascular disease. Interestingly, the authors also observed that MD might be related to a reduction in the intake of certain drugs such as β-blockers, which are involved in blood pressure control. According to the conclusions of the authors, the observed effect might be mediated by a reduction in inflammation and oxidative stress marker levels. Similar results were observed in the Seguimiento Universidad de Navarra (SUN) study regarding the likely role of MD in lowering the risk of cardiovascular disease. Moreover, the findings of this study suggested that adherence to MD might also be related to a lower risk of type-2 diabetes.[87]

The likely impact of MD on cognitive function has also been investigated. After analysing the results from forty-five different studies, all of which included non-demented older (>60) adults, Coelho-Júnior *et al.*[88] concluded that adherence to this dietary pattern might be related to a reduced risk of global cognitive function age-associated decline. The results obtained by Coelho-Júnior *et al.*[88] did not support the idea of the preventive role of MD in dementia; however, other authors consider that the high content of anti-inflammatory and antioxidant compounds might help delay or even cancel the onset of neurodegenerative diseases such as Alzheimer's disease. To date, findings from different studies are inconclusive and further analyses are needed.[89]

Adherence to MD might in addition be related to a lower risk of suffering cancer, according to a recent meta-analysis from Morze *et al.*[90] As aforementioned, chronic inflammation and oxidative stress might display a key role in cancer onset. Therefore, it is reasonable to assume that a great intake of anti-inflammatory and antioxidant compounds could prevent cell malignant transformation. Upon analysis of 117 studies, authors concluded that cancer-related mortality was inversely related to MD adherence and to a lower risk of suffering colorectal cancer. These might be mediated not only by the intake of anti-inflammatory and antioxidant compounds but also by the high consumption of probiotics and prebiotics that maintain 'healthy' gut microbiota.[91] Other authors have focused on the likely potential of MD to prevent the onset of breast cancer. However, mixed results have been obtained so far. Whereas the meta-analysis from Papadaki *et al.*[86] found some evidence in favour of this hypothesis, data obtained from Morze *et al.*[90] were inconclusive. In consequence, further studies with novel methodologies are needed to obtain conclusive results to support or discard the potential of MD in preventing breast cancer.

Some authors have proposed that the health-promoting effect of MD might not be limited to preventing or delaying the onset of chronic disorders. Evidence suggests that adherence to MD during pregnancy might have a positive impact on a child's health. Studies that analyse the impact of MD adherence during pregnancy are based on the Developmental Origins of Health and Disease hypothesis, which suggests that adult health status is strongly influenced by foetal exposure to nutrients, mother's stress, *etc.*[92] Given that MD is characterized by a high intake of high added value compounds, according to the previously mentioned hypothesis this dietary pattern might go beyond mother's health. Biagi *et al.*[93] analysed 29 studies regarding this topic and observed that a mother's adherence to MD might be related to a reduced incidence of foetal growth restriction and risk of newborns small for gestational age, as well as a certain level of protection against preterm delivery. Furthermore, by reviewing 22 articles,[94] Amati *et al.* concluded that adherence to MD during pregnancy might result in less depressive or aggressive behaviours during the early years of life, reduced allergic disorders in childhood and a protective effect toward cardiometabolic and congenital defects.

Taken together, all of the above-discussed results point out that a dietary pattern rich in anti-inflammatory and antioxidant compounds has great potential for human health.

2.2.2 Impact of the Diet on Diseases

In Section 2.2.1, we have discussed the potential of a certain dietary pattern, the Mediterranean diet, toward healthy individuals. This point, 'healthy individuals', is of great relevance, given that we have concluded that an adequate diet might result in maintaining this healthy condition while aging. However, how far does the influence of diet extend once the 'healthy condition' no longer exists? In other words, can diet be considered as an adjuvant treatment for certain disorders? According to some studies, it might be.

One of the best examples of how patients can benefit from dietary intervention for the control of their disorders is the case of blood pressure control. It is well known that high daily intakes of salt (>5) lead to an increase in systolic blood pressure, which can be related to further cardiovascular disorders. Therefore, physicians strongly recommend dietary sodium restriction in order to lower blood pressure in hypertensive patients. Filippini *et al.*[95] analysed 85 studies regarding the effect of dietary salt reduction on blood pressure and found that hypertensive patients can actually benefit from this dietary intervention. The authors also observed that healthy, non-hypertensive persons might also lower their blood pressure by limiting the total amount of salt.

However, achieving a reduction in dietary salt can be difficult at certain dietary patterns such as the Western diet, given the high amounts of salt from ultraprocessed foods. Therefore, Cicero *et al.*[96] suggested that, in order to successfully reduce blood pressure by reducing salt intake, the whole diet must be revised and, if needed, changed. According to their systematic review, authors propose diets with not only a low salt content but also with high levels of plant-based foods, such as the Mediterranean diet or vegetarian diet, as promising tools for lowering blood pressure.

Some authors have focused on the likely potential of single ingredient supplementations for blood pressure lowering, which can be termed as 'nutraceuticals' or 'functional foods'. The terminology is confusing and currently there exists no scientifical consensus.[97] Herein we will use the term 'nutraceutical' as a nutritional supplement, *e.g.*, vitamin C, or as a whole food added to the diet with the only purpose of obtaining health-related benefits. On the other hand, we will consider 'functional foods' as foods that have been added with some ingredients to enhance their anti-inflammatory or antioxidant basal properties. More information can be found in Figure 2.4.

Next, some examples of the potential benefits of using nutraceuticals for blood pressure control are discussed. Beetroot juice is one of the most common nutraceuticals for lowering blood pressure. Bahadoran *et al.*[98] evaluated a series of 22 studies regarding the effectivity of this nutraceutical on blood pressure levels and concluded that this kind of supplementation might indeed lower both systolic and diastolic blood pressure. Interestingly, findings from these authors suggest that the benefits of beetroot juice intake might not be related to its high content of inorganic NO_3, in contrast to the main hypothesis. The authors propose that the capacity of this nutraceutical to lower blood pressure might be the consequence of its high content of anti-inflammatory and antioxidant compounds more than inorganic NO_3 levels. According to this proposal, the combined effect of various plant-derived compounds might be more powerful than a single compound, suggesting some kind of synergy among them.

Other authors have focused on nutraceuticals less popular than beetroot, but also of great interest due to their high content of anti-inflammatory and antioxidant compounds. A meta-analysis of eight clinical trials performed by Moradi *et al.*[99] found that artichoke supplementation might be helpful for

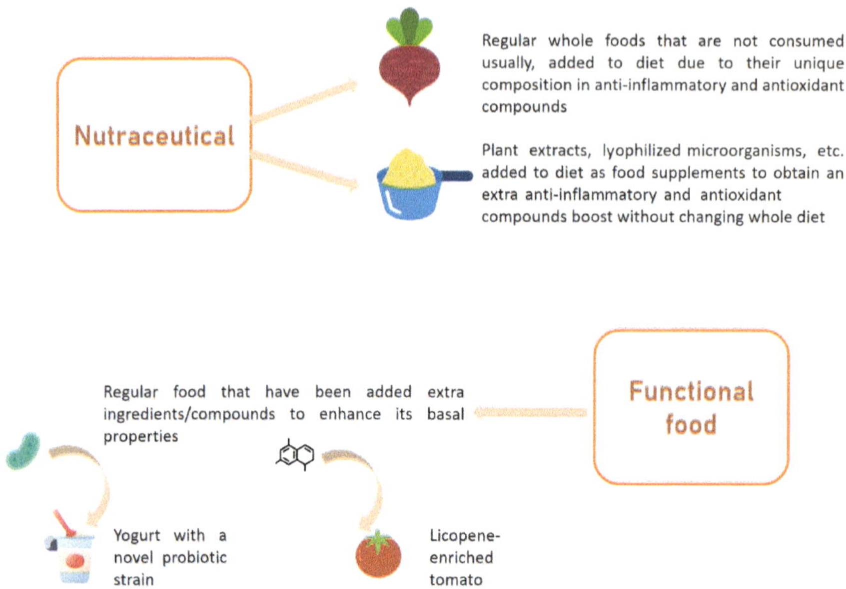

Figure 2.4 Differences among nutraceuticals and functional foods settled for this chapter. Icons obtained from Flaticon. Beetroot designed by Nasyat. Measuring cup and bacteria designed by Smashicons. Yogurt designed by Chanut-is-Industries. Molecule designed by Flowicon. Tomato designed by Icongeek 26.

lowering both systolic and diastolic blood pressure in hypertensive patients. Hawkins *et al.*[100] reviewed 7 studies regarding the potential of chokeberry as a nutraceutical for blood pressure control. The authors observed that daily supplementation with this berry resulted in a reduction in systolic blood pressure and noticed that this effect was stronger in adults older than 50. Hasani *et al.*[101] analysed 6 studies regarding the potential role of ginger in blood pressure control and concluded that it might reduce blood pressure in adults younger than 50.

Schön *et al.*[102] proposed a different and interesting approach toward the use of nutraceuticals for blood pressure control. Instead of supplementing with whole food, the authors evaluated the effect of a standardized grape seed extract, commercially available as Enovita®, on 80 volunteers with mildly elevated blood pressure. 300 mg of the grape extract was orally administered daily for 16 weeks and an effective decrease in blood pressure was noticed, the effect being more significant in men. However, the greatest interest of this work lies in the fact that it might not be necessary to change the overall diet of patients to obtain benefits. Some authors have pointed out the difficulties related to acquiring different, healthier dietary patterns such as the Mediterranean diet for people outside Mediterranean countries.[84] Therefore, the intake of plant extracts might be helpful until a complete diet change is achieved.

2.3 Plant Extracts for the Management of Oxidative Stress-related Disorders: An Opportunity for Agri-food Waste

2.3.1 Influence of the Extraction Method Towards the Biological Activity of Agri-food Waste

Plant extracts (PE) are an interesting source of bioactive compounds. One of the main advantages of this kind of nutraceutical is little amount of PE content with great amounts of anti-inflammatory and antioxidant phytocompounds, given that undesired components might be removed and the bioactive ones are thus enriched. However, the extraction process might require a great amount of the primary source. Therefore, resorting to fruits and/or vegetables suitable for human consumption might be inappropriate. Given the global hunger crisis, using edible parts of plants for pseudo-pharmaceutical purposes is not an ethical option. However, non-edible parts such as bark or peel, which are considered waste in the food industry, display a comparable content of bioactive compounds to edible parts.[103–107] Using agri-food waste as a source for nutraceuticals or bioactive compounds for functional foods would have an additional advantage: this would be a novel way to reduce waste, with the consequent environmental and economic positive repercussions.

A key point in the obtention of plant extracts from agri-food waste is the selected extraction method. Some examples of the potential of the extraction method to influence the final composition of the extract and, consequently, its biological activity, can be found in the scientific literature. Recently, Mármol *et al.*[105] analysed the influence of an extraction solvent on the chemical composition and *in vitro* biological properties of samples from *Pinus pinaster* Aiton subsp. *atlantica* bark. The authors found that the chemical composition was dependent on the extraction solvent and that biological activity changed as a consequence: samples extracted with ethanol or with 50/50 water/ethanol displayed greater antibacterial activity against a panel of Gram-positive bacterial strains, whereas samples extracted with water displayed greater anticancer potential towards a panel of cancer cell models. These results suggest that the extraction solvent for each sample must be carefully selected according to the desired final biological activity of the extract.

Similarly, Bennour *et al.*[108] evaluated the influence of both extraction solvent and extraction method on the composition and properties of extracts from *Moringa oleifera* leaves. In this study, the authors found that extracts obtained by successive maceration and further extraction with butanol displayed the highest total phenolic content and the greatest biological potential in terms of antioxidant capacity, antibacterial potential and cytotoxic effect against a lung cancer cell model. Saptarini *et al.*[109] analysed the antioxidant capacity of the papery skin of onion (*Allium cepa* L. var. *Ascalonicum*) using different extraction methods and observed that it was strongly influenced by them. Taken together, all these studies conclude that agri-food waste is in fact a promising source of bioactive compounds that might be used for nutraceutical purposes. However, the extraction method used in each case must be carefully selected in order to maximize the final properties of the extract.

2.3.2 *In Vivo* Antioxidant Potential of Plant Extracts from Agri-food Waste: State-of-art

To date, most studies using agri-food waste as raw materials for the obtention of bioactive compounds and/or extracts have focused on analysing their antioxidant capacity *in vitro*. Given the promising results obtained, some researchers have evaluated the potential of agri-food-derived plant extracts in *in vitro* preclinical models. In this section, we will discuss some of the most interesting results obtained so far.

Barreto *et al.*[110] evaluated the potential of *Agave sislana* waste for preventing oxidative stress-induced damage on a *Caenorhabditis elegans* model. The main objective of the authors consisted of finding a use for the discarded material from *A. sislana* (*circa* 95% of the gross weight of the leaves), given that a great number of residues are produced in areas like the North region of Brazil, where it is widely cultivated and has a great socioeconomic impact. The extracts showed promising *in vitro* antioxidant capacity that was further confirmed *in vivo*: extracts successfully reduced the intracellular levels of ROS and achieved an increase in lifespan upon exposure to oxidative stress.

Gao *et al.*[111] obtained an extract enriched in polyphenols from tea waste using a superheated steam pre-treatment and a further ultrasound-assisted hot water extraction. The authors evaluated the potential protective effect of this extract toward a zebrafish embryo model upon H_2O_2-induced damage and found that it effectively ameliorated toxicity. Similar to the previous findings from Barreto *et al.*,[110] the main interest of the work from Gao *et al.*[111] lies in the fact that tea consumption, one of the most popular beverages worldwide, results in the production of high amounts of waste. Therefore, the use of these residues would have advantages beyond those related to human health.

Guo *et al.*[112] evaluated the antioxidant potential of green pea (*Pisum sativum* L.) hulls in a rat model. Hulls are produced in great amounts during pea processing for human consumption and currently are considered by-products with low or even no added value. However, findings from these authors showed that green pea hulls displayed great antioxidant capacity both *in vitro* and *in vivo* and should be considered a promising source of bioactive compounds. The authors analysed the bioavailability of the bioactive compounds from green pea hull extracts and concluded that they can be used for antioxidant purposes *in vivo*. In a further study using extracts from the hulls of yellow peas, similar conclusions were obtained,[113] suggesting a promising future for peas hulls.

da Silva-Maia *et al.*[114] analysed the effect of an aqueous extract of the peel from a Brazilian berry, jaboticaba (*Plinia jaboticaba*), on a healthy rat model. The authors found that extract intake resulted in the normal function of the enzymatic antioxidant system and, interestingly, that was able to modulate the composition of gut microbiota: an increase in the *Lactobacillus* genus was observed. Therefore, the authors concluded that this extract is well tolerated and might provide certain benefits for the gut microbiota status, with a consequent positive impact on the overall health.

Mejri *et al.*[115] used non-edible parts of artichoke (*Cynara scolymus* L.) discarded from the food processing industry as a raw material for an added-value plant extract. The likely health-promoting properties of such extract were evaluated in a diabetic mouse model. The authors found that administration of the artichoke by-product extract resulted in amelioration of oxidative stress and overall amelioration of diabetes-related symptoms.

Taken together, all the results discussed in this section suggest that agri-food waste might be considered a promising source of extract with high antioxidant potential that can be used for health-promoting purposes *in vivo*.

2.4 Conclusions

Redox balance has a great impact on human health, with oxidative stress being one of the main causes behind the onset and or progression of serious disorders such as cancer, neurodegenerative diseases or metabolic syndrome. Therefore, a diet rich in plant-based foods, *e.g.*, the Mediterranean diet, might prevent those disorders due to its high content of anti-inflammatory and antioxidant compounds. Moreover, an adequate dietary pattern might also influence disease progression once it has appeared, as in the case of hypertension. When the diet itself is insufficient to control disease, the bioactive compounds from plants can be used as adjuvant treatments to improve the redox status and ameliorate the symptoms. Agri-food waste arises then as a promising source of bioactive compounds, given that it can be obtained in great amounts and is a cheap raw material. Moreover, processing those residues would have a positive environmental impact. According to previous research using *in vivo* models, extracts obtained from agri-food waste display great antioxidant potential. Further research is now needed to escalate these findings to human volunteers.

References

1. H. Sies, Oxidative eustress: On constant alert for redox homeostasis, *Redox Biol.*, 2021, **41**, 101867.
2. K. Sugiura, H. Tanaka, G. Kurisu, K. I. Wakabayashi and T. Hisabori, Multicolor redox sensor proteins can visualize redox changes in various compartments of the living cell, *Biochim. Biophys. Acta, Gen. Subj.*, 2019, **1863**(6), 1098–1107.

3. T. Konno, E. P. Melo, J. E. Chambers and E. Avezov, Intracellular sources of ROS/H_2O_2 in health and neurodegeneration: Spotlight on endoplasmic reticulum, *Cells*, 2021, **10**(2), 1–26.
4. J. Roy, J. M. Galano, T. Durand, J. Y. Le Guennec and J. C. Lee, Physiological role of reactive oxygen species as promoters of natural defenses, *FASEB J.*, 2017, **31**(9), 3729–3745.
5. S. Di Meo, T. T. Reed, P. Venditti and V. M. Victor, Role of ROS and RNS Sources in Physiological and Pathological Conditions, *Oxid. Med. Cell. Longevity*, 2016, **2016**, 1245049.
6. K. Brieger, S. Schiavone, F. J. Miller and K. H. Krause, Reactive oxygen species: From health to disease, *Swiss Med. Wkly.*, 2012, **142**, w13659.
7. M. Canton, R. Sánchez-Rodríguez, I. Spera, F. C. Venegas, M. Favia and A. Viola, *et al.*, Reactive Oxygen Species in Macrophages: Sources and Targets, *Front. Immunol.*, 2021, **12**, 734229.
8. S. Cuevas, V. A. M. Villar and P. A. Jose, Genetic polymorphisms associated with reactive oxygen species and blood pressure regulation, *Pharmacogenomics J.*, 2019, **19**, 315–336.
9. T. Fukai and M. Ushio-Fukai, Cross-Talk between NADPH Oxidase and Mitochondria: Role in ROS Signaling and Angiogenesis, *Cells*, 2020, **9**, 1849.
10. D. P. Jones, Radical-free biology of oxidative stress, *Am. J. Physiol.: Cell Physiol.*, 2008, **295**, C849-68.
11. A. M. Pisoschi and A. Pop, The role of antioxidants in the chemistry of oxidative stress: A review, *Eur. J. Med. Chem.*, 2015, **97**, 55–74.
12. A. Altomare, G. Baron, E. Gianazza, C. Banfi, M. Carini and G. Aldini, Lipid peroxidation derived reactive carbonyl species in free and conjugated forms as an index of lipid peroxidation: limits and perspectives, *Redox Biol.*, 2021, **42**, 101899.
13. A. Negre-Salvayre, N. Auge, V. Ayala, H. Basaga, J. Boada and R. Brenke, *et al.*, Pathological aspects of lipid peroxidation, *Free Radical Res.*, 2010, **44**, 1125–1171.
14. J. U. Dahl, M. J. Gray and U. Jakob, Protein quality control under oxidative stress conditions, *J. Mol. Biol.*, 2015, **427**, 1549–1563.
15. A. Krisko and M. Radman, Protein damage, ageing and age-related diseases, *Open Biol.*, 2019, **9**, 180249.
16. A. Hartley, D. Haskard and R. Khamis, Oxidized LDL and anti-oxidized LDL antibodies in atherosclerosis – Novel insights and future directions in diagnosis and therapy, *Trends Cardiovasc. Med.*, 2019, **29**(1), 22–26.
17. E. C. Kennett, C. Y. Chuang, G. Degendorfer, J. M. Whitelock and M. J. Davies, Mechanisms and consequences of oxidative damage to extracellular matrix, *Biochem. Soc. Trans.*, 2011, **39**, 1279–1287.
18. Y. Yu, Y. Cui, L. J. Niedernhofer and Y. Wang, Occurrence, Biological Consequences, and Human Health Relevance of Oxidative Stress-Induced DNA Damage, *Chem. Res. Toxicol.*, 2016, **29**, 2008–2039.
19. G. Bresciani, I. B. M. da Cruz and J. González-Gallego, Manganese superoxide dismutase and oxidative stress modulation, in *Advances in Clinical Chemistry*, Academic Press Inc., 2015, pp. 87–130.

20. L. Gebicka and J. Krych-Madej, The role of catalases in the prevention/promotion of oxidative stress, *J. Inorg. Biochem.*, 2019, **197**, 110699.
21. A. Nandi, L.-J. Yan, C. Jana Kumar and N. Das, Review Article Role of Catalase in Oxidative Stress-and Age-Associated Degenerative Diseases, *Oxid. Med. Cell. Longevity*, 2019, **2019**, 9613090.
22. E. Lubos, J. Loscalzo and D. E. Handy, Glutathione peroxidase-1 in health and disease: From molecular mechanisms to therapeutic opportunities, *Antioxid. Redox Signaling*, 2011, **15**, 1957–1997.
23. D. Hinerfeld, M. D. Traini, R. P. Weinberger, B. Cochran, S. R. Doctrow and J. Harry, *et al.*, Endogenous mitochondrial oxidative stress: Neurodegeneration, proteomic analysis, specific respiratory chain defects, and efficacious antioxidant therapy in superoxide dismutase 2 null mice, *J. Neurochem.*, 2004, **88**(3), 657–667.
24. M. Fukui and B. T. Zhu, Mitochondrial superoxide dismutase SOD2, but not cytosolic SOD1, plays a critical role in protection against glutamate-induced oxidative stress and cell death in HT22 neuronal cells, *Free Radical Biol. Med.*, 2010, **48**(6), 821–830.
25. S. Y. Kim and S. C. Park, Physiological antioxidative network of the bilirubin system in aging and age-related diseases, *Front. Pharmacol.*, 2012, **3**, 45.
26. S. Jayanti, L. Vítek, C. Tiribelli and S. Gazzin, The role of bilirubin and the other "yellow players" in neurodegenerative diseases, *Antioxidants*, 2020, **9**, 1–26.
27. Y. Kishimoto, K. Kondo and Y. Momiyama, The protective role of heme oxygenase-1 in atherosclerotic diseases, *Int. J. Mol. Sci.*, 2019, **20**, 3628.
28. A. Ayer, A. Zarjou, A. Agarwal and R. Stocker, Heme Oxygenases in Cardiovascular Health and Disease, *Physiol. Rev.*, 2016, **96**(4), 1449–1508.
29. K. Ulrich and U. Jakob, The role of thiols in antioxidant systems, *Free Radicals Biol. Med.*, 2019, **140**, 14–27.
30. E. Kalinina and M. Novichkova, Glutathione in protein redox modulation through s-glutathionylation and s-nitrosylation, *Molecules*, 2021, **26**(2), 435.
31. O. Adeoye, J. Olawumi, A. Opeyemi and O. Christiania, Review on the role of glutathione on oxidative stress and infertility, *JBRA Assist. Reprod.*, 2018, **22**(1), 61–66.
32. X. Li, J. Pan, Q. Liu, E. Xiong, Z. Chen and Z. Zhou, *et al.*, Glutathione S-transferases gene polymorphisms and risk of male idiopathic infertility: a systematic review and meta-analysis, *Mol. Biol. Rep.*, 2013, **40**(3), 2431–2438.
33. C. Y. Hu, D. L. Lu, T. Wu, S. L. Cheng, T. T. Wu and S. Wang, *et al.*, Glutathione-S-transferases M1/T1 gene polymorphisms and male infertility risk in Chinese populations: A meta-analysis, *Medicine*, 2019, **98**(6), e14166.
34. Y. Song, L. Tang, J. Han, Y. Gao, B. Tang and M. Shao, *et al.*, Uric Acid Provides Protective Role in Red Blood Cells by Antioxidant Defense: A Hypothetical Analysis, *Oxid. Med. Cell. Longevity*, 2019, **2019**, 3435174.

35. D. X. Tan, L. C. Manchester, E. Esteban-Zubero, Z. Zhou and R. J. Reiter, Melatonin as a potent and inducible endogenous antioxidant: Synthesis and metabolism, *Molecules*, 2015, **20**, 18886–18906.
36. R. J. Reiter, D. X. Tan, S. Rosales-Corral, A. Galano, X. J. Zhou and B. Xu, Mitochondria: Central organelles for melatonins antioxidant and anti-Aging actions, *Molecules*, 2018, **23**, 509.
37. S. Rajaram, J. Jones and G. J. Lee, Plant-based dietary patterns, plant foods, and age-related cognitive decline, *Adv. Nutr.*, 2019, **10**(4), 422–436.
38. E. Scoditti, M. Massaro, S. Garbarino and D. M. Toraldo, Role of diet in chronic obstructive pulmonary disease prevention and treatment, *Nutrients*, 2019, **11**, 1357.
39. K. V. Gantenbein and C. Kanaka-Gantenbein, Mediterranean diet as an antioxidant: The impact on metabolic health and overall wellbeing, *Nutrients*, 2021, **13**, 1951.
40. J. Quero, I. Mármol, E. Cerrada and M. J. Rodríguez-Yoldi, Insight into the potential application of polyphenol-rich dietary intervention in degenerative disease management, *Food Funct.*, 2020, **11**, 2805–2825.
41. J. Bagyi, V. Sripada, A. M. Aidone, H. Y. Lin, E. H. Ruder and D. R. Crawford, Dietary rational targeting of redox-regulated genes, *Free Radical Biol. Med.*, 2021, **173**, 19–28.
42. K. J. A. Davies and E. Percy, Critical Review Oxidative Stress, Antioxidant Defenses, and Damage Removal, Repair, and Replacement Systems, *IUBMB Life*, 2000, **50**, 279–289.
43. C. T. Aiken, R. M. Kaake, X. Wang and L. Huang, Oxidative stress-mediated regulation of proteasome complexes, *Mol. Cell. Proteomics*, 2011, **10**, R110.006924.
44. F. Shang and A. Taylor, Ubiquitin-proteasome pathway and cellular responses to oxidative stress, *Free Radicals Biol. Med.*, 2011, **51**, 5–16.
45. A. B. Fisher, Peroxiredoxin 6 in the repair of peroxidized cell membranes and cell signaling, *Arch. Biochem. Biophys.*, 2017, **617**, 68–83.
46. A. B. Fisher, J. P. Vasquez-Medina, C. Dodia, E. M. Sorokina, J. Q. Tao and S. I. Feinstein, Peroxiredoxin 6 phospholipid hydroperoxidase activity in the repair of peroxidized cell membranes, *Redox Biol.*, 2018, **14**, 41–46.
47. J. Liu, R. Kang and D. Tang, Signaling pathways and defense mechanisms of ferroptosis, *FEBS J.*, 2021, 7038–7050.
48. X. Chen, C. Yu, R. Kang, G. Kroemer and D. Tang, Cellular degradation systems in ferroptosis, *Cell Death Differ.*, 2021, **28**, 1135–1148.
49. E. Dai, L. Meng, R. Kang, X. Wang and D. Tang, ESCRT-III-dependent membrane repair blocks ferroptosis, *Biochem. Biophys. Res. Commun.*, 2020, **522**(2), 415–421.
50. S. Mitra, I. Boldogh, T. Izumi and T. K. Hazra, Complexities of the DNA Base Excision Repair Pathway for Repair of Oxidative DNA Damage, *Environ. Mol. Mutagen.*, 2001, **38**(2–3), 180–190
51. J. P. M. Melis, H. Van Steeg and M. Luijten, Oxidative DNA damage and nucleotide excision repair, *Antioxid. Redox Signaling*, 2013, **18**, 2409–2419.

52. N. Kumar, S. Raja and B. Van Houten, The involvement of nucleotide excision repair proteins in the removal of oxidative DNA damage, *Nucleic Acids Res.*, 2020, **48**(20), 11227–11243.
53. A. Sassa and M. Odagiri, Understanding the sequence and structural context effects in oxidative DNA damage repair, *DNA Repair*, 2020, **93**, 102906.
54. V. I. Grishko, L. I. Rachek, D. R. Spitz, G. L. Wilson and S. P. LeDoux, Contribution of mitochondrial DNA repair to cell resistance from oxidative stress, *J. Biol. Chem.*, 2005, **280**(10), 8901–8905.
55. Z. Huang, Y. Chen and Y. Zhang, Mitochondrial reactive oxygen species cause major oxidative mitochondrial DNA damages and repair pathways, *J. Biosci.*, 2020, **45**, 84.
56. P. Sharma and H. Sampath, Mitochondrial DNA integrity: Role in health and disease, *Cells*, 2019, **8**, 100.
57. N. V. Paniker, S. K. Srivastava and E. Beutler, Glutathione metabolism of the red cells. Effect of glutathione reductase deficiency on the stimulation of hexose monophosphate shunt under oxidative stress, *Biochim. Biophys. Acta*, 1970, **215**(3), 456–460.
58. H. J. Forman and H. Zhang, Targeting oxidative stress in disease: promise and limitations of antioxidant therapy, *Nat. Rev. Drug Discovery*, 2021, **20**, 689–709.
59. T. Miyazawa, G. C. Burdeos, M. Itaya, K. Nakagawa and T. Miyazawa, Vitamin E: Regulatory Redox Interactions, *IUBMB Life*, 2019, **71**, 430–441.
60. C. Pietro Ghezzi, P. Ghezzi, V. Jaquet, F. Marcucci and H. H. H. W. Schmidt, Themed Section: Redox Biology and Oxidative Stress in Health and Disease, *Br. J. Pharmacol.*, 2017, **174**, 1784. Available from http://onlinelibrary.wiley.com/doi/10.1111/bph.v174.12/issuetoc.
61. S. Reuter, S. C. Gupta, M. M. Chaturvedi and B. B. Aggarwal, Oxidative stress, inflammation, and cancer: How are they linked?, *Free Radicals Biol. Med.*, 2010, **49**, 1603–1616.
62. A. Leone, M. S. Roca, C. Ciardiello, S. Costantini and A. Budillon, Oxidative Stress Gene Expression Profile Correlates with Cancer Patient Poor Prognosis: Identification of Crucial Pathways Might Select Novel Therapeutic Approaches, *Oxid. Med. Cell. Longevity*, 2017, **2017**, 2597581.
63. J. N. Moloney and T. G. Cotter, ROS signalling in the biology of cancer, *Semin. Cell Dev. Biol.*, 2020, **80**, 50–64. Available from https://www.sciencedirect.com/science/article/pii/S1084952116303834?via%3Dihub.
64. M. C. Mentella, F. Scaldaferri, C. Ricci, A. Gasbarrini and G. A. D. Miggiano, Cancer and mediterranean diet: A review, *Nutrients*, 2019, **11**, 2059.
65. L. Soldati, L. Di Renzo, E. Jirillo, P. A. Ascierto, F. M. Marincola and A. De Lorenzo, The influence of diet on anti-cancer immune responsiveness, *J. Transl. Med.*, 2018, **16**, 75.

66. A. Tasdogan, J. M. Ubellacker and S. J. Morrison, Redox regulation in cancer cells during metastasis, *Cancer Discovery*, 2021, **11**, 2682–2692.
67. S. Galadari, A. Rahman, S. Pallichankandy and F. Thayyullathil, Reactive oxygen species and cancer paradox: To promote or to suppress?, *Free Radical Biol. Med.*, 2017, **104**, 144–164.
68. A. W. Caliri, S. Tommasi and A. Besaratinia, Relationships among smoking, oxidative stress, inflammation, macromolecular damage, and cancer, *Mutat. Res., Rev. Mutat. Res.*, 2021, **787**, 108365.
69. M. Badran and I. Laher, Waterpipe (shisha, hookah) smoking, oxidative stress and hidden disease potential, *Redox Biol.*, 2020, **34**, 101455.
70. K. Sommerfeld-Klatta, M. Łukasik-Głębocka and B. Zielińska-Psuja, Oxidative stress and biochemical indicators in blood of patients addicted to alcohol treated for acute ethylene glycol poisoning, *Hum. Exp. Toxicol.*, 2022, **41**, DOI: 10.1177/09603271211061502.
71. M. C. Huang, C. H. Chen, F. C. Peng, S. H. Tang and C. C. Chen, Alterations in Oxidative Stress Status During Early Alcohol Withdrawal in Alcoholic Patients, *J. Formosan Med. Assoc.*, 2009, **108**(7), 560–569.
72. J. A. Hernández, R. C. López-Sánchez and A. Rendón-Ramírez, Lipids and Oxidative Stress Associated with Ethanol-Induced Neurological Damage, *Oxid. Med. Cell. Longevity*, 2016, **2016**, 1543809.
73. F. J. Llorente-Cantarero, F. J. Aguilar-Gómez, R. Leis, G. Bueno, A. I. Rupérez and A. Anguita-Ruiz, *et al.*, Relationship between physical activity, oxidative stress, and total plasma antioxidant capacity in Spanish children from the genobox study, *Antioxidants*, 2021, **10**(2), 1–14.
74. M. A. Sánchez-Rodríguez, M. Zacarías-Flores, E. Correa-Muñoz, A. Arronte-Rosales and V. M. Mendoza-Núñez, Oxidative Stress Risk Is Increased with a Sedentary Lifestyle during Aging in Mexican Women, *Oxid. Med. Cell. Longevity*, 2021, **2021**, 9971765.
75. S. Dato, P. Crocco, P. D'Aquila, F. de Rango, D. Bellizzi and G. Rose, *et al.*, Exploring the role of genetic variability and lifestyle in oxidative stress response for healthy aging and longevity, *Int. J. Mol. Sci.*, 2013, **14**, 16443–16472.
76. A. W. C. Man, H. Li and N. Xia, Impact of Lifestyles (Diet and Exercise) on Vascular Health: Oxidative Stress and Endothelial Function, *Oxid. Med. Cell. Longevity*, 2020, **2020**, 1496462.
77. G. Grande, C. Qiu and L. Fratiglioni, Prevention of dementia in an ageing world: Evidence and biological rationale, *Ageing Res. Rev.*, 2020, **64**, 101045.
78. I. J. Malesza, M. Malesza, J. Walkowiak, N. Mussin, D. Walkowiak and R. Aringazina, *et al.*, High-fat, western-style diet, systemic inflammation, and gut microbiota: A narrative review, *Cells*, 2021, **10**, 3164.
79. A. Christ, M. Lauterbach and E. Latz, Western Diet and the Immune System: An Inflammatory Connection, *Immunity*, 2019, **51**, 794–811.
80. B. M. González Olmo, M. J. Butler and R. M. Barrientos, Evolution of the human diet and its impact on gut microbiota, immune responses, and brain health, *Nutrients*, 2021, **13**, 1–16.

81. A. Więckowska-Gacek, A. Mietelska-Porowska, M. Wydrych and U. Wojda, Western diet as a trigger of Alzheimer's disease: From metabolic syndrome and systemic inflammation to neuroinflammation and neurodegeneration, *Ageing Res. Rev.*, 2021, **70**, 101397.
82. F. Vernia, S. Longo, G. Stefanelli, A. Viscido and G. Latella, Dietary Factors Modulating Colorectal Carcinogenesis, *Nutrients*, 2021, **13**(1), 143.
83. H. Cena and P. C. Calder, Defining a healthy diet: Evidence for the role of contemporary dietary patterns in health and disease, *Nutrients*, 2020, **12**, 334.
84. L. J. Dominguez, G. Di Bella, N. Veronese and M. Barbagallo, Impact of mediterranean diet on chronic non-communicable diseases and longevity, *Nutrients*, 2021, **13**(6), 2028.
85. C. Davis, J. Bryan, J. Hodgson and K. Murphy, Definition of the mediterranean diet: A literature review, *Nutrients*, 2015, **7**, 9139–9153.
86. A. Papadaki, E. Nolen-Doerr and C. S. Mantzoros, The effect of the mediterranean diet on metabolic health: A systematic review and meta-analysis of controlled trials in adults, *Nutrients*, 2020, **12**, 1–21.
87. J. Godos and F. Galvano, Insights on mediterranean diet from the sun cohort: Cardiovascular and cognitive health, *Nutrients*, 2020, **12**, 1332.
88. H. J. Coelho-Júnior, A. Trichopoulou and F. Panza, Cross-sectional and longitudinal associations between adherence to Mediterranean diet with physical performance and cognitive function in older adults: A systematic review and meta-analysis, *Ageing Res. Rev.*, 2021, **70**, 101395.
89. M. Siervo, O. M. Shannon, D. J. Llewellyn, B. C. M. Stephan and L. Fontana, Mediterranean diet and cognitive function: From methodology to mechanisms of action, *Free Radical Biol. Med.*, 2021, **176**, 105–117. Available from https://www.sciencedirect.com/science/article/pii/S0891584921007383.
90. J. Morze, A. Danielewicz, K. Przybyłowicz, H. Zeng, G. Hoffmann and L. Schwingshackl, An updated systematic review and meta-analysis on adherence to mediterranean diet and risk of cancer, *Eur. J. Nutr.*, 2021, **60**(3), 1561–1586.
91. G. Merra, A. Noce, G. Marrone, M. Cintoni, M. G. Tarsitano and A. Capacci, *et al.*, Influence of mediterranean diet on human gut microbiota, *Nutrients*, 2021, **13**, 1–12.
92. H. Danielewicz, G. Myszczyszyn, A. Dębińska, A. Myszkal, A. Boznański and L. Hirnle, Diet in pregnancy—more than food, *Eur. J. Pediatr.*, 2017, **176**, 1573–1579.
93. C. Biagi, M. Di Nunzio, A. Bordoni, D. Gori and M. Lanari, Effect of adherence to mediterranean diet during pregnancy on children's health: A systematic review, *Nutrients*, 2019, **11**, 997.
94. F. Amati, S. Hassounah and A. Swaka, The impact of mediterranean dietary patterns during pregnancy on maternal and offspring health, *Nutrients*, 2019, **11**, 1098.

95. T. Filippini, M. Malavolti, P. K. Whelton, A. Naska, N. Orsini and M. Vinceti, Blood Pressure Effects of Sodium Reduction, *Circulation*, 2021, **143**(16), 1542–1567.
96. A. F. G. Cicero, M. Veronesi and F. Fogacci, Dietary Intervention to Improve Blood Pressure Control: Beyond Salt Restriction, *High Blood Pressure Cardiovasc. Prev.*, 2021, **28**, 547–553.
97. J. K. Aronson, Defining 'nutraceuticals': neither nutritious nor pharmaceutical, *Br. J. Clin. Pharmacol.*, 2017, **83**, 8–19.
98. Z. Bahadoran, P. Mirmiran, A. Kabir, F. Azizi and A. Ghasemi, The nitrate-independent blood pressure-lowering effect of beetroot juice: A systematic review and meta-analysis, *Adv. Nutr.*, 2017, **8**, 830–838.
99. M. Moradi, G. Sohrabi, M. Golbidi, S. Yarmohammadi, N. Hemati and M. S. Campbell, *et al.*, Effects of artichoke on blood pressure: A systematic review and meta-analysis, *Complementary Therapies Med.*, 2021, **57**, 102668.
100. J. Hawkins, C. Hires, C. Baker, L. Keenan and M. Bush, Daily supplementation with aronia melanocarpa (chokeberry) reduces blood pressure and cholesterol: a meta analysis of controlled clinical trials, *J. Diet. Suppl.*, 2021, **18**(5), 517–530.
101. H. Hasani, A. Arab, A. Hadi, M. Pourmasoumi, A. Ghavami and M. Miraghajani, Does ginger supplementation lower blood pressure? A systematic review and meta-analysis of clinical trials, *Phytotherapy Res.*, 2019, **33**, 1639–1647.
102. C. Schön, P. Allegrini, K. Engelhart-Jentzsch, A. Riva and G. Petrangolini, Grape seed extract positively modulates blood pressure and perceived stress: A randomized, double-blind, placebo-controlled study in healthy volunteers, *Nutrients*, 2021, **13**(2), 1–18.
103. Z. Amri, F. Zaouay, H. Lazreg-Aref, H. Soltana, A. Mneri and M. Mars, *et al.*, Phytochemical content, Fatty acids composition and antioxidant potential of different pomegranate parts: Comparison between edible and non edible varieties grown in Tunisia, *Int. J. Biol. Macromol.*, 2017, **104**, 274–280. Available from https://www.sciencedirect.com/science/article/pii/S0141813017305706.
104. N. Drabińska, M. Jeż and M. Nogueira, Variation in the accumulation of phytochemicals and their bioactive properties among the aerial parts of cauliflower, *Antioxidants*, 2021, **10**(10), 1597.
105. I. Mármol, C. Vieito, V. Andreu, A. Levert, A. Amiot and C. Bertrand, *et al.*, Influence of extraction solvent on the biological properties of maritime pine bark (Pinus pinaster), *Int. J. Food Stud.*, 2022, **11**(1), 51–62.
106. G. Mannino, G. Serio, C. M. Bertea, R. Chiarelli, A. Lauria and C. Gentile, Phytochemical profile and antioxidant properties of the edible and non-edible portions of black sapote (Diospyros digyna Jacq.), *Food Chem.*, 2022, **380**, 132137. Available from https://www.sciencedirect.com/science/article/pii/S030881462200098X.

107. J. Chang, M. Wang, Y. Jian, F. Zhang, J. Zhu and Q. Wang, *et al.*, Health-promoting phytochemicals and antioxidant capacity in different organs from six varieties of Chinese kale, *Sci. Rep.*, 2019, **9**(1), 20344.
108. N. Bennour, H. Mighri, T. Bouhamda, M. Mabrouk, E. Apohan and O. Yesilada, *et al.*, Moringa oleifera leaves: could solvent and extraction method affect phenolic composition and bioactivities?, *Prep. Biochem. Biotechnol.*, 2021, **51**(10), 1018–1025.
109. N. M. Saptarini and Y. Wardati, Effect of Extraction Methods on Antioxidant Activity of Papery Skin Extracts and Fractions of Maja Cipanas Onion (Allium cepa L. var. ascalonicum), *Sci. World J.*, 2020, **2020**, 3280534.
110. S. Maria Andrade Gomes Barreto, C. Orlando Muñoz Cadavid, R. Amir De Oliveira Moura, G. Melo Martins Silva, S. Vitória Ferreira De Araújo and J. Antônio Aderaldo Da Silva Filho, *et al.*, In vitro and in vivo antioxidant activity of agave sisalana agro-industrial residue, *Biomolecules*, 2020, **10**(10), 1–17.
111. T. Gao, Y. Shi, Y. Xue, F. Yan, D. Huang and Y. Wu, *et al.*, Polyphenol extract from superheated steam processed tea waste attenuates the oxidative damage in vivo and in vitro, *J. Food Biochem.*, 2020, **44**(1), e13096.
112. F. Guo, H. Xiong, X. Wang, L. Jiang, N. Yu and Z. Hu, *et al.*, Phenolics of Green Pea (Pisum sativum L.) Hulls, Their Plasma and Urinary Metabolites, Bioavailability, and in Vivo Antioxidant Activities in a Rat Model, *J. Agric. Food Chem.*, 2019, 11955–11968.
113. F. Guo, R. Tsao, X. Wang, L. Jiang, Y. Sun and H. Xiong, Phenolics of Yellow Pea (Pisum sativum L.) Hulls, Their Plasma and Urinary Metabolites, Organ Distribution, and in Vivo Antioxidant Activities, *J. Agric. Food Chem.*, 2021, **69**(17), 5013–5025.
114. J. K. da Silva-Maia, A. G. Batista, L. C. Correa, G. C. Lima, S. Bogusz Junior and M. R. Maróstica Junior, Aqueous extract of berry (Plinia jaboticaba) byproduct modulates gut microbiota and maintains the balance on antioxidant defense system in rats, *J. Food Biochem.*, 2019, **43**(2), e12705.
115. F. Mejri, T. Baati, A. Martins, S. Selmi, M. Luisa Serralheiro and P. L. Falé, *et al.*, Phytochemical analysis and in vitro and in vivo evaluation of biological activities of artichoke (Cynara scolymus L.) floral stems: Towards the valorization of food by-products, *Food Chem.*, 2020, **333**, 127506.

CHAPTER 3

Sustainable Environmental Remediation by Valorization of Agro-food Industrial Waste and By-products

SONIA MORYA, (0000-0002-8137-2759)*[a]
DEEPIKA SANDHU, (0000-0003-4476-4523)[a]
ARNO NEUMANN (0000-0002-8816-0636)[b] AND
JAMES SIMON CHACHA (0000-0002-6147-0458)[c]

[a] Department of Food Technology and Nutrition, School of Agriculture, Lovely Professional University, Phagwara-144411, Punjab, India, Email: deepika.sandhu12@gmail.com; [b] BET Bioscience Extraction Technologies Inc., P.O. Box 2212, Abbotsford, B.C., V2T 3X8, Canada, Email: arnoneumanncanada@gmail.com; [c] Department of Food Science and Agroprocessing, School of Engineering and Technology, Sokoine University of Agriculture, P.O. Box 3006, Morogoro, Tanzania, Email: chachajs26@gmail.com
*Email: sonia.morya8911@gmail.com

3.1 Introduction

Massive amounts of biodegradable solid or liquid waste, which are organic remnants of processed raw materials, are generated by the agro-food sectors. According to Baiano (2014), the drinks industry generates approximately 26% of food wastes, followed by the dairy industry (21%), cereal processing and manufacturing (12.9%), fruit and vegetable production and processing

Green Chemistry Series No. 78
Agri-food Waste Valorisation
Edited by Pankaj Chowdhary and Abhay Raj

Published by the Royal Society of Chemistry, www.rsc.org

(14.8%), meat product processing and preservation (8%), manufacturing and processing of animal and vegetable oils (3.9%), fish product processing and preservation (0.4%), and others.[1,2] One-third of all food produced for human use (1.3 billion tonnes) is lost or wasted each year. Fruits and vegetables, as well as tubers and roots, have the highest food waste rates of any food, accounting for a total of 40–50% of global quantitative food losses and waste per year (520–650 million tonnes). Food waste in the EU totals 89 million tonnes per year (39% of which occurs during industrial processes), while total agricultural residue production in the EU totals 367 million tonnes per year (crop residues or parts of crop plants that are not consumed as food).[3] Annually, billions of tonnes of agricultural leftovers are generated in the agricultural and various food processing chains, making exploitation of any given agro-waste an important need. The most common destination for these materials, disposal, requires the implementation of valorization recycling options that help to reduce the negative impact of disposal. Chronic undernutrition, which occurs in various developing regions around the world, has focused the international community's emphasis on increasing food supply. Paradoxically, this increase necessitates the disposal of massive volumes of solid leftovers, which is still another major issue in developing countries.[4,5] One of the valorization goals for food processing byproducts, waste, and effluents is the recovery of fine and beneficial chemicals and the generation of valuable metabolites using chemical and biotechnological processes.[6,7] Due to the loss of natural resources as a result of increased technological development, it is necessary to take decisive and coordinated action to treat waste as a valuable resource that may be reused or recycled.[8,9] Agricultural co-products/by-products and waste other than being considered as an important energy source are also indeed important factors for restoring the level of organic matter in the soil.[10,11] The agro-food industry and agricultural production generate large quantities of residues, solid wastes, and various by-products, which are generated in the primary agro-forestry production (crop-based) or by secondary processing industries. The majority of organic matter becomes a source of environmental problems. However, if residues are used to improve food production, they are no longer considered waste but rather new resources. Residues such as corn cobs, cereals straw, cotton stalks, reed stems and various grasses, maize and sorghum stover, sugarcane, vine pruning and tequila bagasse, banana and coconut residues, coffee pulp, corn husks and coffee husk, sunflower and cotton seed hulls, rice husks, peanut shells, sunflower seed hulls, wood sawdust, waste paper, and chips are some examples of residues and by-products that can be valorized and upgraded to higher value useful products.[12,13] Globally present food agro-industrial by-product disposal and management practices have had a significant impact on the environment, as well as the social and economic sectors. In this situation, greenhouse gas (GHG) emissions pollute the environment because organic materials are frequently dumped in landfills or burned to avoid the growth of parasites, bacteria, and pests.[14] Food industry agro-industrial products generate a considerable volume of fruit and vegetable trash which affects municipal landfills. Despite being highly biodegradable, these generate leachate and are also responsible for methane emission.[15–17] Waste from vegetable and fruit manufacturing often includes significant levels of proteins, carbohydrates, and lipids as well as specific aromatic and aliphatic molecules, and they could be inexpensive and abundant sources of bioactives and biomaterials. A major component of agro-industrial waste is lignocellulose. This type of waste contains cellulose, hemicelluloses, and lignin and is collectively known as "lignocellulosic materials." The most abundant component is cellulose followed by hemicellulose and lignin. The composition of cellulose, hemicellulose, and lignin in agro-industrial waste depends upon the species, tissue type, and structure and also the maturity of the plant.[18,19] As the agriculture industry becomes more industrialised, the resulting waste generation poses a serious environmental concern. Agriculture generates 5 million metric tonnes of biomass each year. Most wastes created by agro-based food enterprises are abundant in nutrients and if left unprocessed and untreated can serve as breeding grounds for disease-causing bacteria. Surprisingly, these wastes can be used as raw materials to make value-added products or also as a source of renewable energy. Many wastes generated by agro-industries are lignocellulosic in nature which means they are generally high in polysaccharides such as cellulose and hemicellulose and lignin (an aromatic polymer) in addition to containing other nutrients such as proteins, pectin, lipids, and polyphenols.[3,20]

3.2 Types of Agro-food Industrial Waste/Bioproduct

Agro-industrial wastes are wastes formed during the industrial processing of agricultural or animal products, or wastes obtained from agricultural activities in the form of stem, straw, stalk, husk, leaves, shell, lint, seed, peel, pulp, grains or legumes, bagasse from sweet sorghum milling or sugarcane, brewer's spent grains, wasted coffee grounds, and many others (Figure 3.1). These wastes are mostly made up of fibres, carbohydrates, proteins, and minerals. The main elements of such agro-industrial wastes are cellulose, hemicelluloses, and lignin, which are collectively referred to as "lignocellulosic materials."

3.3 Chemical Composition of Agro-industrial Waste

Lignocellulosic biomass leftovers are gaining popularity as a waste product from agricultural wastes. The principal components of lignocellulosic biomass include cellulose, hemicellulose, and lignin with minor components including pectin, protein, wax, and extractives. The proportion of cellulose, hemicellulose, and lignin varies depending on the type of biomass or waste (Figure 3.2).

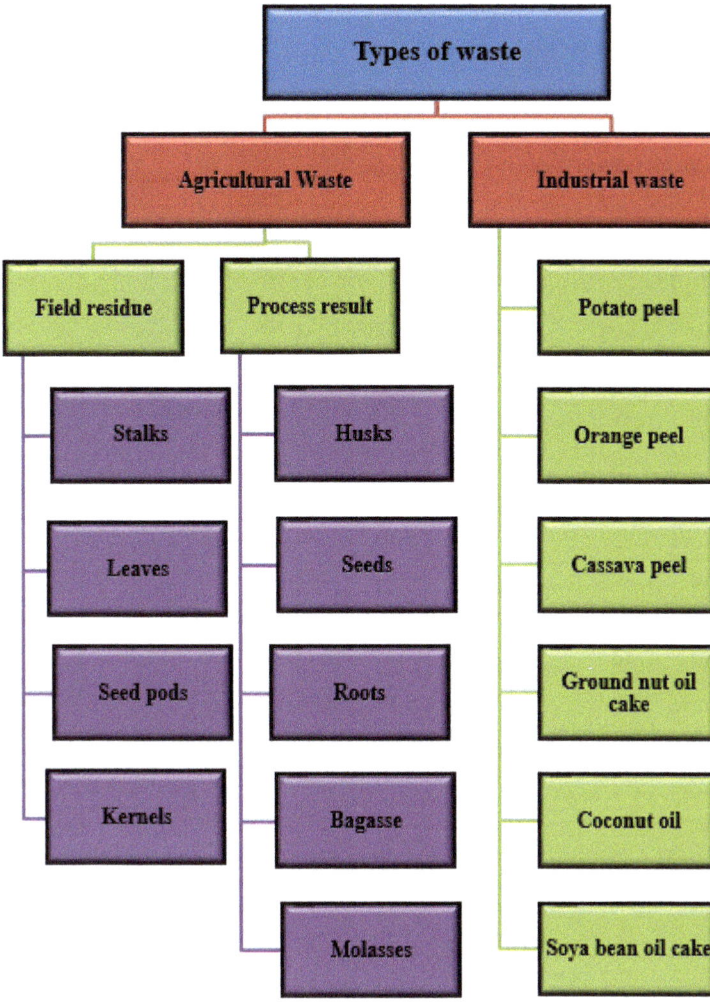

Figure 3.1 Types of agro-industrial waste.

3.3.1 Cellulose

Plants produce around 180 billion tonnes of cellulose annually and constitute the world's greatest reserve of organic carbon. Cellulose is the most plentiful and renewable polymer feedstock accessible today. Cellulose is formed from D-glucose units that condense *via* $\beta(1 \rightarrow 4)$-glycosidic linkages.[23] Cellulose is tasteless, odourless, hydrophilic with a contact angle of 20°–30°, insoluble in water and most of the organic solvents, chiral, and biodegradable. It can be chemically broken down into its glucose units by treating it with concentrated acids at higher temperatures. Cellulose can be extracted from a variety of raw materials, including pea hulls, soybean hulls, maize bran, oat hull, and dried beet pulp. The use of agro-wastes and

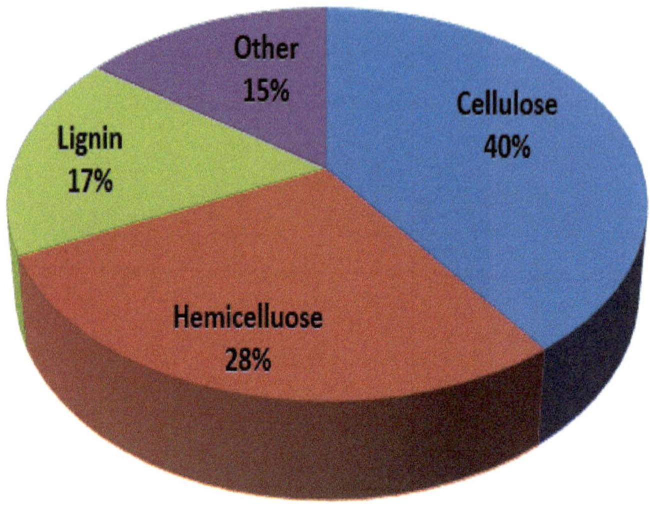

Figure 3.2 Chemical composition of agro-industrial wastes.[21,22]

byproducts as raw materials has practical implications for developing the material components as an alternative to traditional manufacturing supplies, food, and pharmaceutical uses that are also environmentally benign.[24] The sources of cellulose can be divided into the primary, secondary, tertiary, and quaternary levels. The conventional or primary sources of cellulose are fibres used for textile or paper, wood, and industrial crops for building and bio-ethanol. The secondary source of cellulose is the nonprocessed waste directly obtained from the food industry or as a residue from agricultural/forestry activities such as bark, straw, leaves, and husks, whereas the tertiary source comprises the processed waste from the use, transformation, and conversion of cellulosic biomass, for example, pulp, bagasse, and food residues.[25]

3.3.2 Hemicellulose

Hemicelluloses, the second most common polysaccharides in nature, represent about 20–35% of lignocellulosic biomass. Xylans are the most abundant hemicelluloses. Hemicelluloses are heterogeneous polymers of pentoses (xylose, arabinose), hexoses (mannose, glucose, galactose), and sugar acids. Unlike cellulose, hemicelluloses are not chemically homogeneous. The hemicellulose proportion varies with the type of feedstock. It can be categorized into four major groups, namely, (1) hardwood such as poplar (26.2–28.7%) and oak (35.9%); (2) softwood such as pine (24.0–27.0%) and spruce (22.9%); (3) agricultural waste such as barley hull (36.0%), corn cob (31.9–36.0%), and sugarcane bagasse (28.0–32.0%); and (4) grasses such as switch grass (25.0–50.0%). Its practical applications in various agro-industrial processes include efficient conversion of hemicellulosic biomass into fuels and chemicals, delignification of paper pulp, digestibility enhancement of animal feedstock, clarification of juices, and improvement in the consistency of beer.[26,27]

3.3.3 Lignin

Lignin is the most abundant natural phenolic polymer in the world. In - nature, lignin polymer usually forms ether or ester linkages with hemicellulose which is also associated with cellulose. Therefore, these nature polymers construct a complicated and valuable lignocellulose polymer.[28,29] Lignin has a complicated cross-linking structure and contains several functional groups within the molecule, including aliphatic hydroxyl, phenolic hydroxyl, and methoxyl groups. These functional groups affect the reactivity and chemical properties of lignin, especially the hydroxyl groups and aromatic structure are the most critical functional groups to determine the characteristics of the polymers. The three major precursors of the lignin polymer are p-coumaryl alcohol, coniferyl alcohol, and sinapyl alcohol.[30] Lignin can be classified into native lignin and technical lignin. Native lignin is referred to the original lignin structure in the lignocellulose without any modification. As a natural polymer, native lignin does not exist solely in nature. It always exists as part of the lignocellulose. Technical lignin from the industrial by-product can be directly used as raw materials for the production of other chemicals. Aliphatic and aromatic hydroxyl groups are major constituents and active sites in technical lignin; thus, it can also be directly used as polyols for producing polyurethane and replacing 30% of petroleum-based polyols during polyurethane production.[31] Potential applications of lignin include energy production and its use in dispersants, biodegradable polymers, protective UV-absorbents, nanoparticles, and phenolic resins. It can be further used in drug delivery and heavy metal absorption in the environment.[32]

The different compositions of cellulose, hemicellulose, lignin, moisture, ash, carbon, nitrogen, *etc.* and their constituents have the potential to be biochemically digested to produce useful products such as biogas, bioethanol, and other commercially useful examples. Approximately, 20% of the production of fruits and vegetables in India is wasted every year. Most of these wastes are left unutilized or untreated, which causes adverse effects on the environment as well as human and animal health but the composition of these wastes includes a large number of organic compounds that produced a variety of value-added products and also reduced the cost of production.

3.4 Applications

Agricultural residues are rich in bioactive compounds. These residues can be used as an alternate source for the production of different products such as biogas, biofuel, mushroom, and tempeh as raw materials in various research studies and industries. The use of agro-industrial wastes as raw materials

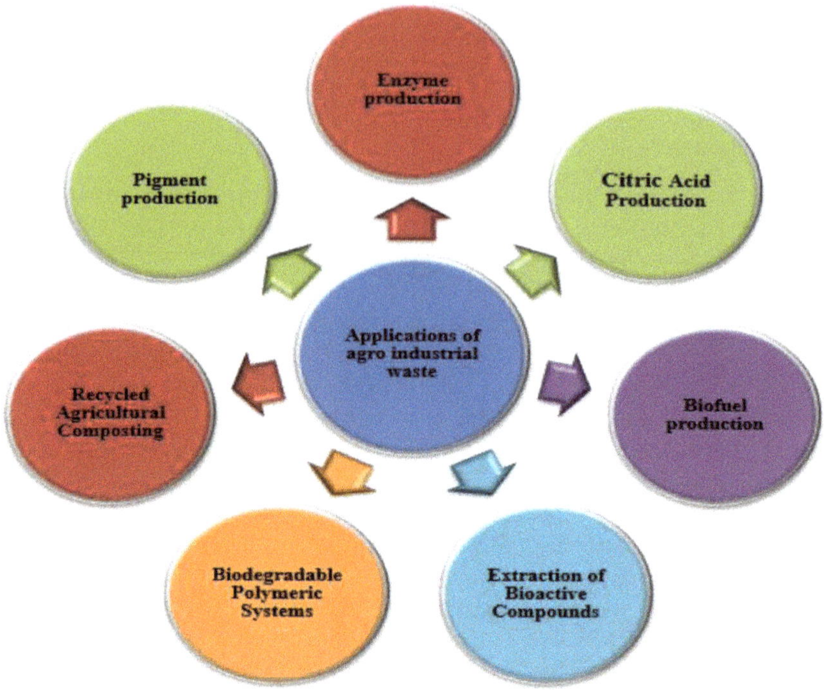

Figure 3.3 Applications of agro-industrial waste.

can help reduce the production cost and also reduce the pollution load from the environment.[33–35] Agro-industrial wastes are used for manufacturing enzymes, biofuels, vitamins, animal feed, antioxidants, antibiotics, and other chemicals through various methods of production (Figure 3.3).

3.4.1 Enzyme Production

Enzymes are biological catalysts that have plenty of applications in several industries that range from baking and brewing to paper, pulp, and the detergent industry. Due to their high degree of substrate specificity and stringent yet robust operational parameters, they are preferred over chemical catalysis in many situations.[3] Agro-industrial wastes consist of variable compositions that support the growth of microorganisms as a result of fermentation produced different valuable enzymes. The growth rate of fungi is enhanced by the use of these substrates which resulted in the conversion of lignocellulosic substrates into less complicated ones by degrading the action of several enzymes. One of the important enzymes, *i.e.*, amylase, was used in starch processing industries for the degradation of polysaccharides into sugar components.[26,36] Recent advances in the valorization of agro-industrial wastes by the production of lignocellulolytic enzymes under solid-state fermentation (SSF) are reviewed. SSF is a promising technology for the production of lignocellulolytic enzymes; however, the large-scale feasibility is the main challenge of SSF being the control of operational parameters and adequate reactor design of the first locks.[37] The microbial production of enzymes can be carried out both by submerged (SMF) and solid-state fermentation (SSF); the latter being the preferred method of production in the industrial sector. However, the trend has begun to shift towards SSF owing to the utilization of different agro-industrial wastes as low-cost carbon and nitrogen sources, thereby reducing the cost of enzyme production. Different organisms are responsible for the production of enzymes that may vary in properties such as hydrolyzing, oxidizing, or reducing enzymes.[38] Agro-industrial wastes such as wheat straw, sugarcane bagasse, rice bran, wheat bran, corn cob, *etc.* are the cheapest and have plentifully available natural carbon sources which have been successfully utilized recently for the production of industrially important enzymes.[39,40]

3.4.2 Citric Acid Production

Citric acid (CA) is an organic acid found in a variety of fruits such as limes, lemons, oranges, pineapples, and grapefruits. It is a natural ingredient that aids in detoxification, maintaining the energy levels and supporting healthy digestion and kidney function. It has a slightly tart and refreshing flavor and is employed for balancing the sweetness in soft drinks, juices, and other beverages. Citric acid is used in the food and beverage (F&B) industry due to its antioxidant properties to preserve the food or as an acidifier to enhance the flavors and aromas of fruit juices, ice creams, and marmalades.[41,42] It is accepted worldwide as a GRAS (generally recognized as safe), as approved by the Joint FAO/WHO Expert Committee on Food Additives. CA and its salts (primarily sodium and potassium) are used in many industrial applications, as a chelating agent, buffer, pH adjustment, and derivatization agent. Applications include laundry detergents, shampoos, cosmetics, enhanced oil recovery, and chemical cleaning.[43] It is produced mainly by submerged fermentation using *Aspergillus niger* or *Candida* sp. from different sources of carbohydrates, such as molasses and starch-based media. In view of surges in demand and growing markets, there is always a need for the discovery and development of better production techniques and solutions to improve the production yields and the efficiency of product recovery. To support the enormous scale of production, it is necessary and important for the production process to be environmentally friendly by utilizing readily available and inexpensive agro-industrial waste products, while maintaining high production yields.[44]

3.4.3 Biofuel Production

The global scientific community has a leading interest to search for new technologies for the production of energy from lignocellulosic biomass, while attempting to mitigate environmental concerns. Biofuels, including bioethanol, biohydrogen, biogas, biodiesel, bio-oil, vegetable oils, biosynthetic gas, bio-char, and Fischer–Tropsch liquids are all produced from raw materials. Several alternatives for bioethanol production from agro-industrial wastes (AIW) are discussed, such as separate hydrolysis and fermentation (SHF), separate hydrolysis and co-fermentation (SHCF), simultaneous saccharification and fermentation (SSF), simultaneous saccharification and co-fermentation (SSCF), pre-saccharification followed by simultaneous saccharification and fermentation (PSSF), and consolidated bioprocessing (CBP).[45,46] There is a high current utilization in the form of domestic fuel. These are chickpea stalks, cotton stalks, groundnut stalks, maize stover, mustard stalks, paddy husk, and pigeon pea stalks. For these residues combined, an estimated total of 47 Mton per year is used as a domestic fuel. Considering that improved cooking stoves can lead to a 35% savings in fuel (Household Energy Network, 2010), approximately 17 Mton per year of these residues could be saved from use as a domestic fuel just by implementing such stoves and could thus be freed up for other uses, including conversion to more efficient fuel types such synthesis gas produced by gasification (*e.g.*, the gasification efficiency of rice husk is around 65%).[47] Bioethanol and related liquid-fuel-based production are the only commercial operational processes that utilize agro-industry waste as raw materials. This is achieved by following fermentation strategies such as hydrolysis of waste followed by fermentation or simultaneous saccharification and fermentation.[3]

3.4.4 Extraction of Food Flavoring and Preservative Components

Agro-industrial wastes are a low-cost and sustainable resource for the production of renewable value-added products. Many natural flavoring agents can be produced from wastes *via* microbial conversions. Vanillin is one of the most important flavoring compounds in the food industry. Several by-products from the cereal industry, *i.e.*, maize, rice, wheat, and sugar beet pulp, have been examined as a source of ferulic acid that has been extracted at a high yield from wastes. The production of vanillin from agro-food industry by-products represents an opportunity to produce this flavor in a new, economically, and environmentally sustainable way, which also allows for the valorization of waste matrixes. Ascorbic acid or vitamin C, a natural compound obtained from several plant tissues, is the best example of the potential use in the food industry.[39,40] Different antimicrobial packaging systems including lemon extracts have been used to preserve mozzarella cheese. Results showed an increase in the shelf life of all active packaged mozzarella cheeses, confirming that lemon extract may exert an inhibitory effect on the microorganisms responsible for spoilage phenomena without affecting the functional microbiota of the product. Pomegranate peel extract (PE) showed excellent antioxidant activity while the seed extract did not have any significant activity, probably to the difference in the type and number of bioactive compounds present in both tissues. Pomegranate peel extract showed good antimicrobial activity against *Staphylococcus aureus* and *Bacillus cereus*. In general, the addition of pomegranate peel extract to popular chicken and meat products enhanced their shelf life by 2–3 weeks.[48]

3.4.5 Extraction of Bioactive Compounds

Bioactive compounds are the phytochemicals that are present naturally in lesser extents in plants as well as foodstuffs and have the potential to amend metabolic processes for the promotion of better health.[33,49] Bioactive compounds are demanded for the market, and they could be included in AIW, for example, pectin, dietary fibre, polyunsaturated lipids, essential oils, flavonoids, and peptides, among others. Also structuring agents, mainly referring to colloidal polymers with interesting gelling or viscosant properties, can be selectively extracted from AIW. Some recent studies relevant to bioactive extraction (*e.g.*, carotenoids, essential oils, polyphenols, and anthocyanins) from AIW using novel technologies include the use of ultrasounds, supercritical carbon dioxide, microwaves, and pulsed electric fields. For these reasons, the extraction of specific compounds from AIW could be an affordable, sustainable, and profitable strategy for industries.[50,51] The environmentally friendly method was considered a promising, technical, and economically viable solution in the recovery of bioactive compounds such as cocoa lipids, caffeine, and theobromine, suggesting the application in pharmaceutical products, adding value to the commonly discarded waste. The technology of supercritical and subcritical fluid for the hydrolysis of biomass and extraction of bioactive compounds from waste is already a reality.[52,53]

3.4.6 Production of Biodegradable Polymeric Systems

Biodegradable polymers are a unique class of polymers that are ecologically benign (biodegradable and biocompatible). The production process for these biopolymers is grouped into four different classifications depending on the desired products and the available materials/precursors. Carbon-rich precursors are used in the production of bio-based polymers through microbial, biopolymer blending, and chemical methods. Bio-based polymers have been synthesized from plant-based precursors containing lignocellulose fibres, cellulose esters, polylactic acid, and polyhydroxy alkanoates (PHA). The Food and Agriculture Organization (FAO) estimates that 20–30% of fruits and vegetables are discarded as waste during post-harvest handling. The development of bio-based polymers is essential, considering the scale of global environmental pollution that is directly linked to the production of synthetic plastics such as polypropylene (PP) and polyethylene (PET).[40] Globally, 400 million tons of synthetic plastics are produced each year, and less than 9% are recycled. The optical, mechanical, and chemical properties such as ultraviolet (UV) absorbance, tensile strength, and water permeability are influenced by the synthetic route. The production of bio-based polymers from renewable sources and microbial synthesis are scalable and facile and pose a minimal impact on the environment compared to chemical synthesis methods that rely on alkali and acid treatment or co-polymer blending. Sustainability is a primary criterion that influences the choice of the precursor (the type of agro-wastes). The production of biopolymers requires commercially viable quantities of agro-waste, a key challenge considering that the wastes occur at the retail and household levels, and there is no mechanism for sorting and disposal of the wastes. Additionally, there is a global variability in the availability of agro-wastes, a factor that influences the mechanical and optical properties of the polymers developed.[54,55]

3.4.7 Recycled Agricultural Composting

Composting is a self-heating aerobic biodegradative process of organic matter that is carried out under controlled conditions *via* microbial activities to produce valuable, stable, humus-rich products useful for the cultivation of crops and the soil. In addition, composting is an alternative to the landfill and incineration approaches for waste disposal.[56] Composting is a managed process that utilizes natural microorganisms available in organic matter and soil to decompose organic wastes. The addition of microorganisms could enhance the composting process such as by accelerating the organic matter degradation, mineralization, and microbial enzyme activities and the quality of the end-products such as a high germination index. These microorganisms require sufficient basic nutrients, oxygen, and water for decomposition to occur at an accelerated pace. The raw materials going into the compost are often referred to as feedstock. The end product, compost, is a dark brown, humus-like material which can be easily and safely stored, handled, and applied to land as a valuable soil conditioner. Various inoculation strategies, such as inoculation of a single or mixed microbial culture, and the addition of inoculants at different stages, can be used to improve the agricultural waste composting process. These strategies have a substantial influence on the composting productivity such as organic matter degradation, humification temperature, enzyme activity, microbial population, and the quality of its end products such as the germination index.[57,58]

3.4.8 Pigment Production

Microorganisms such as bacteria, mold, yeast, and fungi produce different types of pigments depending on their source. Microorganisms belonging to genera *Monascus, Rhodotorula, Aspergillus,* and *Penicillium* have been reported as potential producers of biopigments. A wide range of bacterial strains such as *Serratia marcescens*, *Pseudomonas magneslorubra*, *Vibrio psychroerythrus*, *S. rubidaea*, *V. gazogenes*, *Alteromonas rubra*, *Rugamonas rubra*, *Streptoverticillium rubrireticuli*, and *Streptomyces longisporus* have shown their potential in pigment production. Yeast cultures are considered a good source of microbial pigments due to their unicellular nature and high growth rate by utilizing low-cost media as compared to bacteria and fungi.[59,60] Fruit byproducts have become an important source of those pigments and colors, mainly because they present high color stability and purity. Further criteria for new viable sources of natural pigments are good availability, a low price, and a high-yielding material. All those characteristics can easily be presented in tropical exotic fruit byproducts. The great activity in this field is supported by data from a survey, which forecasts an expansion of the European color market by 1% per year till 2008, whereas coloring foodstuff is expected to grow even by 10–15% in the same period.[48]

3.5 Techniques Used for Waste Valorization

3.5.1 Extraction

In order to address the drawbacks of conventional extraction techniques, novel extraction strategies such as ultrasound-assisted extraction (UAE), pressurised liquid extraction (PLE), supercritical fluid extraction (SFE), accelerated solvent extraction (ASE), microwave-assisted solvent extraction (MASE), and solid-phase microextraction (SPME) are considered promising.[61] The increased yield and shorter extraction durations seen during high-pressure extraction of colorants from food waste have been attributed to changes in the diffusivity coefficient generated by denaturation of the colorant binding protein as a result of high pressure. The level of positive pressure, solid-to-liquid ratio, duration, and/or temperature all play crucial roles in the extraction process because solubility also rises as pressure rises. Because changes in the diffusivity coefficient promote the transfer of chemicals from food waste, using high pressure shortens processing times and improves the extraction. High pressure increases the plant cell membrane's diffusivity coefficient, which in turn increases the membrane's permeability and, consequently, the extraction solvent.[62]

It has been suggested that a combination process of cavitation, turbulence, and collision accounts for an increased yield, shorter extraction time, and reduced energy consumption during negative pressure extraction of natural components. Colorants, polyphenols, polysaccharides, flavonoids, and alkaloids can all be extracted from plants, fruits, and vegetables using negative pressure extraction. It releases intracellular components into the extraction solvent through cavitation, which is caused by negative pressure creating bubbles that increase turbulence, collision, and mass transfer between the solid matrix and solvent.[63] The greater yield during the SFE of colorants has reportedly been attributed to supercritical conditions increasing the fluid's diffusivity. The total extraction process is significantly influenced by the pressure and temperature. With respect to the physicochemical properties of supercritical fluids, associated with temperature, pressure, and the addition of modifier(s), distinct fractions of rich and highly pure bioactive substances, such as pigments, can be successively extracted under consecutive conditions. The SFE technology was used to recover carotenoids from a variety of fruit and vegetable wastes that were high in carotenoids.[64]

In the MEF extraction process, the voltage placed across the food material ruptures the cell and damages the cell membrane's permeability. By electroporating tissues, MEF increases the extraction yield by allowing cells to be permeable to foreign substances. Therefore, throughout the MEF extraction process, factors such as temperature, electrical conductivity of the solid and solvent, level of voltage gradient or electric field strength, frequency, the form of the current wave (sine, triangular, and square), and frequency all play a vital role. MEF extraction shows promising results as compared to the traditional (steam distillation) approach considering the color, yield, and concentration.[65] Ultrasonication-assisted extraction (UAE) entails moving a substance with the aid of sound waves from any matrix to a liquid phase. Ultrasonication extraction is widely used in the food industry to recover ingredients from the trash. By breaking up the food's particles and enhancing foaming, ultrasound can improve food's ability to emulsify. Due to the effective separation of the pigments by UAE, this potent method is utilized to extract colorants from food waste. Carotenoids were recovered from pomegranate peels using US in the presence of various oils. Regarding the extraction duration, temperature, amplitude level, and solid/oil ratio, the carotenoid yield was assessed.[66]

3.5.2 Fermentation

Streptococcus thermophilus, *Leuconostoc* spp., *Lactococcus lactis*, and *Lactobacillus* spp. are the LAB cultures that are frequently employed in regulated food preparation. The three bacteria *Oenococcus, Pediococcus,* and *Weissella* are crucial to the fermentation of plant-based products.[67] Leading techniques are employed to enhance these industrial bacteria's qualities, but natural resources are equally important for discovering new strains with biotechnological uses. Therefore, the same methods can be applied to choosing the best LAB candidates to ferment plant material leftovers.[68,69] The generation of lactic acid, which might be reintegrated into the food chain, and improving the protein digestibility and sensory qualities of these vegetable by-products, which may be employed as food ingredients, are all examples of value-adding methods that utilize LAB.[67–69]

To ferment products made from rice, soy, oak, fruit, and sorghum, the genus *Bacillus* with species *B. coagulans*, *B. licheniformis*, *B. amyloliquefaciens,* and *B. subtilis* has been utilized alone or in combination with other bacteria and/or fungi. Following enzymatic hydrolysis, rice bran is a plentiful by-product stream created during the processing of rice that can be fermented by *B. coagulans* in order to produce extremely pure (>99%) L-lactic acid.[66] Surprisingly, the *B. coagulans* A107 strain was able to produce 90% of the lactic acid needed to break down the hydrolyzate's carbohydrates into lactic acid without the use of any extra nutrients. On the other hand, alternative valorization techniques can also be used by *Bacillus* strains to ferment soybean meal, soybean dregs, and okara.[70] A mixture of *B. coagulans* and *Lactobacillus johnsonii* can ferment soybean meal, improving its antioxidant capabilities following fermentation. To develop soil biostimulants with enhanced β-glucosidase, phosphatase, and dehydrogenase activity, *B. licheniformis* was also able to ferment enzymatically hydrolyzed okara, a waste product of the production of soy milk. These three examples demonstrate a range of valorization processes mediated by the *Bacillus* species that ought to be further investigated in order to increase the number of uses for soy-derived products.[71]

Fungal fermentation has also been used to improve the nutritional qualities including protein and fibre contents. Examples of fungal enzymes which have significant technological applications include: (1) xylanases (produced through the fermentation of brewer's spent grains), (2) phytases (produced through the fermentation of triticale residues), (3) alpha-amylases (produced through the fermentation of wheat bran, soy bran and other starchy residues), (4) proteases (produced through the fermentation of soybean and wheat bran), and (5) lipases (produced through the fermentation).[72]

3.6 Application of Techniques for Waste Valorization

3.6.1 Bio-methane Production

Anaerobic digestion has been identified as a widely-accepted method in which biodegradable organic food wastes and wastewater sludges are degraded anaerobically under regulated temperature and pH conditions and in the presence of bacterial consortia for the creation of biogas. A renewable energy source is the bio-gas, which is mostly composed of carbon dioxide, methane, and minute amounts of other pollutants such as nitrogen, hydrogen sulphide, oxygen, methyl mercaptans, and hydrogen.[73,74] The energy value of the biofuel is increased by the removal of contaminants using procedures such as chemical absorption, membrane separation, pressure swing absorption, *etc.* A nutrient-rich digestate that may be utilized as a soil fertiliser is also generated along with the bio-gas. The technology of anaerobic digestion is widely accepted, mostly because it is inexpensive to operate, produces little leftover waste, and may be used to generate renewable energy.[75,76] By reducing the amount or bulk of garbage, it also aids in the management of solid waste. Large complex polymers found naturally in waste matrices, such as proteins and carbohydrates, are first hydrolyzed, followed by fermentation to produce CO_2 and hydrogen acetate and final conversion to methane. The processes of hydrolysis, acetogenesis, acidogenesis, and methanogenesis are all carried out by certain classes of bacteria. Anaerobic digestion may reduce waste volumes and also contribute to energy recovery in a cost-effective way, as evidenced by the high degradability of food wastes and the increased output of bio-gas when run under regulated circumstances. The biggest drawback of this approach is its lengthier treatment durations by microbial breakdown, which range from 20 to 40 days. Food waste's chemical makeup prevents it from degrading as quickly as it should, which causes the digestion process to frequently have a long solid residence time and poor conversion efficiency. Numerous chemical, physical, enzymatic, and thermal pre-treatments have been recorded to enhance the solubility of the food substrate and speed the decomposition of solid organic waste since hydrolysis has been shown to be the rate-limiting step for anaerobic degradation.[3,40]

3.6.2 Recovery of Value-added Compounds for Food Additives

Sugars, proteins, pectin, lipids, fibre, polysaccharides, flavorings, and phytochemicals abound in the waste products produced during the production of diverse plant-based meals. If these items with added value are extracted, they can be used as cosmetics, nutraceuticals, therapeutics, and food additives, among other things. If these items with added value are extracted, they can be used as cosmetics, nutraceuticals, therapeutics, and food additives, among other things.[77,78] The carotenoids and polyphenols that are extracted from vegetable and fruit wastes are employed as natural food preservatives because they boost the antioxidant capacity of the finished product's shelf life. Through successive solvent extraction, essential oils, sugar, flavonoids, and pectin were extracted from orange peel wastes, and the resulting product was sold as a natural sweetener for food. High-value components have also been extracted from other wastewaters, such as the effluent from an olive mill where hydroxytyrosol, a substance needed for food preservation as well as a functional element in bread, was recovered. De-pectinated apple pomace was used to make soluble dietary fibre granules that were sold as nutritional supplements.[3] From the pomace of apples and grapes, a wide range of compounds, including ethanol, tartrates, citric acid, grape seed oil, and nutritional fibres, were effectively extracted. It is known that a number of polyphenols, including catechin, epicatechin, anthocyanins, flavonols, and others, present in grape skin and seeds can prevent human low-density lipoproteins from oxidizing.[79,80]

3.6.3 Bio-alcohol Production

Agricultural and biomass-based sugars are fermented by microorganisms to produce bio-alcohol, the most prevalent of which is ethanol. The major reason ethanol has become popular as an alternative liquid fuel in the face of rapidly dwindling petroleum supplies is that it shares many of the same properties as petroleum.[81] It is well known that ethanol has a high-octane rating, which enables its blend to have the same anti-knock effect as petroleum. In order to produce polyethylene and other polymers, ethanol is a crucial substrate. The first generation feedstock, which is composed of agricultural cereals and sugar-based crops, has limited uses for the manufacturing of ethanol and is unsustainable primarily due to problems with food security and land usage.[3] It is true that the manufacturing of alcohol from food and agricultural wastes offers two key benefits. First, using these wastes as substrates for the manufacture of bio-ethanol lowers the cost of waste disposal. Second, it lowers the cost of producing ethanol, which was previously generated using first-generation feedstocks. Its capacity to regenerate, long-term sustainability, minimal net carbon emissions, and excellent energy efficiency are further advantages. An attractive alternative is the efficient conversion of food waste to bioethanol since it reduces the accumulation of such waste and is ecologically benign.[82]

3.6.4 Bio-absorbent Production

There are numerous agricultural wastes including rice husk and bran, wheat husk and bran, sawdust, coconut shells, groundnut shells, hazelnut shells, cotton seed hulls, walnut shells, maize corn cob, banana, apple, orange peels, grape pomace/stalks, soybean hulls, olive pomace, water hyacinth, sunflower stalks, sugar beet pulp, cotton stalks, coffee beans, *etc.* This is mostly due to the cellulose and lignin content, which are abundant in various functional groups including alcohols, ketones, aldehydes, acids, ethers, and phenols.[83] They are also inexpensive and widely available. The low-cost precursors are first heated at lower temperatures to form chars (carbonization) and then heated at higher temperatures in the presence of an oxidizing agent, such as steam or CO_2. The chemical process comprises heating the precursor in an electrical furnace with inert atmospheres at lower temperatures after impregnating it with an activating agent. The activating agents used by different workers may be divided into three categories: basic solutions, acid solutions made up of minerals and organic materials, and oxidizing agents.[3,84] In addition to resolving the issue of managing solid waste, the development of such bio-based wastes into efficient adsorbents would help treat wastewater in an economical and sustainable manner. Improvements in the surface characteristics of activated carbons created from inexpensive bio-based wastes as well as improvements in the adsorption capacity for the removal of various types of pollutants have been documented. However, there are still significant obstacles to overcome in order to improve the stability of the produced activated carbons as well as for economical and eco-friendly regeneration techniques.[85]

3.6.5 Bio-oil Production

Effective energy recovery from food waste can assist save expenses associated with using conventional fuels or with trash disposal. According to a review by Arena, because of the unfavourable features of the food waste composition, pyrolysis and gasification technologies need dried-up, pre-treated food wastes which raises the overall cost due to the energy needed for drying.[86,87]

Hydrothermal carbonization, hydrothermal gasification, and hydrothermal liquefaction are thermal processes that do not require a pre-treatment phase of drying. Food wastes are successfully transformed into high-energy liquid (bio-oil) and solid bio-fuels (hydro-char) during a short residence time of 5–60 minutes when done under subcritical water conditions of high pressure (7–30 MPa) and moderate temperatures (250–380 °C). As a by-product, one also obtains a gaseous phase and an aqueous phase with dissolved organics. The hydrothermal liquefaction technique has proven to have significant benefits over gasification and pyrolysis. First, the method shows adaptability by handling a variety of food waste kinds with a high moisture content. In comparison with bio-oil obtained from pyrolysis (16–19 MJ kg^{-1}), the hydrothermal liquefied bio-oil has demonstrated greater heating values (35 MJ kg^{-1}). Additionally, its heating value was discovered to be equivalent to that of standard petroleum fuels. Compared to the bio-oil produced by pyrolysis, which contains 40% O_2, the hydrothermal liquefaction bio-oil has a much lower (10%) O_2 level. The bio-oil has a greater calorific value as a result.[87–89]

The method is advantageous in terms of energy usage due to the modest temperature need compared to gasification and pyrolysis and the absence of a drying stage. The main issue is that the bulk of the organic by-products are dissolved in the aqueous phase during operation. This calls for the aqueous phase's downstream treatment before it is released into the environment in order to recover the organics.[90] According to a techno-economic analysis (TEA), the loss of organics into the aqueous phase has an impact on the process's economic performance. The hydrothermal carbonization process is the least expensive of the three hydrothermal conversion methods since it uses the lowest temperatures and uses less energy. The carbonization technologies and hydrothermal liquefaction provide a number of benefits over anaerobic methods, including better waste volume reduction, fewer treatment stages, quicker treatment periods, and no hazardous emissions.[3] Both procedures are becoming more significant in the conversion to energy primarily from food wastes rich in moisture content due to the limits of the pyrolysis and gasification techniques, which need substantial moisture contents found in food wastes to be removed. Additionally, part of the carbon content in such waste is lost to the atmosphere during anaerobic digestion or fermentation processes through the production of carbon dioxide, but the majority of the organic carbon in the waste is preserved in the final hydro-char product.[91,92]

3.7 Conclusion

Waste utilization without a doubt not only reduces disposal issues but also handles pollution-related issues. In response to the need for environmental sustainability and conservation, global opinion of agro-industrial waste is fast shifting. Agro-industrial trash or leftovers are high in nutrients and bioactive compounds. Agro-industry wastes, notably food industrial wastes, contain a large amount of organic matter that is high in nutrient content and has higher COD, BOD, and TSS values. If agro-industrial wastes are not properly processed, they might pose serious pollution issues. Furthermore, they imply a loss of vital biomass and nutrients. As a result of the heterogeneity in the content of such wastes such as minerals, sugars and proteins should be regarded as "raw materials" rather than "wastes" for other industrial operations. Agro-industrial waste is no longer regarded as a problem, but as an opportunity. The challenges of employing biorefinery principles to extract biomass and suppress microorganisms are also investigated. Bio-refineries can mimic the operation of an oil refinery by using renewable biomass to create a variety of products. The potential future of technological breakthroughs for improving people's lives and preserving our environment lies in the use of clean bioprocess technology.

Conflicts of Interest

The authors declare no conflict of interest.

Acknowledgements

The authors are thankful to Lovely Professional University, Punjab, India for providing the research facilities used for this study.

References

1. A. Baiano, Recovery of biomolecules from food wastes—A review, *Molecules*, 2014, **19**(9), 14821–14842.
2. A. Nayak and B. Bhushan, An overview of the recent trends on the waste valorization techniques for food wastes, *J. Environ. Manage.*, 2019, **233**, 352–370.
3. R. Ravindran, S. S. Hassan, G. A. Williams and A. K. Jaiswal, A review on bioconversion of agro-industrial wastes to industrially important enzymes, *Bioengineering*, 2018, **5**(4), 93.
4. R. Domínguez-Perles, D. A. Moreno and C. García-Viguera, Waking up from four decades' long dream of valorizing agro-food byproducts: toward practical applications of the gained knowledge, *J. Agric. Food Chem.*, 2018, **66**(12), 3069–3073.
5. L. F. Călinoiu, L. Mitrea, G. Precup, M. Bindea, B. Rusu, K. Szabo, F. V. Dulf, B. E. Ştefănescu and D. C. Vodnar, *76. Sustainable use of agro-industrial wastes for feeding 10 billion people by 2050*, Wageningen Academic Publishers, 2018, p. 482.
6. F. Federici, F. Fava, N. Kalogerakis and D. Mantzavinos, Valorisation of agro-industrial by-products, effluents and waste: concept, opportunities and the case of olive mill wastewaters, *J. Chem. Technol. Biotechnol.*, 2009, **84**(6), 895–900.

7. C. Marzo, A. B. Díaz, I. Caro and A. Blandino, Valorization of agro-industrial wastes to produce hydrolytic enzymes by fungal solid-state fermentation, *Waste Manage. Res.*, 2019, **37**(2), 149–156.
8. M. Kieliszek, K. Piwowarek, A. M. Kot and K. Pobiega, The aspects of microbial biomass use in the utilization of selected waste from the agro-food industry, *Open Life Sci.*, 2020, **15**(1), 787–796.
9. W. Chen, T. L. Oldfield, S. I. Patsios and N. M. Holden, Hybrid life cycle assessment of agro-industrial wastewater valorisation, *Water Res.*, 2020, **170**, 115275.
10. D. Statuto and P. Picuno, Planning the energy valorization of agricultural co-products, by-products and waste in a landscape context, in *Proceedings of the 11th International AIIA Conference on: "Biosystems Engineering addressing the human challenges of the 21st century"*, Bari (Italy), 2017, pp. 206–210.
11. C. S. Lin, A. A. Koutinas, K. Stamatelatou, E. B. Mubofu, A. S. Matharu, N. Kopsahelis, L. A. Pfaltzgraff, J. H. Clark, S. Papanikolaou, T. H. Kwan and R. Luque, Current and future trends in food waste valorization for the production of chemicals, materials and fuels: a global perspective, *Biofuels, Bioprod. Biorefin.*, 2014, **8**(5), 686–715.
12. A. N. Philippoussis and P. A. Diamantopoulou, Agro-food industry wastes and agricultural residues conversion into high value products by mushroom cultivation, in *Proceedings of the 7th international conference on mushroom biology and mushroom products (ICMBMP7)*, France, 2011, pp. 4–7.
13. N. Mirabella, V. Castellani and S. Sala, Current options for the valorization of food manufacturing waste: a review, *J. Cleaner Prod.*, 2014, **65**, 28–41.
14. K. Winans, A. Kendall and H. Deng, The history and current applications of the circular economy concept, *Renewable Sustainable Energy Rev.*, 2017, **68**, 825–833.
15. L. D. Shirahigue and S. R. Ceccato-Antonini, Agro-industrial wastes as sources of bioactive compounds for food and fermentation industries, *Ciência Rural*, 2020, **50**(4), 1–17.
16. M. E. Álvarez-Cao, A. Rico-Díaz, M. E. Cerdán, M. Becerra and M. I. González-Siso, Valuation of agro-industrial wastes as substrates for heterologous production of α-galactosidase, *Microb. Cell Fact.*, 2018, **17**(1), 1–3.
17. X. Wang, C. Li, C. H. Lam, K. Subramanian, Z. H. Qin, J. H. Mou, M. Jin, S. S. Chopra, V. Singh, Y. S. Ok and J. Yan, Emerging waste valorisation techniques to moderate the hazardous impacts, and their path towards sustainability, *J. Hazard. Mater.*, 2022, **423**, 127023.
18. J. Kumla, N. Suwannarach, K. Sujarit, W. Penkhrue, P. Kakumyan, K. Jatuwong, S. Vadthanarat and S. Lumyong, Cultivation of mushrooms and their lignocellulolytic enzyme production through the utilization of agro-industrial waste, *Molecules*, 2020, **25**(12), 2811.
19. F. Girotto, L. Alibardi and R. Cossu, Food waste generation and industrial uses: A review, *Waste Manage.*, 2015, **45**, 32–41.

20. R. Gómez-García, D. A. Campos, C. N. Aguilar, A. R. Madureira and M. Pintado, Valorisation of food agro-industrial by-products: From the past to the present and perspectives, *J. Environ. Manage.*, 2021, **299**, 113571.
21. T. S. El-Tayeb, A. A. Abdelhafez, S. H. Ali and E. M. Ramadan, Effect of acid hydrolysis and fungal biotreatment on agro-industrial wastes for obtainment of free sugars for bioethanol production, *Braz. J. Microbiol.*, 2012, **43**, 1523–1535.
22. J. C. Motte, E. Trably, R. Escudié, J. Hamelin, J. P. Steyer, N. Bernet, J. P. Delgenes and C. Dumas, Total solids content: a key parameter of metabolic pathways in dry anaerobic digestion, *Biotechnol. Biofuels*, 2013, **6**(1), 1–9.
23. D. Klemm, B. Heublein, H. P. Fink and A. Bohn, Cellulose: fascinating biopolymer and sustainable raw material, *Angew. Chem., Int. Ed.*, 2005, **44**(22), 3358–3393.
24. A. A. Sundarraj and T. V. Ranganathan, A review on cellulose and its utilization from agro-industrial waste, *Drug Invent. Today*, 2018, **10**(1), 89–94.
25. A. N. Ahmad Khorairi, N. S. Sofian-Seng, R. Othaman, H. Abdul Rahman, N. S. Mohd Razali, S. J. Lim and W. A. Wan Mustapha, A review on agro-industrial waste as cellulose and nanocellulose source and their potentials in food applications, *Food Rev. Int.*, 2021, 1–26.
26. B. C. Saha, Hemicellulose bioconversion, *J. Ind. Microbiol. Biotechnol.*, 2003, **30**(5), 279–291.
27. S. Banerjee, A. F. Patti, V. Ranganathan and A. Arora, Hemicellulose based biorefinery from pineapple peel waste: xylan extraction and its conversion into xylooligosaccharides, *Food Bioprod. Process.*, 2019, **117**, 38–50.
28. J. Ralph, C. Lapierre and W. Boerjan, Lignin structure and its engineering, *Curr. Opin. Biotechnol*, 2019, **56**, 240–249.
29. D. Kumar, K. K. Yadav, M. Singh and N. Garg, Biochemical utilization of agro-industrial lignocelluloses rich waste for cellulase production, *Res. J. Agric. Sci.*, 2012, **44**(2), 184–191.
30. C. Chio, M. Sain and W. Qin, Lignin utilization: a review of lignin depolymerization from various aspects, *Renewable Sustainable Energy Rev.*, 2019, **107**, 232–249.
31. C. A. Cateto, M. F. Barreiro and A. E. Rodrigues, Monitoring of lignin-based polyurethane synthesis by FTIR-ATR, *Ind. Crops Prod.*, 2008, **27**(2), 168–174.
32. B. M. Upton and A. M. Kasko, Strategies for the conversion of lignin to high-value polymeric materials: review and perspective, *Chem. Rev.*, 2016, **116**(4), 2275–2306.
33. P. K. Sadh, S. Duhan and J. S. Duhan, Agro-industrial wastes and their utilization using solid state fermentation: a review, *Bioresour. Bioprocess.*, 2018, **5**(1), 1–5.
34. S. Kusch, C. C. Udenigwe, C. Cavinato, M. Gottardo and F. Micolucci, Value-added utilization of agro-industrial residues, *Advances in Food Biotechnology*, 2015, p. 415.

35. G. C. Premier, J. R. Kim, J. Massanet-Nicolau, G. Kyazze, S. R. Esteves, B. K. Penumathsa, J. Rodríguez, J. Maddy, R. M. Dinsdale and A. J. Guwy, Integration of biohydrogen, biomethane and bioelectrochemical systems, *Renewable Energy*, 2013, **49**, 188–192.
36. M. Kapri, U. Singh, S. M. Behera, P. P. Srivastav and S. Sharma, Nutraceutical augmentation of agro-industrial waste through submerged fermentation using Calocybe indica, *LWT*, 2020, **134**, 110156.
37. P. Leite, D. Sousa, H. Fernandes, M. Ferreira, A. R. Costa, D. Filipe, M. Gonçalves, H. Peres, I. Belo and J. M. Salgado, Recent advances in production of lignocellulolytic enzymes by solid-state fermentation of agro-industrial wastes, *Curr. Opin. Green Sustainable Chem.*, 2021, **27**, 100407.
38. P. S. Panesar, R. Kaur, G. Singla and R. S. Sangwan, Bio-processing of agro-industrial wastes for production of food-grade enzymes: progress and prospects, *Appl. Food Biotechnol.*, 2016, **3**(4), 208–227.
39. M. Barcelos, C. L. Ramos, M. Kuddus, S. Rodriguez-Couto, N. Srivastava, P. W. Ramteke, P. K. Mishra and G. Molina, Enzymatic potential for the valorization of agro-industrial by-products, *Biotechnol. Lett.*, 2020, **42**(10), 1799–1827.
40. M. Yusuf, Agro-industrial waste materials and their recycled value-added applications, *Handbook of Ecomaterials*, 2017, vol. 1, p. 1.
41. B. C. Behera, R. Mishra and S. Mohapatra, Microbial citric acid: Production, properties, application, and future perspectives, *Food Front.*, 2021, **2**(1), 62–76.
42. A. M. Canete-Rodriguez, I. M. Santos-Duenas, J. E. Jimenez-Hornero, A. Ehrenreich, W. Liebl and I. Garcia-Garcia, Gluconic acid: Properties, production methods and applications—An excellent opportunity for agro-industrial by-products and waste bio-valorization, *Process Biochem.*, 2016, **51**(12), 1891–1903.
43. P. L. Show, K. O. Oladele, Q. Y. Siew, F. A. Aziz Zakry, J. C. Lan and T. C. Ling, Overview of citric acid production from Aspergillus niger, *Front. Life Sci.*, 2015, **8**(3), 271–283.
44. S. V. Lende, H. Karemore and M. J. Umekar, Review on production of citric acid by fermentation technology, *GSC Biol. Pharm. Sci.*, 2021, **17**(3), 85–93.
45. D. Carrillo-Nieves, M. J. Alanís, R. de la Cruz Quiroz, H. A. Ruiz, H. M. Iqbal and R. Parra-Saldívar, Current status and future trends of bioethanol production from agro-industrial wastes in Mexico, *Renewable Sustainable Energy Rev.*, 2019, **102**, 63–74.
46. D. Basso, F. Patuzzi, D. Castello, M. Baratieri, E. C. Rada, E. Weiss-Hortala and L. Fiori, Agro-industrial waste to solid biofuel through hydrothermal carbonization, *Waste Manage.*, 2016, **47**, 114–121.
47. D. Cardoen, P. Joshi, L. Diels, P. M. Sarma and D. Pant, Agriculture biomass in India: Part 2. Post-harvest losses, cost and environmental impacts, *Resour., Conserv. Recycl.*, 2015, **101**, 143–153.
48. J. F. Ayala-Zavala, V. Vega-Vega, C. Rosas-Domínguez, H. Palafox-Carlos, J. A. Villa-Rodriguez, M. W. Siddiqui, J. E. Dávila-Aviña and

G. A. González-Aguilar, Agro-industrial potential of exotic fruit byproducts as a source of food additives, *Food Res. Int.*, 2011, **44**(7), 1866–1874.
49. Z. A. Pollard and J. L. Goldfarb, Valorization of cherry pits: Great Lakes agro-industrial waste to mediate Great Lakes water quality, *Environ. Pollut.*, 2021, **270**, 116073.
50. E. Caballero and C. Soto, Valorization of agro-industrial waste into bioactive compounds: techno-economic considerations, *Biorefinery*, 2019, 235–252.
51. M. A. de Oliveira, L. Y. Madruga, B. L. de Lima, M. A. Villetti, M. S. de Souza Filho, M. J. Kipper, N. N. Marques and R. C. Balaban, Agro-industrial waste valorization: Transformation of starch from mango kernel into biocompatible, thermoresponsive and high swelling nanogels, *J. Braz. Chem. Soc.*, 2021, **32**, 1607–1616.
52. L. C. Freitas, J. R. Barbosa, A. L. da Costa, F. W. Bezerra, R. H. Pinto and R. N. de Carvalho Junior, From waste to sustainable industry: How can agro-industrial wastes help in the development of new products?, *Resour., Conserv. Recycl.*, 2021, **169**, 105466.
53. V. Taurisano, G. Anzelmo, A. Poli, B. Nicolaus and P. Di Donato, Re-use of agro-industrial waste: Recovery of valuable compounds by eco-friendly techniques, *Int. J. Performability Eng.*, 2014, **10**(4), 419.
54. C. Maraveas, Production of sustainable and biodegradable polymers from agricultural waste, *Polymers*, 2020, **12**(5), 1127.
55. E. Aziable, S. Tchegueni, M. B. Bodjona, A. K. Degbe, M. Zamama, M. Hafidi, M. El Meray and K. A. Kili, Valorization of agro-industrial waste by bio-process aerobic "composting", *JMES*, 2017, **8**(4), 1277–1283.
56. H. Abu Qdais and M. Al-Widyan, Evaluating composting and co-composting kinetics of various agro-industrial wastes, *Int. J. Recycling Org. Waste Agriculture*, 2016, **5**(3), 273–280.
57. M. H. Zainudin, A. Zulkarnain, A. S. Azmi, S. Muniandy, K. Sakai, Y. Shirai and M. A. Hassan, Enhancement of agro-industrial waste composting process via the microbial inoculation: a brief review, *Agronomy*, 2022, **12**(1), 198.
58. E. Lizundia, F. Luzi and D. Puglia, Organic waste valorisation towards circular and sustainable biocomposites, *Green Chem.*, 2022, **24**, 5429–5459.
59. R. Panesar, S. Kaur and P. S. Panesar, Production of microbial pigments utilizing agro-industrial waste: a review, *Curr. Opin. Food Sci.*, 2015, **1**, 70–76.
60. *Sustainable Microbial Technologies for Valorization of Agro-Industrial Wastes*, ed. J. K. Saini, S. Singh and L. Nain, CRC Press, 2022.
61. S. Otles and C. Kartal, Food waste valorization, in *Sustainable food systems from agriculture to industry*, Academic Press, 2018, pp. 371–399.
62. R. J. Sadus, *High pressure phase behaviour of multicomponent fluid mixtures*, Elsevier, 2012.
63. S. A. Khan, R. Aslam and H. A. Makroo, High pressure extraction and its application in the extraction of bio-active compounds: A review, *J. Food Process Eng.*, 2019, **42**(1), e12896.

64. T. Lefebvre, E. Destandau and E. Lesellier, Sequential extraction of carnosic acid, rosmarinic acid and pigments (carotenoids and chlorophylls) from Rosemary by online supercritical fluid extraction-supercritical fluid chromatography, *J. Chromatogr. A*, 2021, **1639**, 461709.
65. H. A. Makroo, N. K. Rastogi and B. Srivastava, Ohmic heating assisted inactivation of enzymes and microorganisms in foods: A review, *Trends Food Sci. Technol.*, 2020, **97**, 451–465.
66. I. G. Moorthy, J. P. Maran, S. Muneeswari, S. Naganyashree and C. S. Shivamathi, Response surface optimization of ultrasound assisted extraction of pectin from pomegranate peel, *Int. J. Biol. Macromol.*, 2015, **72**, 1323–1328.
67. C. Sabater, L. Ruiz, S. Delgado, P. Ruas-Madiedo and A. Margolles, Valorization of vegetable food waste and by-products through fermentation processes, *Front. Microbiol.*, 2020, **11**, 581997.
68. A. A. Koutinas, A. Vlysidis, D. Pleissner, N. Kopsahelis, I. L. Garcia, I. K. Kookos, S. Papanikolaou, T. H. Kwan and C. S. Lin, Valorization of industrial waste and by-product streams via fermentation for the production of chemicals and biopolymers, *Chem. Soc. Rev.*, 2014, **43**(8), 2587–2627.
69. P. Petrova and K. Petrov, Lactic acid fermentation of cereals and pseudocereals: Ancient nutritional biotechnologies with modern applications, *Nutrients*, 2020, **12**(4), 1118.
70. Á. Orts, M. Tejada, J. Parrado, P. Paneque, C. García, T. Hernández and I. Gómez-Parrales, Production of biostimulants from okara through enzymatic hydrolysis and fermentation with Bacillus licheniformis: comparative effect on soil biological properties, *Environ. Technol.*, 2019, **40**(16), 2073–2084.
71. M. V. Tacin, F. P. Massi, M. H. Fungaro, M. F. Teixeira, A. V. de Paula and V. de Carvalho Santos-Ebinuma, Biotechnological valorization of oils from agro-industrial wastes to produce lipase using Aspergillus sp. from Amazon, *Biocatal. Agricultural Biotechnol.*, 2019, **17**, 369–378.
72. S. Jirasatid, K. Limroongreungrat, M. Nopharatana and K. Monacolin, pigments and citrinin of rice pasta by-products fermented by Monascus purpureus, *Int. Food Res. J.*, 2019, **26**(4), 1279–1284.
73. Q. Zhang, J. Hu and D. J. Lee, Biogas from anaerobic digestion processes: Research updates, *Renewable Energy*, 2016, **98**, 108–119.
74. R. Liguori, A. Amore and V. Faraco, Waste valorization by biotechnological conversion into added value products, *Appl. Microbiol. Biotechnol.*, 2013, **97**(14), 6129–6147.
75. M. S. Kim, J. G. Na, M. K. Lee, H. Ryu, Y. K. Chang, J. M. Triolo, Y. M. Yun and D. H. Kim, More value from food waste: Lactic acid and biogas recovery, *Water Res.*, 2016, **96**, 208–216.
76. R. Iglesias-Iglesias, M. M. Fernandez-Feal, C. Kennes and M. C. Veiga, Valorization of agro-industrial wastes to produce volatile fatty acids: combined effect of substrate/inoculum ratio and initial alkalinity, *Environ. Technol.*, 2021, **42**(25), 3889–3899.

77. C. M. Galanakis, P. Tsatalas and I. M. Galanakis, Implementation of phenols recovered from olive mill wastewater as UV booster in cosmetics, *Ind. Crops Prod.*, 2018, **111**, 30–37.
78. R. D. Farias, C. M. García, T. C. Palomino, F. Andreola, I. Lancellotti and L. Barbieri, Valorization of agro-industrial wastes in lightweight aggregates for agronomic use: Preliminary study, *Environ. Eng. Manage. J.*, 2017, **16**(8), 1691–1700.
79. A. Nayak, B. Bhushan, A. Rosales, L. R. Turienzo and J. L. Cortina, Valorisation potential of Cabernet grape pomace for the recovery of polyphenols: Process intensification, optimisation and study of kinetics, *Food Bioprod. Process.*, 2018, **109**, 74–85.
80. J. V. Madeira Junior, C. B. Teixeira and G. A. Macedo, Biotransformation and bioconversion of phenolic compounds obtainment: an overview, *Crit. Rev. Biotechnol.*, 2015, **35**(1), 75–81.
81. C. Kourmentza, C. N. Economou, P. Tsafrakidou and M. Kornaros, Spent coffee grounds make much more than waste: Exploring recent advances and future exploitation strategies for the valorization of an emerging food waste stream, *J. Cleaner Prod.*, 2018, **172**, 980–992.
82. H. S. Hafid, U. K. Shah, A. S. Baharuddin and A. B. Ariff, Feasibility of using kitchen waste as future substrate for bioethanol production: a review, *Renewable Sustainable Energy Rev.*, 2017, **74**, 671–686.
83. V. K. Gupta, A. Nayak, B. Bhushan and S. Agarwal, A critical analysis on the efficiency of activated carbons from low-cost precursors for heavy metals remediation, *Crit. Rev. Environ. Sci. Technol.*, 2015, **45**(6), 613–668.
84. T. Mehmood, F. Nadeem, S. A. Qamar, M. Bilal and H. Iqbal, Bioconversion of agro-industrial waste into value-added compounds, in *Sustainable bioconversion of waste to value added products*, Springer, Cham, 2021, pp. 349–368.
85. M. Alexandri, J. P. López-Gómez, A. Olszewska-Widdrat and J. Venus, Valorising agro-industrial wastes within the circular bioeconomy concept: The case of defatted rice bran with emphasis on bioconversion strategies, *Fermentation*, 2020, **6**(2), 42.
86. T. P. Pham, R. Kaushik, G. K. Parshetti, R. Mahmood and R. Balasubramanian, Food waste-to-energy conversion technologies: Current status and future directions, *Waste Manage.*, 2015, **38**, 399–408.
87. J. Berbel and A. Posadillo, Review and analysis of alternatives for the valorisation of agro-industrial olive oil by-products, *Sustainability*, 2018, **10**(1), 237.
88. A. Gallifuoco, L. Taglieri, F. Scimia, A. A. Papa and G. Di Giacomo, Hydrothermal carbonization of Biomass: New experimental procedures for improving the industrial Processes, *Bioresour. Technol.*, 2017, **244**, 160–165.
89. V. K. Gaur, P. Gupta, V. Tripathi, R. S. Thakur, R. K. Regar, D. K. Patel and N. Manickam, Valorization of agro-industrial waste for rhamnolipid production, its role in crude oil solubilization and resensitizing bacterial pathogens, *Environ. Technol. Innovation*, 2022, **25**, 102108.

90. S. B. Tradler, S. Mayr, M. Himmelsbach, R. Priewasser, W. Baumgartner and A. T. Stadler, Hydrothermal carbonization as an all-inclusive process for food-waste conversion, *Bioresour. Technol. Rep.*, 2018, **2**, 77–83.
91. N. A. Qambrani, M. M. Rahman, S. Won, S. Shim and C. Ra, Biochar properties and eco-friendly applications for climate change mitigation, waste management, and wastewater treatment: A review, *Renewable Sustainable Energy Rev.*, 2017, **79**, 255–273.
92. N. Klai, B. Yadav, O. El Hachimi, A. Pandey, B. Sellamuthu and R. D. Tyagi, Agro-industrial waste valorization for biopolymer production and life-cycle assessment toward circular bioeconomy, in *Biomass, Biofuels, Biochemicals*, Elsevier, 2021, pp. 515–555.

CHAPTER 4

Valorisation of Coproducts and By-products Obtained from Nuts

JANE MARA BLOCK,*[a] GABRIELA POLMANN,[a] MARIA ISABEL LANDIM NEVES,[b] JOSE DAVID SÁNCHEZ-MARTÍNEZ,[c] ALEJANDRO CIFUENTES[c] AND ELENA IBAÑEZ[c]

[a] Food Science and Technology Department – Federal University of Santa Catarina, Rod. Admar Gonzaga 1346, 88034-001, Florianopolis, SC, Brazil; [b] Faculty of Food Engineering, Estadual University of Campinas, Rua Monteiro Lobato, 13083-745, Campinas, SP, Brazil; [c] Institute of Food Science Research (CIAL), Calle Nicolás Cabrera, 9, 28049, Madrid, Spain
*Email: janeblock@gmail.com

4.1 Introduction

The effects of nut consumption on health have been extensively reported since the publishing of the first studies in the early 90s.[1–3] The nuts present high contents of macronutrients such as lipids, proteins, and fibers. They also are rich in micronutrients such as vitamins and minerals and bioactive compounds such as tocopherols, phytosterols, polyphenols, and phenolic acids.[4–7] Other bioactive compounds that have been reported recently in nuts include bioactive peptides, betaine, choline, and isorhamnetin.[8,9]

The daily consumption of nuts has been associated with the prevention of non-communicable chronic diseases (NCCDs) such as cardiovascular and neurodegenerative diseases, diabetes, and cancer. These effects on health have been attributed to the bioactive compounds present.[10] Although nuts have a high energy value, they have also been recommended for weight control. The control of satiety, an increase of thermogenesis, displacement

Green Chemistry Series No. 78
Agri-food Waste Valorisation
Edited by Pankaj Chowdhary and Abhay Raj

Published by the Royal Society of Chemistry, www.rsc.org

of unfavorable nutrients, availability for metabolizable energy, composition in terms of bioactive compounds with anti-obesity action, and the improvement of the gut microbiome have been associated with this effect.[11] These findings led the nuts to be classified as functional foods.

The positive effects of nuts on health, which are widely established within the scientific community, the recommendation of the World Health Organization[12] for its incorporation in the diet, and the media coverage, have probably contributed to the increase in their consumption in the last few decades. The report of the International Nut and Dried Fruit Council showed that the production of nuts reached 5.3 million metric tons in 2020/2021. The global nut production increased by 15% when compared to 2019/20 and 65% when compared to the previous decade (2011/2012).[13]

A large quantity of co-products and by-products are generated during the processing of foods. The goods generated in the processing with added value that can be sold or reused are considered co-products. The residual outputs, which present low commercial value and need to be disposed of at a financial or environmental cost, are considered by-products or wastes. The valorisation of coproducts, by-products, and wastes from food processing has been described in the literature as adding value to the original raw materials. Simultaneously, this strategy can improve the economic and environmental sustainability of the food production chain. The full use of these raw materials can eliminate waste and pollution, which is the first principle of the circular economy. The integrated value chains have been described as an efficient approach to achieving the zero-waste goal and accelerating the transition of the food industry to a circular bioeconomy.[14–16]

The processing of nuts generates between 40 and 60% of by-products and, considering that the production of nuts has increased in the last few decades, a large volume of biomass is generated annually. The almond industry in the USA represented by the Almond Board of California has committed to generating zero waste by 2025. This challenge can be achieved by converting by-products and high volumes of feedstock into useful commercial products.[17]

Even considering the high potential of these co-products and by-products, which are rich in macro and micronutrients and bioactive compounds, they are currently underutilized. The specialty oils obtained after the pressing of nuts can be used in foods and cosmetics. The defatted flour obtained from the cake and the wholemeal flour are co-products rich in nutrients and bioactive compounds. These flours can improve the nutritional composition or functionality in bakery and pastry products, sweets, sauces, chocolates, and ice creams. In addition, they can be used for producing plant-based milk and yogurts.[18–20] Moreover, green technologies such as microwave-assisted extraction (MAE), pressurized liquid extraction (PLE), supercritical fluid extraction (SFE), and ultrasonic-assisted extraction (UAE) can be used to convert by-products such as nutshells and skins to functional ingredients for food, cosmetic, and pharmaceutical industries turning by-products and co-products into products with added value (Figure 4.1).

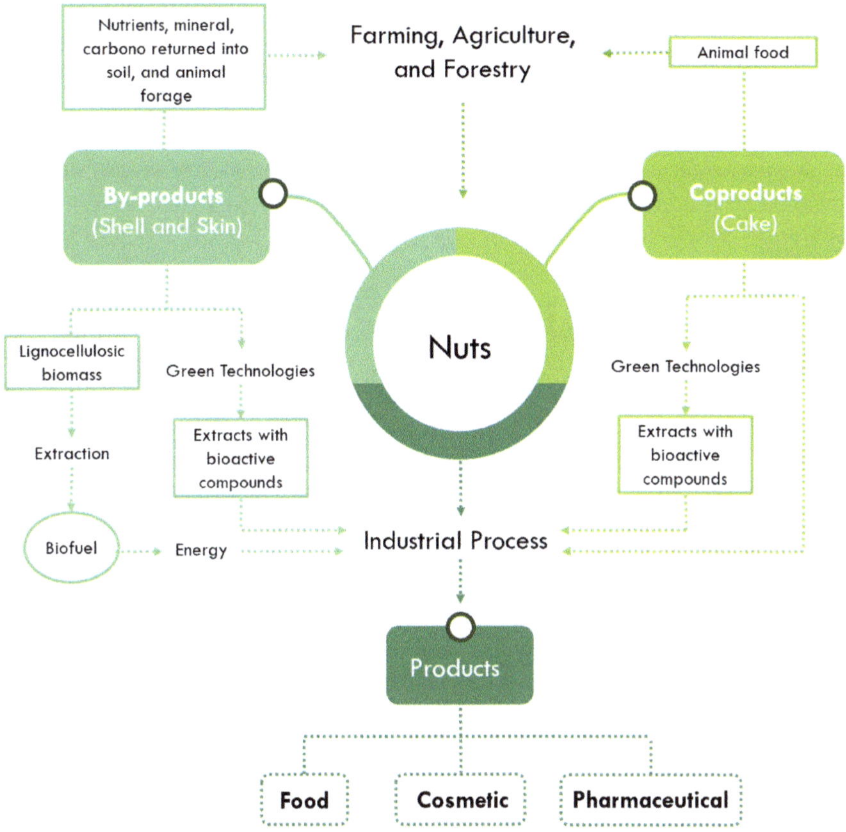

Figure 4.1 Integrated chain for processing of nuts.

In this chapter, the composition of nuts and their co-products and by-products, the extraction of bioactive compounds using green technologies and their applications, and the effects of nuts and their co-products and by-products on health will be presented and discussed. Some of the under-exploited nuts from the Amazon biome that are currently little known and explored will also be described. The knowledge of these raw materials is important to increase their consumption and commercialization worldwide. The full exploitation of these raw materials could allow the production of zero waste in the industries that process these raw materials in the near future.

4.2 Chemical Composition of Nuts

It is widely known that nuts present a high concentration of lipids and a low concentration of sugars.[9] However, even with their high fat concentration, nuts are considered healthy because they are rich in macronutrients, micronutrients, and bioactive compounds such as tocopherols, phytosterols, and phenolic compounds.[6,7] The consumption of nuts has been associated with positive effects on human health such as the prevention of cardiovascular and inflammatory diseases, obesity, and type 2 diabetes mellitus. These health benefits have been attributed to their nutritional and phytochemical composition. Table 4.1 shows the nutritional composition of the most consumed nuts worldwide.

The fat content among the most consumed nuts ranges from around 40% (almonds, pistachios, and cashews) to around 70% (macadamia, pine, and Brazil nuts), which represents an energy value between 477 and 791 kcal 100 g^{-1}, respectively. The protein content in nuts ranges from 7.9% to 36.3% with the highest concentration reported for almonds (19.5–26 g 100 g^{-1}) and cashews (18.2–36.3 g 100 g^{-1}). On the other hand, the lowest content in protein has been reported for macadamia (7.9 to 8.4 g 100 g^{-1}) and pecan (9.2 to 10 g 100 g^{-1}) nuts.

Although low levels of carbohydrates are usually observed in nuts, a higher concentration of this nutrient has been reported for cashews and pistachios (1.5–27.2 and 7.9–30.1 g 100 g^{-1}, respectively). Pecan nuts, pistachios, and almonds can present concentrations higher than 10 g 100 g^{-1} of fibers. The low moisture content reported for nuts is important to maintain their shelf life, preventing microbial growth, premature seed germination, and many undesirable chemicals and biological alterations.[9,21]

Nuts are rich in bioactive lipids such as mono and polyunsaturated fatty acids, tocopherols, and phytosterols (Table 4.2). The saturated fatty acids reported for the most consumed nuts are at lower concentrations when compared to the unsaturated fatty acids.

Nuts are a good source of tocopherols, which are widely known for their capacity to donate a hydrogen atom of the hydroxyl group to the free radicals retarding the oxidation processes[29] and for their vitamin E activity. The higher total tocopherol content has been reported for cashews (1714.8 $\mu g\, g^{-1}$), followed by almonds (452.0 $\mu g\, g^{-1}$). In addition, phytosterols present in nuts (from 11 to 558.6 mg 100 g^{-1}) have been associated with a decrease in plasma LDL-cholesterol levels. This bioactivity is related to its chemical structure which is similar to cholesterol. The transport of phytosterols, which present higher hydrophobicity when compared to cholesterol, is improved by micelles in the human body. On the other hand, the absorption of phytosterols is much lower when compared to cholesterol (around 5 and 45%, respectively) promoting the inhibition of cholesterol absorption.[30]

Recently, bioactive peptides have been reported in nuts such as walnuts, pine nuts, and cashews.[9] Zhao *et al.*[31] reported activity from peptides from walnuts against the SARS-CoV-2 virus. In addition, Wang *et al.*[32] reported the antioxidant and anti-diabetic activity of peptides obtained from the same raw material. Jahanbani *et al.*[33] showed the antioxidant and anticancer activity of peptides obtained from pine nuts. Moreover, Bisinotto *et al.*[34] reported the antioxidant and anti-allergy activities of peptides obtained from cashew nuts. The maintenance of homeostasis and antihypertensive, antimicrobial, antioxidant, and anti-inflammatory activities are the mechanisms that have been attributed to the bioactivity of the peptides from nuts.[9]

Table 4.1 Nutritional composition of the most consumed nuts worldwide.

Nuts[21–24]	Lipids (g 100 g^{-1})	Proteins (g 100 g^{-1})	Carbohydrates (g 100 g^{-1})	Total fiber (g 100 g^{-1})	Moisture content (g 100 g^{-1})	Energy value (kcal 100 g^{-1})
Almonds 	43.4–67.0	19.5–26.0	2.1–12.0	2.5–14.0	3.0–9.5	477–755
Brazil nuts 	64.9–67.2	14.3–16.2	6.3–7.6	8.0–8.5	3.1–3.2	666–710
Cashew nuts 	43.8–51.2	18.2–36.3	7.9–30.1	2.5–4.2	5.2–7.2	499–707
Hazelnuts 	60.8–61.5	14.1–15.0	1.4–16.7	9.7	4.2–5.3	609–680
Macadamias 	66.2–75.8	7.9–9.5	4.4–13.8	8.6	1.4–2.1	633–791
Pecans 	59.5–62.2	9.2–10	13.0–13.4	10.0–13.0	3.0–3.3	626–649
Pine nuts 	61.7–68.4	13.1–13.7	1.8–13.1	3.7	1.5–2.3	615–722
Pistachios 	45.1–45.3	10.8–20.2	1.5–27.2	10.6	4.4–5.7	491–597
Walnuts 	64.5–65.2	13.5–15.2	2.1–13.7	6.7	2.7–4.1	643–702

Table 4.2 Lipid composition of the most consumed nuts worldwide.

Nuts[21,22,25–28]	MUFAs[a] (g 100 g^{-1} of total lipids)	PUFAs[b] (g 100 g^{-1} of total lipids)	UFAs[c] (MUFAs + PUFAs)	SFAs[d] (g 100 g^{-1} of total lipids)	Tocopherols (μg g^{-1})	Phytosterols (mg 100 g^{-1})
Almonds	31.3–69.2	12.3–21.7	43.6–90.2	3.7–8.6	452.0	217.8
Brazil nuts	28.5–36.8	36.1–37.7	64.6–74.5	24.9–28.9	152.8	47–148
Cashew nuts	23.8–61.1	7.8–17.2	31.6–78.3	7.5–21.3	1714.8	11.8
Hazelnuts	45.5–79.6	7.8–10.9	53.3–90.5	4.4–8.8	371.3	11.0
Macadamias	56.8–82.5	1.4–2.4	58.2–84.9	1.6–15.0	122.3	161.8
Pecans	40.6–71.8	21.0–21.6	61.6–93.4	6.1–7	28.7	—
Pine nuts	23.8–40.8	15.6–46.0	39.4–86.8	4.9–8.3	229.5	273.7
Pistachios	23.2–59.5	14.4–30.7	37.6–90.2	5.8–9.3	291.0	558.6
Walnuts	21.2	69.1	90.3	9.3	321.1	123.6

[a] MUFAs: monounsaturated fatty acids.
[b] PUFAs: polyunsaturated fatty acids.
[c] UFAs: unsaturated fatty acids.
[d] SFAs: saturated fatty acids.

The main vitamins reported in nuts are vitamin E and vitamins of the B complex such as folic acid (B9), niacin (B3), pantothenic acid (B3), pyridoxine (B6), riboflavin (B2), and thiamine (B1). Vitamins C, A, and K have also been reported in nuts.[9,35] Nuts are rich in minerals such as potassium (K) followed by phosphorus (P), calcium (Ca), and magnesium (Mg).[36] Moreover, it has been reported that nuts have betaine, a modified amino acid consisting of glycine with three methyl groups involved in different metabolic pathways. In addition, choline, an amine precursor of acetylcholine and phosphatidylcholine, has been reported in almonds, hazelnuts, and walnuts. Acetylcholine is a neurotransmitter involved in memory and muscle control, and phosphatidylcholine, present in cell membranes, is involved in membrane signaling and transport of lipids through lipoproteins.[9,37]

In addition to macro and micronutrients, nuts are rich in phenolic compounds such as flavonoids and phenolic acids.[36] Stilbene derivatives, coumarin lignans, phenolic aldehydes, phytates, tyrosols, phytoestrogens, and caffeic acid phenethyl esters (CAPEs) have also been reported in the literature.[21,36,38–41] Flavonoids are usually found as soluble conjugated glycosides bounded through an OH group (O-glycosides) or through carbon–carbon bonds (C-glycosides). The insoluble form of phenolic acids, which are the main form in nuts, showed a significantly higher antioxidant capacity compared to free and soluble conjugated phenolics.[42–45] An extensive review of the bound phenolics in foods and their activities has been published by Acosta-Estrada *et al.*[46]

The total phenolic content of nuts ranges from 32 to 1657 mg of gallic acid equivalents (mg GAE 100 g^{-1}) (Table 4.3). Pecans, walnuts, and pistachios have the highest values reported for the TPC (2016, 1556, and 1657 mg GAE 100 g^{-1}, respectively).[47]

Almonds, hazelnuts, pecans, pine nuts, pistachios, and walnuts present the most diverse phenolic profiles. The highest concentration of phenolic acids is found in pecans (3374.3 μg 100 g^{-1}), and the lowest in almonds (9.6 μg 100 g^{-1}). Pistachios, hazelnuts, and almonds present the highest number of flavonoids. Isoflavones, lignans, and phytoestrogens have been reported mainly in pistachios. Proanthocyanidins are present in higher concentrations in hazelnuts when compared to other nuts.[36]

4.3 Products, Coproducts, and By-products Obtained from Nuts

Nuts are consumed as a snack in their natural form, roasted, salted, or caramelized. Consumers usually prefer roasted nuts since roasting improves nuts' flavor, color, crunchiness, and texture. Brown pigments and pyrazine, associated with the typical roasted flavors, are formed during roasting through Maillard reactions.[52] In addition, the roasting process causes changes in lipids by oxidation, decreases the moisture, and increases the concentration of volatile compounds.[20]

Figure 4.2 shows that the processing of nuts results in products, coproducts, and by-products that can be used either as ingredients in foodstuffs or

Table 4.3 Phenolic compounds of the most consumed nuts worldwide.[a]

Nuts[36,40,41,47–51]	Total phenolic content (mg GAE[a] 100 g^{-1})	Flavonoids (mg 100 g^{-1})	Phenolic acids[b] (μg 100 g^{-1})	Proanthocyanidins (mg 100 g^{-1})	Isoflavones (μg 100 g^{-1})	Lignans (μg 100 g^{-1})	Phytoestrogens (μg 100 g^{-1})
Almonds	418	11.0	9.6	153	18	112	131
Brazil nuts	310	na[c]	193.2	na	na	na	na
Cashew nuts	274	1.98	456.8	2.02	22.1	99.4	122
Hazelnuts	835	11.96	457.0	491	30.2	77.1	108
Macadamias	156	na	675.8	na	na	na	na
Pecans	2016	34	3374.3	477	na	21.0	na
Pine nuts	68	0.49	810.0	na	na	na	na
Pistachios	1657	15.24	693.3	226	177	199	383
Walnuts	1556	2.71	890.9	60.32	53.3	85.7	140

[a] GAE: gallic acid equivalents.
[b] Fresh weight.
[c] na: not available.

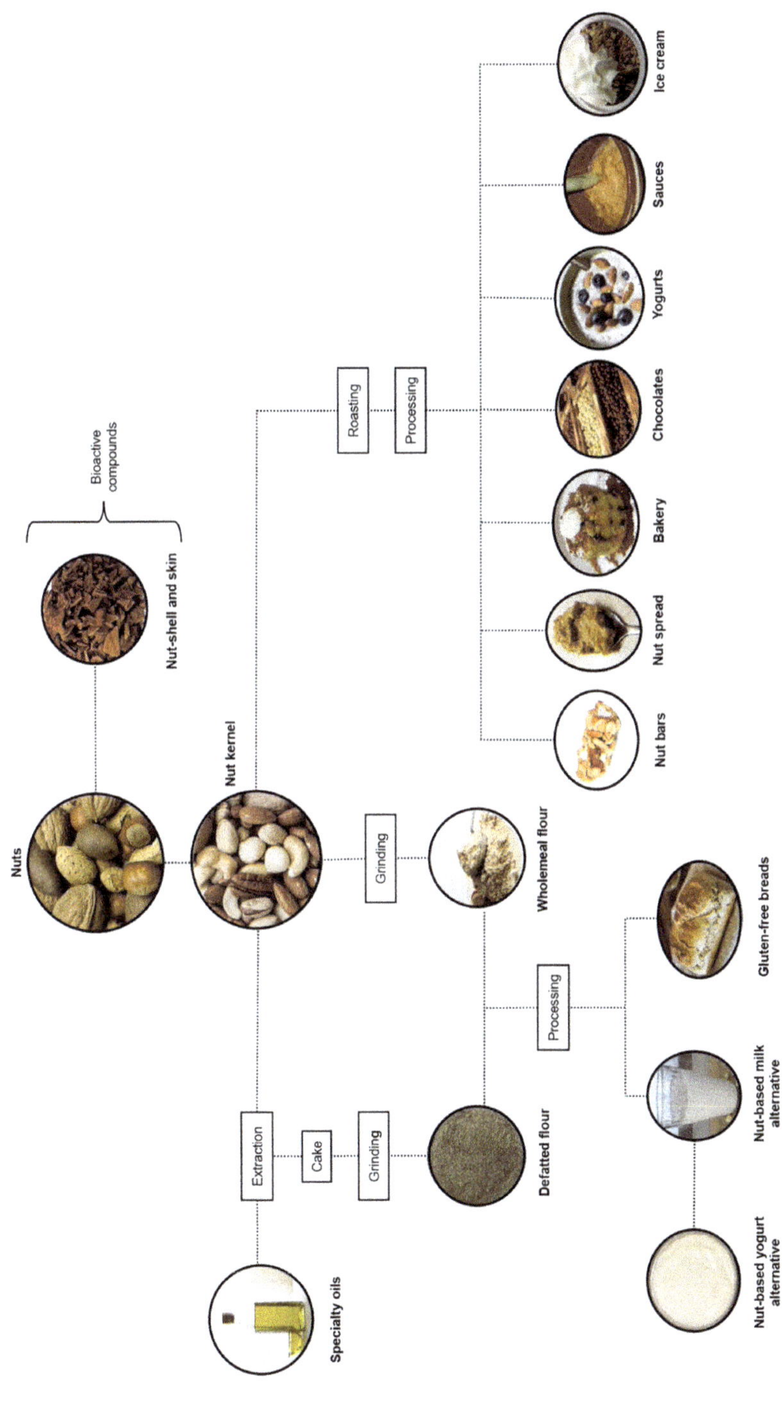

Figure 4.2 Products, coproducts, and by-products obtained from the processing of nuts and their application in foodstuffs.

to obtain extracts rich in bioactive compounds. Bakery and pastry products, sweets, sauces, chocolates, ice creams, spreads, and nut-based milk can be formulated with different nuts usually after the roasting process. Oils obtained from nuts after pressing are considered specialty oils and are used in foods, cosmetics, soaps, shampoos and conditioners, skin moisturizers, and lip balms.[20]

The defatted flour obtained from the cake after the pressing of the nuts for obtaining the oil and the wholemeal flour are coproducts rich in nutrients and bioactive compounds. These coproducts can be used to produce bakery, confectionery products, plant-based milk, and yogurts.[18,19] Moreover, the literature reports that extracts obtained from nutshells with water, ethanol, and their mixtures are rich in polyphenols.[49,53,54] The products, coproducts, and by-products from the nut processing with potential for use in the food, cosmetic, and pharmaceutical industries will be presented in Sections 4.3.1–4.3.5, and their characteristics and uses will be discussed in detail.

4.3.1 Oils

Oils obtained from nuts are classified as specialty oils due to their unique sensory and nutritional attributes. The price of these oils is usually high and can vary according to the availability and the processing used to obtain them.[55,56] Figure 4.3 shows that the main steps for obtaining oil from nuts are pre-drying, peeling, drying, oil extraction, and centrifugation.[57]

The pre-cleaning process includes the selection of the raw material considering its maturation stage. In this step, nuts with injuries or dirt materials such as leaves, stones, and nuts that are not properly matured are removed. Then the nuts are peeled and shelled before the oil extraction process. Oils from nuts can be obtained either by mechanical or hydraulic pressing. They are usually classified as extra virgin as their processing does not include the refining steps used for oils such as palm, soybean, canola, and other vegetable oils.[57] Mechanical pressing is very effective for recovering oil from oilseeds with an oil content above 20%. In addition, this technology uses mild conditions of temperature and pressure, is simple, solvent-free, and does not require much energy and investment.[58] These conditions preserve flavors and thermolabile compounds such as tocopherols and phenolic compounds.[59–61] The use of cold press has been reported for obtaining oil from almonds,[62] walnuts,[63] Brazil nuts,[59] cashew nuts,[64] pecans,[65,66] and macadamias.[67] The major drawback described for this process is the low oil yield obtained, which can be solved using combined extraction methods.[59,68,69] On the other hand, the extraction of oil using organic solvents such as hexane is a conventional method widely used alone or in combination with different raw materials. High temperatures are used, and step evaporation is required. In addition, organic solvents are flammable and toxic, and the nutritional, sensorial, and oxidative quality of the raw material is affected by this process.

The literature has reported that the pressing can be combined with alternative methods to recover specialty oils and high-value-added products

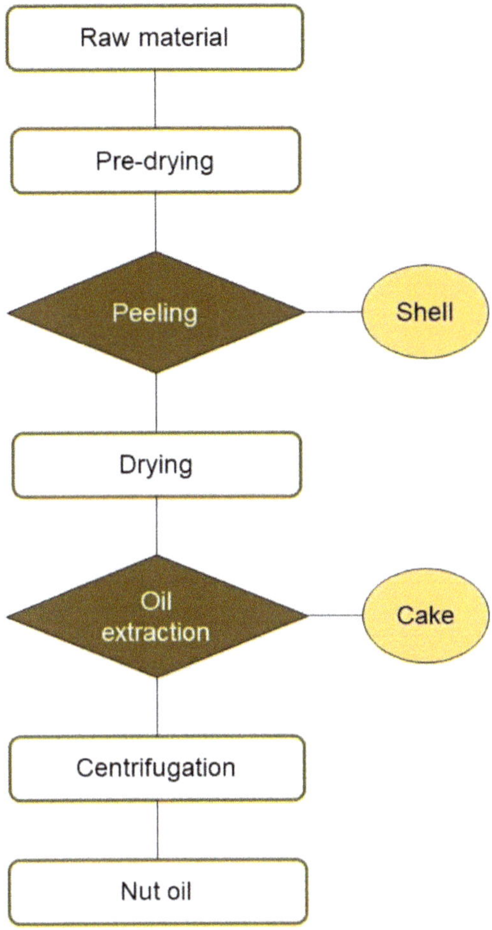

Figure 4.3 Processing steps to obtain oils, coproducts, and by-products from nuts.

from nuts. The emerging green extraction technologies such as supercritical fluid extraction (SFE), pressurized liquid extraction (PLE), subcritical fluid extraction (SubFE), ultrasound-assisted extraction (UAE), microwave-assisted extraction (MAE), and enzyme-assisted extraction (EAE) have been successfully developed and applied for this objective. The description of these methods and which raw materials they have been applied to will be presented in Section 4.4.

Table 4.4 shows that the oils obtained from nuts are rich in monounsaturated fatty acids, mainly oleic acid (C18:1, ω9). The concentration of oleic acid in nuts is similar to that of olive oil and these oils are classified as monounsaturated. The main polyunsaturated and saturated fatty acids are linoleic acid (C18:2, ω6) and palmitic acid (C16:0), respectively. Nuts have a balanced composition of fatty acids containing significant concentrations of

Table 4.4 Fatty acid, tocopherol, and phytosterol profiles in oils obtained from nuts.[a]

Fatty acids (%)	Almonds[56,70,71]	Brazil nuts[35,70]	Cashews[70,71]	Hazelnuts[70,71]	Macadamias[70,71]	Pecans[25,28]	Pistachios[56,70,71]	Walnuts[56,70,71]
Miristic (C14:0)	0–0.1	0–0.1	—	0–0.1	0.7–0.9	—	0–0.1	0.1
Palmitc (C16:0)	3.0–6.9	14.9–16.7	3.9–10.3	3.1–5.8	6.1–8.4	4.4–5.4	5.3–7.4	6.7
Palmitoleic (C16:1 *cis* 9)	0–0.2	0–0.4	0.1–0.3	0.1–0.3	13.0–17.3	—	0.5–0.7	0.2
Margaric (C17:0)	—	—	0–0.1	—	—	—	—	—
Stearic (C18:0)	0.7–1.3	9.9–11.9	3.2–9.8	1.3–2.7	2.3–3.2	1.7–1.4	0.5–0.9	2.3
Oleic (C18:1 *cis* 9)	31.3–69.2	28.5–36.3	23.5–60.6	45.4–79.3	43.8–65.2	40.6–71.8	22.7–58.2	21.0
Linoleic (C18:2 *cis* 9, 12)	12.3–21.5	36.0–37.5	7.8–17.0	7.8–10.4	1.3–2.3	20.2–20.6	14.1–30.3	57.5
Linolenic (C18:3 *cis* 9, 12, 15)	0–0.2	0.1–0.2	0–0.2	0–0.5	0.1–0.2	0.8–1.0	0.3–0.4	11.6
Arachidic (C20:0)	0–0.2	0–0.2	0.2–0.7	0–0.2	1.9–2.3	0–0.1	0–0.6	0.1
Gondoic (C20:1 *cis* 11)	—	0–0.1	0–0.2	—	—	0–0.2	0–0.6	—
Behenic (C22:0)	0–0.1	—	0.2–0.1	—	0.6–0.2	—	0–0.3	0.1
∑Saturated	3.7–8.6	24.9–28.9	7.5–21.3	4.4–8.8	11.6–15.0	6.1–7.0	5.8–9.3	9.3
∑Monounsaturated	31.3–69.2	28.5–36.8	23.8–61.1	45.5–79.6	56.8–82.5	40.6–71.8	23.2–59.5	21.2
∑Polyunsaturated	12.3–21.7	36.1–37.7	7.8–17.2	7.8–10.9	1.4–2.4	21.0–21.6	14.4–30.7	69.1
Tocopherol (μg g^{-1})								
α-Tocopherol	439.5	72.5	78.4	310.1	122.3	1.7	15.6	20.6
β-Tocopherol	—	—	1329.8	—	—	—	—	—
γ-Tocopherol	12.5	74.4	300.3	61.2	—	26.8	275.4	300.5
δ-Tocopherol	—	5.9	6.3	—	—		—	—
Total	452.0	152.8	1714.8	371.3	122.3	28.7	291.0	321.1
Phytosterols (mg 100 g^{-1})								
Brassicasterol	—	1.5	10.1	—	—	—	—	—
β-Sitosterol	207.2	39.5	0.8	99.1	150.7	—	468.6	113.0
Campesterol	5.5	4.0	—	6.7	7.3	—	23.7	5.1
Stigmasterol	5.2	11.3	—	3.8	3.8	—	66.3	5.6
Sitostanol	—	39.5	0.9	—	—	—	—	—
δ5-Avenasterol + δ7-stigmasterol	—	6.7	—	—	—	—	—	—
Total	217.8	47–148	11.8	11.0	161.8	0.2–0.3	558.6	123.6

[a] —: not presented.

MUFAs and saturated fatty acids (SFAs).[21,26] The oil obtained from nuts are also rich in tocopherols and phytosterols. Cardoso *et al.*[35] reported a total tocopherol content of 152.8 $\mu g\ g^{-1}$ for Brazil nuts, mainly α-tocopherol (72.6 $\mu g\ g^{-1}$) and γ-tocopherol (74.4 $\mu g\ g^{-1}$). Cashews followed by almonds have the highest tocopherol content (1714.8 and 452 $\mu g\ g^{-1}$, respectively). The main isomers in these raw materials are β- and α-tocopherols (1329.8 and 439.5 $\mu g\ g^{-1}$, respectively). The highest phytosterol content has been reported for pistachios (558.6 mg 100 g^{-1}) and almonds (217.8 mg 100 g^{-1}). The main phytosterols reported for nuts are β-sitosterol and brassicasterol.

4.3.2 Nut Flours

Almond,[72] Brazil nut,[73] pistachio,[74,75] chestnut,[76] pecan,[77] and cashew flours[78] have been reported as alternatives in restrictive diets. Under some conditions such as celiac disease, gluten intolerance, or gluten allergy the consumption of wheat is not allowed. These flours can also be an alternative to low-carb diets imparting a nutty flavor in bakery, confectionery, fillings, and nut spreads.[79–82] The flour obtained from nuts can be whole or defatted, which results from the defatted cake generated after the extraction of the nut oil.[72,73,83] Figure 4.4 shows the processing for obtaining nut flour.

The manufacturing of gluten-free products requires the replacement of wheat, oat, rye, and barley flours and grains. The viscoelasticity of the dough, gas retention, and water absorption in gluten-free products also need to be improved. Guar gum, xanthan gum, and alginates can be added to improve properties such as the texture and stability of the dough. In addition, the combination of different flours can be used in gluten-free baked goods to achieve the desired texture.[84]

The development of cookies with almond flour[72] and muffins with pecan flour[77] has been reported in the literature. The lipids in the nut flours contributed to the homogenization and emulsification of the dough, improving the nutritional value and the texture of these foodstuffs.[84] Table 4.5 shows that the nut flours present high contents of lipids (18–64 g 100 g^{-1}) and proteins (15–44 g 100 g^{-1}).

Bakery products such as bread, cookies, muffins, cakes, and cream crackers formulated with nut flours have been well accepted by consumers.[77,79,83] Garcia-Mendoza *et al.*[90] reported a total phenolic content between 10.4 and 13.3 mg GAE g^{-1} and antioxidant capacity by DPPH between 83 and 107 μmol Trolox equiv. g^{-1} for walnut cake flour. Bisinotto *et al.*[34] reported that the phenolic compounds of defatted cashew nut flour showed high bioaccessibility in simulated gastrointestinal digestion.

4.3.3 Shells and Skins of Nuts

The shells and skins (peel) of nuts, which represent between 40 and 50% of the raw material, are by-products usually discarded. Almond,[91,92] macadamia,[93] and hazelnut shells[94] can be used to produce lignocellulosic biomass

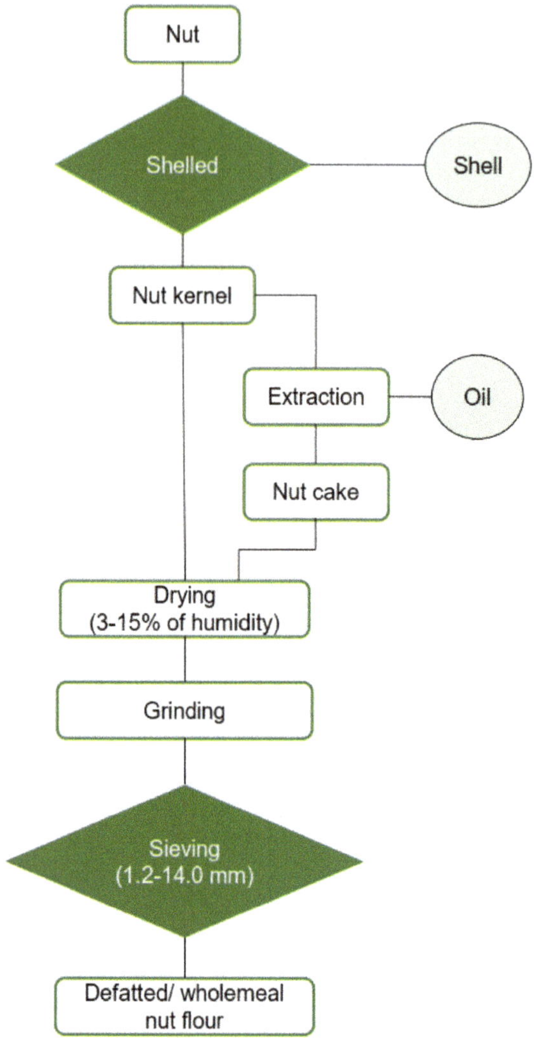

Figure 4.4 Processing steps to obtain nut flours.

since they have high concentrations of cellulose, hemicellulose, and lignin. Almond and pistachio shells have been studied to obtain biochar.[95,96]

The prebiotic potential of natural and blanched almond skins was demonstrated using a model of gastrointestinal digestion and a model of *in vitro* colonic fermentation.[97] The authors showed that the almond skin powder increased the growth of the *Bifidobacteria* and *Clostridium coccoides* group and decreased the numbers of the *Clostridium hystolyticum* group. The prebiotic activity was attributed to the dietary fiber in almond skins. In an intervention study in humans, the consumption of almond skin powder increased significantly the number of *Bifidobacterium* spp. and *Lactobacillus* spp. in faecal samples.[98]

Table 4.5 Composition of nut flours from different raw materials on a wet basis.

Nuts flour	Moisture (g 100 g^{-1})	Ash (g 100 g^{-1})	Proteins (g 100 g^{-1})	Lipids (g 100 g^{-1})	Carbohydrates (g 100 g^{-1})	Energy value (kcal g^{-1})
Almonds[a] [85]	4.67	3.32	22.24	58.70	—	—
Brazil nuts[a] [73]	—	—	22.24–24.69	51.14–58.70	—	—
Cashew Nuts[a] [86]	5.70	4.40	25.30	36.70	28.00	543.50
Hazelnuts[a] [87]	4.64	2.06	15.35	64.04	13.92	693.44
Pecans[b] [88]	10.70	5.90	42.10	1.60	39.80	342.00
Pistachios[b] [82]	6.01	4.50	36.01	21.90	36.61	487.58
Walnuts[b] [89]	3.5–11.00	4.00–5.00	39.00–44.00	18.00–26.00	8.00–15.00	350–470

[a] Wholemeal nut flour.
[b] Defatted nut flour.

Sharma *et al.*[99] reported that the testa, a thin layer of reddish-brown color that covers the cashew nut, is rich in hydrolyzable tannins, polymeric proanthocyanidins, catechin, epicatechin, and syringic, gallic, and *p*-coumaric acids.[99] John and Shahidi[49] reported a higher concentration of phenolic compounds in the shell and brown skin of Brazil nuts when compared to the kernels. The authors reported high antioxidant activity for Brazil nuts using the Trolox equivalent antioxidant capacity (TEAC), 1,1-diphenyl-2-picrylhydrazyl (DPPH) radical, hydroxyl radical scavenging activities by electron paramagnetic resonance (EPR), and oxygen radical scavenging capacity (ORAC). The phenolic compounds identified were gallic acid, gallocatechin, protocatechuic acid, catechin, vanillic acid, taxifolin, myricetin, ellagic acid, and quercetin.

Pecan nutshells are commercialized to make tea and used in folk medicine in southern Brazil for the treatment of intestinal constipation, gastritis, and as a diuretic. These therapeutic effects have been associated with the phenolic acids, flavonoids, and proanthocyanidins in the pecan nutshell.[100] do Prado *et al.*[100] reported a high content of total phenolics and antioxidant activity in pecan nutshell infusion. Moreover, Hilbig *et al.*[53] reported cytotoxic effects, induction of cell death by apoptosis, and cell cycle arrest in MCF-7 cells for pecan nutshell extracts. An increase in the survival time of 67% in mice with Ehrlich ascites tumors fed with the extracts was also observed. In addition, DNA damage was observed in the CT-DNA, plasmid DNA, and comet assays. The authors indicated that the mechanism involved in the antitumor effect of pecan nutshell extracts might be related to the activation of essential proteins involved in apoptotic cell death and cell cycle regulation. These results were attributed to the phenolic compounds in the extracts such as gallic, 4-hydroxybenzoic, chlorogenic, vanillic, caffeic, and ellagic acid, catechin, epicatechin, epigallocatechin, and epicatechin gallate. The results indicated that pecan nutshell extracts are effective against tumor cell growth and may be considered an alternative to cancer treatment.[53,54] Barreira *et al.*[79] also reported antimicrobial and antitumor activity for extracts obtained from pecan peels.

4.3.4 Nut-based Milk Alternatives

Plant-based milk alternatives can be formulated with nuts (almonds, coconut, hazelnuts, pistachios, walnuts, Brazil nuts, and cashew nuts), oilseeds (sesame, flax, hemp, and sunflower), vegetables (soy, peanuts, lupines, and cowpea), cereals (oats, rice, corn, and spelt), and pseudocereals (quinoa, teff, and amaranth).[101,102] The consumption of this type of beverage has been observed worldwide because of conditions such as cow's milk allergy, lactose intolerance, caloric concern, and high cholesterol prevalence. Veganism, vegetarianism, concerns with the environmental impact caused by milk production, and fears about the presence of hormones or antibiotic residues in dairy cows also stimulate the sales and increase of this market.[103–105]

The first step in the processing of nut-based milk is the crushing of the raw material. The resulting particle size distribution should range from 5 to 20 μm to mimic the cow's milk appearance and consistency. Then, the crushed raw material is homogenized in water in different proportions depending on its composition, especially the percentage of lipids. The steps for producing nut-based milk alternatives from Brazil nuts,[106] almonds,[107] and walnuts[108] include cracking, deshelling, peeling, grain grinding, homogenization, decanting, and sterilization (Figure 4.5).

Table 4.6 shows the composition of nut-based milk alternatives obtained from different raw materials. The protein content in plant-based milk

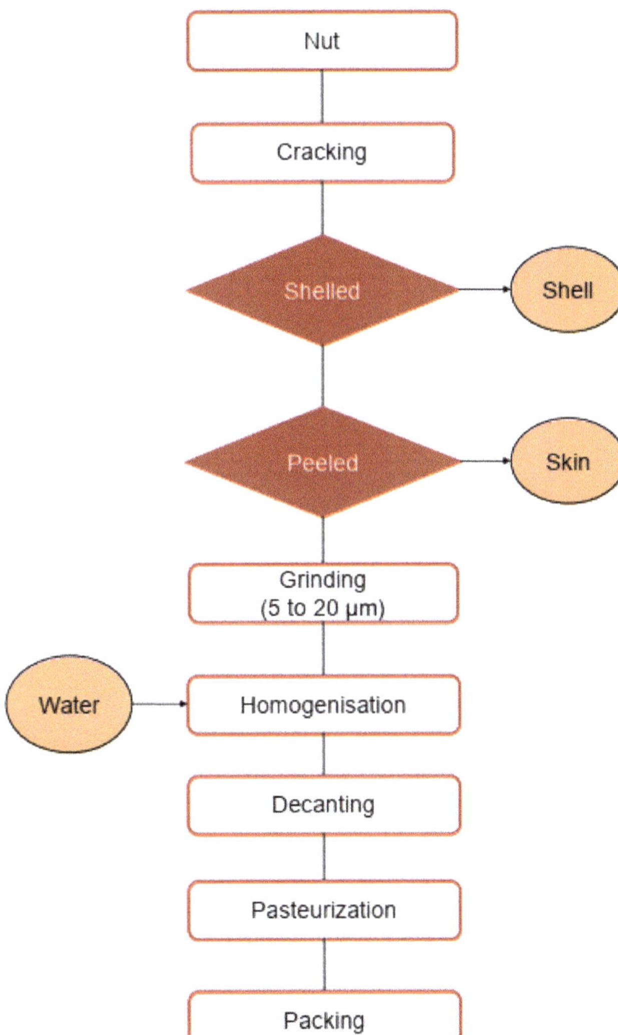

Figure 4.5 Processing of nut-based milk alternative.

Table 4.6 Nutritional composition per 100 mL of nut-based milk alternative obtained from different raw materials.[a]

Nut beverages	Energy (kcal)	Proteins (g)	Lipids (g)	Carbohydrates (g)	Ca (mg)	Fe (mg)	Mg (mg)	Na (mg)
Almonds[109,112]	13–55	0.3–2.0	1.1–5.5	0.2–4.5	42–188	1.4	42.1	6.4
Cashew nuts[112,113]	25–79	0.4–2.2	1.0–5.3	3.6–5.7	9–188	—	—	—
Hazelnuts[112]	29	0.40	1.60	3.1	—	—	—	—
Macadamias[109,113]	38	0.3	2.7	2.8	—	—	—	—

[a] —: not presented.

(0.3–2.2 g 100 mL^{-1}) is lower than that of commercial milk (3.4 g 100 mL^{-1}). Almonds and cashews usually present a higher protein level than other nut-based milk alternatives. Scholz-Ahrens *et al.*[109] showed that cow's milk proteins have a Digestible Indispensable Amino Acid (DIAAS) score of 114–118, higher than plant proteins. The DIAAS value reported for almond extract is 40 which is considered low. Proteins provide amino acids that perform structural, defense, transport, and regulatory functions in the human body. The deficiency of essential amino acids in the diet can reduce protein synthesis leading to physiological and biochemical changes. Thus, the replacement of milk with plant-based milk needs to be planned to ensure adequate consumption of essential amino acids.[110] On the other hand, these beverages are richer in fibers and unsaturated fatty acids and they are cholesterol-free. The lipid content ranges from 1.0 to 5.5 g 100 mL^{-1}, which is higher than in whole milk, 3.0–3.5%.[102,111] The calcium values of nut-based beverages range from 9 to 188 g 100 mL^{-1} which is lower than those of cow's milk (122.21 mg 100 mL^{-1}).[110] The iron content of almond-based beverages (1.4 g 100 mL^{-1}) is higher compared to that of cow's milk (0.05 mg 100 mL^{-1}).[110,112]

Based on the differences in the composition between plant-based and bovine milk, consumers should be advised that plant-based milk cannot be the only alternative to substitute bovine milk. This is especially important for calcium since a diet rich in other sources of calcium or supplementation with this mineral is necessary. This is strongly recommended for babies and children who do not consume the same variety of foods as adults. Labeling and public education are also necessary to ensure that the public understands the nutritional differences between plant-based and bovine milk.[112] The positive effects of the consumption of nut-based milk alternatives are related mainly to their high content of unsaturated fatty acids and phenolic compounds. However, there is a lack of information on the composition of sterols, vitamins, amino acids, dietary fiber, and other bioactive substances present in nut-based milk alternatives.[104]

Prebiotic beverages have in their composition compounds that probiotic microorganisms can ferment in the large intestine conferring beneficial effects on health. Prebiotics include oligosaccharides, human milk oligosaccharides, conjugated linoleic acid, polyunsaturated fatty acids, fermentable dietary fiber, phenolics, and other phytochemicals.[114] The International Scientific Association for Probiotics and Prebiotics (ISAPP) has defined symbiotics, which combine prebiotics and probiotics, as "a mixture comprising live microorganisms and substrate(s) selectively utilized by host microorganisms that confers a health benefit on the host".[115]

Nuts are nutrient-dense, sensorial pleasant and they have polymerized polysaccharides and polyphenols with a prebiotic effect. These characteristics are ideal for the formulation of dairy-free probiotic beverages and yogurt alternatives.[116,117] The presence of prebiotics has been reported in the skins of hazelnuts,[94,118] pistachios,[119] and almonds.[120] The formulations of prebiotic beverages with cashew nuts,[121] and cashew nuts and passion fruit juice[122] have also been reported.

The development of yogurt alternatives formulated with cashew nuts and almonds with texture, flavor, and probiotic properties similar to yogurt has been reported. Yogurt is defined "as the product manufactured from cow milk with a gel structure that results from the coagulation of the milk proteins, due to the lactic acid produced by defined species of bacteria during fermentation".[123] *Streptococcus salivarius* subsp. *thermophilus* and *Lactobacillus delbrueckii* subsp. *bulgaricus*, which have an optimal growth temperature of 40–44 °C, are used as bacterial cultures because of their synergistic action.[123,124] The same initial cultures used in yogurts are used to produce nut-based yogurts. On the other hand, in probiotic beverages, which are more liquid and drinkable when compared to yogurts, other microorganisms and temperatures are used for the fermentation process.[124]

4.3.5 Nut Spreads and Snack Bars

Nut spreads are usually made with 40% of roasted and ground nuts and they have between 10% and 90% of fat.[125] The oily fraction is formed by the mixture of vegetable fats and/or a dispersed phase of the roasted nuts. Sugar, cocoa powder, milk, whey powder, lecithin, or other emulsifiers are usually added. The development of nut spreads has been reported as an alternative to increasing the value-addition to nuts and their coproducts.[125,126] Lima *et al.*[126] obtained cashew nut spread with good results for physicochemical, microbiological, and sensory properties. Oxidative stability of 20 days was reported for pistachio spread kept at 60 °C.[125] Mônaco *et al.*[127] reported that flavor and texture are the main attributes for acceptance of the hazelnut spread.

Cereal bars have been developed as an alternative to replacing the consumption of sweet products without nutritional value and as a source of energy and protein for athletes.[128,129] Nuts such as almonds, cashews, walnuts, and Brazil nuts are used in the formulation of cereal bars and they are widely accepted by consumers.[128,130,131]

4.4 Extraction of Bioactive Compounds from Nuts and Their Applications

There is no universal extraction procedure suitable to promote the best extraction yield for bioactive compounds. The recovery and composition of the extracts obtained depend on the solvent and the method of extraction used. It happens because the polyphenols, polysaccharides, and proteins from the raw material can form complexes.[132] The phenolic compounds have a high number of hydrogen donor and acceptor sites (HDS and HAS) which can establish intermolecular interactions through electrostatic attractions and hydrogen bonds with the HDS and HAS of the solvents.[40,132] Thus, the phenolic compounds can be solubilized in solvents that can form hydrogen bonds such as water, methanol, ethanol, acetone, and ethyl acetate.

Therefore, the most suitable solvent and technique for the extraction of phenolic compounds depend on the raw material and the composition of the extract desired. The water solubility of phenolic compounds increases with the number of hydroxyl groups present, making the phenolic glycosides more soluble in water than their corresponding aglycones. On the other hand, methanol is effective for obtaining anthocyanin-rich phenolic extracts since it can denature cell membranes, dissolving and stabilizing the anthocyanins.[134]

Albuquerque Mendes *et al.*[135] obtained extracts of phenolic compounds from cashew nuts using acetone, water, ethanol, and methanol. The highest values of total phenolic compounds using the Folin Ciocalteu method were obtained using ethanol (5.93 mg GAE g^{-1}). In addition, ESI-Q/TOF-MS analyses revealed the presence of fatty acids, phospholipids, and sugars in the extract obtained with the ternary mixture of acetone–methanol–ethanol using different proportions. On the other hand, the authors reported that for the ethanolic extract only phenolic compounds were extracted. Erşan *et al.*[134] obtained acidified (0.1% HCl, v/v) aqueous methanol (80 : 20 v/v) extracts from ground red and green pistachio hulls. They identified, by HPLC-DAD-ESI-MS, 66 phenolic compounds in the extracts. However, the interest in using safe and biodegradable solvents for separation processes is growing since the organic solvents from petroleum resources are volatile, toxic, flammable, and associated with several health and environmental risks.[7,133] These safety and environmental concerns have encouraged the use of water or water–ethanol mixtures for the extraction of phenolic compounds from different raw materials. Recently, green solvents, which are produced from renewable sources by fermentation or chemical transformation processes, have been classified into four main solvent classes: (a) aqueous solvent systems, (b) bio-based solvents, (c) ionic liquids (ILs), and (d) deep eutectic solvents (DESs), and more recently (e) natural deep eutectic solvents (NADES). Parameters such as solubility, boiling point, toxicity index, and the energy required for solvent evaporation/separation by the target compound need to be considered for the selection of an alternative solvent.[7,136,137]

Bio-based solvents are obtained from sugar, starch, lignocellulose, protein, and oil, usually from forestry and food wastes. The green solvents have high solvent power and low toxicity and in addition are non-flammable, environmentally safe, and biodegradable.[138] Ethanol is the most common bio-based solvent and is considered as generally recognized as safe (GRAS).[139] The use of water or ethanol and their combination has been reported to extract bioactive compounds from nuts such as almonds;[48] pine nuts;[140] macadamias;[141] walnuts;[142] pistachios;[143] hazelnuts;[144] and hazelnuts, walnuts, almonds, and peanuts,[145] evidencing the possibility to use this type of solvents to obtain an economically viable and environmentally safe extract.

Ionic liquids (ILs) are synthetic organic salts with melting points below 100 °C. ILs are nonvolatile solvents and some of them have low viscosity, facilitating pumping and agitation operations on a large scale. In addition, they present low vapour pressure, high thermal and chemical stability, high conductivity, and capacity for solubilizing various organic and inorganic compounds.[146] On the other hand, ILs present poor biodegradability and biocompatibility and high cost.[147] Because of their limited application, no reports can be found in the literature regarding ILs to recover bioactive compounds from nuts.

Sugars, organic acids, amino acids, or organic bases can be mixed in certain molar ratios resulting in NADES, with a lower melting point at room temperature than their single components alone. NADES showed higher efficiency to recover the target compounds.[148,149] The use of 100% natural ingredients, mainly primary metabolites of vegetables and fruits, and the low price are considered the main advantage of NADES when compared to ILs.[150] NADES composed of choline chloride : 1,2-propylene glycol (1 : 4 w/w) was used to extract phenolic compounds from hazelnuts. The extraction was assisted by microwave and the antioxidant activity of the extracts was 2–3 times higher than those obtained with ethanol.[136] The extracts obtained from the nut *Carya cathayensis* using ultrasound and choline chloride : malic acid (1.5 : 1 w/w) exhibited the highest phenolic compound content, antioxidant capacity, and α-glucosidase and α-amylase inhibition activities when compared to other NADES and with water, ethanol, methanol, and their combination with water (1 : 1 v/v).[151]

Conventional extraction techniques such as maceration and Soxhlet extraction can provide high extraction yields. On the other hand, for both methods, the extraction is time consuming, and the temperature used in Soxhlet extraction can cause oxidation of phenolics, degradation, and decrease of bioactivity. Moreover, some organic solvents are prohibited for use in foods and the elimination of solvent residues is difficult. The interest in the development of green, cheaper, more efficient, and less polluting processes to obtain bioactive compounds from natural sources is increasing.[132] The use of safe solvents, the decrease of extraction time and temperature, the improvement of mass transfer during extraction, and the increase of the extraction yield are the main goals of these alternative technologies. Microwave-assisted extraction (MAE), pressurized liquid extraction (PLE), supercritical fluid extraction (SFE), and ultrasonic-assisted extraction (UAE) are the most studied green alternative processes for the extraction of phenolics from different raw materials.[40,137] The extraction of phenolic compounds using MAE, PLE, and UAE with different proportions of ethanol–water has been reported in several studies (see Table 4.7). Although MAE and UAE usually are related to shorter extraction times, a lower temperature (40 °C) can be used with PLE and SFE. Moreover, the extraction yields can be improved by optimizing and combining different methods.[152] Table 4.7 summarizes the studies reported using different technologies and solvent systems for extracting bioactive compounds or oil from nuts and these technologies will be discussed in more detail in Sections 4.4.1–4.4.4.

Table 4.7 Bioactive compounds recovered from nuts and their coproducts using different extraction techniques.[a]

Method of extraction	Nut	Solvents	Conditions of extraction	Bioactive compound recovered	Ref.
Maceration	Cashew nuts	Ethanol	Stirred at 210 rpm for 2 h and then centrifuged at 3500 rpm for 15 min	Phenolics	Albuquerque Mendes *et al.* (2019)[135]
Maceration	Almonds	Water	Not informed	Flavonoids and phenolics	Milbury *et al.* (2006)[48]
Maceration	Walnuts	Water and ethanol	Not informed	Phenolics	Labuckas *et al.* (2016)[142]
Maceration	Hazelnuts, walnuts, almonds, and peanuts	Ethanol	Not informed	Flavonoids and phenolics	Salem *et al.* (2022)[145]
Water bath	Macadamias	Water	Not informed	Phenolics	Dailey and Vuong (2015)[153]
MAE	Almonds	Ethanol 70% (v/v)	100 W, 60 s	Phenolics	Valdés *et al.* (2015)[154]
MAE	Hazelnuts	NADES (choline chloride: 1,2-propylene glycol)	24% water, 38 min, 92 °C	Phenolics	Bener *et al.* (2022)[136]
MAE	Peanuts	Water/ethanol (70 : 30, v/v)	90% microwave power, 30 s	Phenolics	Ballard *et al.* (2010)[155]
UAE	*Carya cathayensis* Sarg	NADES (choline chloride:malic acid)	Probe sonication 400 W, 10 min, at 65 °C	Several bioactive compounds	Fu *et al.* (2022)[51]
UAE	Walnuts	Ethanol 50% (v/v)	Probe sonication, 1500 W, 15 min, at 25 °C	Phenolics	Han *et al.* (2018)[56]

Table 4.7 *(Continued)*

Method of extraction	Nut	Solvents	Conditions of extraction	Bioactive compound recovered	Ref.
UAE	Pistachios	Acidified (0.1% HCl, v/v) aqueous methanol (80%, v/v)	Probe sonication at 70% amplitude, 30 s	Phenolics	Erşna *et al.* (2016)[134]
UAE	Hazelnuts	Ethanol	Probe sonication (130 W), 90% amplitude for 51 min, at 29 °C	Oil	Wong *et al.* (2019)[144]
SFE	Pecan nuts	Carbon dioxide	300 bar and 58.4 °C	Residual oil	Salvador *et al.* (2016)[157]
SFE	Pecan nuts	Carbon dioxide	250 bar and 20 °C	Residual oil	dos Santos Alves *et al.* (2019)[158]
SFE	Almonds	Carbon dioxide	40 MPa, 40 °C, 10% of ethanol as co-solvent	Oil	Balvardi *et al.* (2015)[159]
SFE	Pistachios	Carbon dioxide	15 MPa, 200 °C	Several bioactive compounds	Bodoira *et al.* (2019)[40]
PLE	Peanuts	Ethanol 60% (v/v)	7 MPa, 220 °C, 7 g min^{-1}	Phenolics	Rossi *et al.* (2020)[160]
PLE	Pine nuts	Water	10 MPa, 50 °C, 3 cycles of 1.5 mL	Extracts enriched in inositols	Ruiz-Aceituno *et al.* (2014)[140]
PLE	Brazil nuts	Isopropyl alcohol	10.35 MPa, 50 °C, 40 min	Oil	Cornelio-Santiago *et al.* (2019)[161]

[a] MAE: microwave-assisted extraction; UAE: ultrasound-assisted extraction; SFE: supercritical fluid extraction; PLE: pressurized liquid extraction.

4.4.1 Microwave-assisted Extraction (MAE)

MAE is performed using high frequency (300 MHz to 300 GHz) non-ionizing electromagnetic radiation which is converted to heat by ionic conduction and dipole rotation.[162] The internal and external heating of the matrix without a thermal gradient increases the solvent diffusion and improves the recovery of the target compounds from the raw material.[163]

The systems for the extraction using MAE usually have an open or a closed vessel. In an open vessel, the extraction is carried out at atmospheric pressure with the reflux of the solvent. On the other hand, in closed vessels, the compounds are extracted at high temperatures and high pressures.[7,155] The selective and homogeneous heating of the sample/solvent mixture, fast energy transfer, low energy consumption, shorter times, low solvent volumes, less production of waste, and higher volume of raw material processed are the main advantages reported for MAE. However, MAE needs an additional stage after extraction to remove solid residues and the efficiency of the extraction can be poor when the target compounds are nonpolar, volatile, or thermolabile.

MAE has been used for the extraction of resveratrol, proanthocyanidins, and flavonoids from nuts.[136,141,154] Valdés *et al.*[154] obtained an extract rich in flavonoids (86.4 mg of QE g^{-1}) from blanched almond skins using MAE and water/ethanol (30 : 70, v/v). In addition, Dailey and Vuong[141] showed that the extraction of proanthocyanins from macadamia skins using MAE and water was effective.

4.4.2 Ultrasound-assisted Extraction (UAE)

UAE has been reported as an efficient emerging technology to recover phytochemical compounds from nuts. The short processing time and high extraction yields are the main advantages of this technique.[162,164] UAE, which is based on the cavitation phenomenon, causes intense shear stress associated with turbulence at extreme levels.[164] The cavitation phenomena change the microstructure of the raw material since it causes ruptures in the cell walls facilitating the penetration of the solvent and improving the mass transfer. UAE is also an environmentally friendly technique since it can be used for extracting bioactive compounds from coproducts and by-products with less residue generation and a shorter processing time.[165] UAE requires low energy consumption and low maintenance cost to obtain high-quality phytochemical extracts. Therefore, UAE has economic potential for large-scale application in the food and pharmaceutical industries.[164,166]

Frequencies above the audible range (10 Hz–20 kHz) are used in low-frequency ultrasound. The energy generated makes the medium vibrate and then the energy is transferred through the ultrasound waves to the particles in the sample.[167] The power, frequency, and amplitude are the main physical parameters that need to be controlled in this process. The energy level at which ultrasound propagates through the medium can be expressed as ultrasound power (W), ultrasound intensity (W cm^{-2}), or acoustic energy density (W cm^3 or W mL^{-1}).[165,167] The UAE system can be designed as a water bath or a probe. UAE with a probe usually increases the mass transfer. It has been attributed to the absence of obstacles during the ultrasound propagation process since it has a direct effect on the sample.[168] The disadvantage of this method is that the ultrasonic probe may lose its effectiveness due to prolonged use. In addition, during the extraction process, a loss of volatile compounds may be observed since the system is open.

Using a water bath is known as indirect sonication since the ultrasound waves need first to pass through the liquid in the ultrasonic device and then through the wall of the container. This results in a lower ultrasound intensity in the sample with low extraction selectivity. Moreover, the solvent cannot be renewed during the process requiring filtration and evaporation of the solvent.[7,168]

The extraction of phenolic compounds from different nuts by UAE has been reported in the literature.[151,156,169–171] Ballistreri *et al.*[169] obtained total phenolic compounds in pistachios ranging from 1.84 to 3.49 mg of GAE g^{-1} of nut using UAE and acidified methanol. Ma *et al.*[172] obtained extracts rich in proanthocyanidins and chlorogenic acid from almond skins using water and polyethylene glycol (50 : 50 v/v). Özcan *et al.*[171] found 106.25 mg of GAE 100 g^{-1} of phenolic compounds in almond kernels using UAE for 10 min and a solution of methanol:water acidified with HCl. Han *et al.*[156] showed that the extraction of phenolic compounds from walnuts using water : ethanol (50 : 50, v/v) and a UAE probe was two times higher when compared to that using a UAE-bath (52.6 and 25.8 mg of GAE g^{-1} of dry weight, respectively).

4.4.3 Pressurized Liquid Extraction (PLE)

PLE has been reported as a promising green extraction method mainly for recovering phenolic compounds from different raw materials using water, ethanol, or their mixtures at high pressures (3.3 to 20.3 MPa) and different temperatures (40–200 °C). Generally, pure water under subcritical conditions can dissolve phenolic compounds with a wide range of polarity.[162] The high pressure and controlled temperatures can maintain the solvents in their liquid state, even at temperatures much higher than their corresponding boiling points.[164] Under these conditions, a decrease in the viscosity and surface tension of water is observed increasing the diffusivity and facilitating the desorption and solubility of the target compounds into the solvent. The efficiency of the extraction under these conditions is increased.[162,173,174] PLE can be applied in a static or batch mode where the extractor is pressurized while the outlet valve is kept closed. The valve is opened and the extract is collected after the time set for the extraction. On the other hand, in the dynamic mode, the outlet valve is kept open and the solvent is pumped continuously through the extractor.[152] PLE has been reported for obtaining aqueous extract rich in inositol from pine nuts[140] and phenolic compounds (350.29 mg GAE 100 g^{-1}) from munguba[174] and from peanut skin using water:ethanol (40 : 60, v/v).[160]

The limitations reported for PLE include the low selectivity of the analyte during extraction, the high level of extract dilution, especially when using several numbers of cycles or continuous flow, and the need for advanced instrumentation increasing the cost of the process.[164]

4.4.4 Supercritical Fluid Extraction (SFE)

SFE is a fast, efficient, and clean extraction method based on a high-pressure extraction technique. The solute is separated from a solid matrix using a supercritical fluid obtained by working at pressures and temperatures above their critical point. In this state, the density is similar to liquids while viscosity is similar to gases, leading to the diffusivity between gases and liquids.[175,176] These properties allow supercritical fluids to penetrate cell matrices more effectively than solvents in the liquid state, improving the extraction yields and the quality of the extract.[152] SFE is also considered a clean and environmentally friendly technique that can be used for the extraction of bioactive compounds from nuts.[143] The temperature and pressure are often relatively low, avoiding damage to the properties of thermolabile compounds. In addition, these variables can be adjusted to optimize the extraction conditions and/or achieve a selective extraction of the target compound.[152]

The main solvent used in SFE is carbon dioxide (CO_2) which is non-toxic, environmentally friendly, non-flammable, easy to remove, and of low cost. Moreover, in this process, it is possible to use moderate pressures and room temperature since the critical point of CO_2 is 7.4 MPa and 31.1 °C. SFE can be run in static or dynamic modes; in static mode, the outlet valve is closed and the sample is kept in contact with the solvent during the extraction time. On the other hand, in the dynamic mode, the supercritical fluid is continuously fed into the extraction vessel, flowing through the sample and to the collection system. The advantages of SFE are its high selectivity, elimination of clean-up steps, low extraction time, and the possibility of obtaining extracts free of residues.[177]

The process may be affected by the mass transfer rate between the fluid and the solute depending on the raw material. Heavy solutes can also clog the exit valve. In addition, CO_2 has low polarity and is more suitable for the extraction of non-polar compounds. This limitation can be solved using a co-solvent with higher polarity such as ethanol, allowing the extraction of more polar compounds such as polyphenols improving the selectivity and extraction yield of the process.[152] SFE has been reported as an effective method to recover the oil from pecan nut cake.[157]

4.5 Effects of Nuts on Health

In the last few years, several studies have reported an inverse relationship between the consumption of nuts and non-communicable chronic diseases (NCCD) such as cardiovascular diseases, diabetes, neurodegenerative diseases, and cancer.[10] It has been reported in the literature that the coproducts and by-products from nuts are also rich in bioactive compounds related to beneficial effects on health. The use of this inexpensive and still underexplored biomass as ingredients in functional foodstuffs may improve the diet, and consequently, the health of the population.[42] Despite their high caloric content, nuts are rich in phytochemicals with bioactive properties. In fact, several studies reported that the by-products of nuts (skins, shells, cakes, and hulls among others) are richer in phenolic compounds, phytosterols, and vitamins. The almond skin contains 70–100% of the total phenolics present in the whole nut.[178] Interestingly, some phenolic compounds have been exclusively found in almond skin (8 out of a total of 19 identified flavonoids).[48] Hazelnuts also present more phenolic compounds in their skin (4 of a total of 7 identified polyphenols).[179] The effects on health of the coproducts and by-products from nuts will be presented and discussed in Sections 4.5.1–4.5.4.

4.5.1 Cardioprotection

Cardiovascular diseases (CVDs) are considered multifactorial diseases that affect the heart and/or the blood vessels. They include coronary artery disease, cerebrovascular disease, and peripheral vascular disease. The principal risk factors for CVDs are well-known and include high levels of low-density lipoprotein (LDL), cholesterol, hypertension, diabetes, abdominal obesity, psychosocial factors, smoking, poor consumption of vegetables, excess alcohol intake, and low physical activity. All these factors lead CVDs to be the main cause of death worldwide.[180]

CVDs are related to excessive cellular oxidative stress and inflammatory response.[181] Inflammatory process generates reactive oxygen and nitrogen species (ROS and RNS) at the site of inflammation faster than oxidative stress. Likewise, this number of reactive species also triggers an intracellular signaling cascade that promotes pro-inflammatory gene expression. Thus, inflammation and oxidative stress are present in many pathophysiological chronic diseases linked with each other such as diabetes, cardiovascular diseases, and neurodegenerative diseases.[182] The failures of the antioxidant/anti-inflammatory clinical trials can be related to the selection of agents that target specifically inflammation or oxidative stress. The use of both antioxidant and anti-inflammatory agents simultaneously that can retard or avoid oxidative and inflammatory pathways can be more effective.[182]

Dietary patterns such as the Mediterranean diet which is rich in vegetables, olive oil, fish, and nuts have been strongly correlated with a lower risk of CVDs as reported by the large nutritional intervention PREDIMED.[183] The cardioprotective mechanism of phytochemicals from nuts and their coproducts was linked to the improvement of the serum lipid profile. LDL cholesterol (LDL-c), especially its oxidized form, has been reported as the main marker of CVD risk[184] and its pathogenesis.[185] Several authors have highlighted the capacity of reduction on LDL-c oxidation by nut coproduct extracts. Wijeratne *et al.*[186] showed that *in vitro* cupric ion induced oxidation of human cholesterol was also effectively inhibited by flavonoid-rich almond skin extracts. Wijeratne *et al.*[186] and Chen *et al.*[187] corroborated these results against LDL-c oxidation in humans after almond skin consumption. These authors also found the synergic effect of α-tocopherol addition.[188] Indeed, nuts and their coproducts have been described as a high source of tocopherols mainly α-tocopherol and γ-tocopherol.[50,180,188] LDL-c oxidation may be prevented by the high antioxidant capacity of tocopherols combined with phenolic compounds such as flavonoids. These antioxidant molecules can act by donating a hydrogen atom to free radicals and forming stable compounds in order to stop the oxidation reaction cascade.[23] The Brazilian nuts are one of the most important sources of selenium in the diet and can be also found in large amounts in Brazilian cake.[188] This trace element is part of the selenoproteins which present antioxidant activity for protection against oxidative stress initiated by excess ROS and NOS.[189] Selenium also has the ability to increase the excretion of mercury (Hg), preventing the formation of reactive oxygen species induced by Hg.[190] Moreover, it has been reported that the ingestion of less than 40 μg of selenium a day can increase the risk of CVDs.[191]

A decrease in LDL and total cholesterol in the plasma of hamsters fed with a high-fat diet and hazelnut skin extracts during 8 weeks has been reported.[192] The reduction in LDL-c plasma levels has been attributed to the phytosterols in the coproducts of nuts. The phytosterols compete with cholesterol in the micelles in the intestinal lumen, reducing the amount of absorbed cholesterol.[193] Another proposed mechanism is the modification in the expression of genes that codify sterol carrier proteins (*e.g.* the Niemann-Pick C1-like 1 protein). Phytosterols may also reduce the rate of cholesterol esterification in the enterocyte increasing the cholesterol transintestinal excretion.[194]

Nuts and their coproducts are also rich in fiber and monounsaturated fatty acids (MUFAs). Fibers can help in total cholesterol and LDL-c downregulation by the increment in the faecal bile acid excretion.[195] On the other hand, consumption of MUFAs has shown benefits on CVD risk, reducing the total and LDL-c.[196] The consumption of nuts enriches the seric lipoproteins with MUFA, which seems to not be a substrate for oxidation reactions, possibly decreasing their susceptibility to oxidation and protecting against CVD.[5]

Salem *et al.*[145] reported the anti-inflammatory effect of peanut, almond, hazelnut, and walnut shell extracts using an *in vitro* THP-1 macrophage stimulated by LPS cell model.[145] Results showed downregulation in pro-inflammatory cytokines such as tumor necrosis factor-alpha (TNF-α) and interleukin 8 (IL-8) and inflammatory mediator enzymes such as nitric oxide synthase (iNOS) and cyclooxygenase (COX-2). The best results were obtained for peanut shells attributed to their high concentrations of flavonoid and phenolic contents.[145]

4.5.2 Neuroprotection

Alzheimer's disease (AD), Parkinson's disease, depression, and schizophrenia are neurodegenerative disorders characterized by neuronal loss and dysfunction with chronic and progressive impairment of brain tissue.[197] AD, which is considered the main neurodegenerative disorder, affected 47.47 million people in 2015. It has been reported that 75.63 and 135.46 million cases are expected in 2030 and 2050, respectively. These data show the great social and economic impact of AD on the population.[197] The etiology of AD remains unclear and the current palliative treatments are based on cholinesterase inhibitors that delay the progression and symptoms associated with AD.[198] The main neuropathological characteristics of AD are oxidative stress, deterioration of the cholinergic system, neuroinflammation, aggregation of amyloid beta (Aβ) plaques, and hyperphosphorylation of tau proteins. These mechanisms are associated with damage and neuron death.[199]

It has been reported that nuts and their coproducts can present neuroprotective effects in complex and multifactorial AD.[37] The aqueous extract of hazelnuts added to rats' diet improved the response of the animals in memory tests. In addition, a reduction in neuroinflammation markers and apoptosis after Aβ plaques induction was observed.[200]

The pecan nut acetone–water extracts showed antioxidant and neuroprotective effects in cells and in animals with neurological dysfunctions.[201] In addition, hydroethanolic walnut hull extract showed neuroprotection against oxidative stress and pro-inflammatory effect induced by isoprenaline administration in Wistar rats.[202] Brain induction using isoprenaline showed histopathological symptoms of neuronal degeneration in the rats. On the other hand, the pretreatment with walnut hull extract showed an improvement in the brain of the rats with neuropathological alterations. It was attributed to the regulation of some enzymes involved in neurodegeneration such as acetylcholinesterase and glutathione peroxidase. The authors attributed the protective effects to the high concentration of phenolic compounds present in walnut hulls. Animals treated with phenolic acids such as caffeic acid,[203] gallic acid,[204] or ellagic acid[205] showed attenuation of neurocognitive dysfunction induced by Aβ plaques aggregation. Nuts and their coproducts rich in flavonoids inhibited the metabolism of Aβ, one of the main biomarkers for AD. The supplementation of the mice diet with catechins showed to be effective against Aβ-42 production. In addition, it also improved the Aβ clearance due to their antioxidant and antiaggregation activity.[206,207]

The isorhamnetin, which is a 3′-methoxylated derivative of quercetin reported in almonds, showed a reduction in oxidative damage and memory abnormalities after Aβ-25–35 induction in rats.[8] Different cell models corroborated the effect of flavonoids such as myricetin,[208] proanthocyanidin,[209] quercetin,[210] and kaempferol[211] for attenuating the neurotoxicity of Aβ plaques. Furthermore, anthocyanins showed neuroprotection against Aβ-induced toxicity in cells and reduced cognitive impairment in an AD mouse model. The neuroprotective mechanism proposed by the authors was the interaction of anthocyanins with the GABA receptor.[209] The non-flavonoid lignans showed potential for inhibiting the inflammatory signaling pathways and cholinesterase enzymes. They also presented antioxidant properties decreasing the neurodegeneration induced by Aβ in mouse and mouse neuronal cells.[212] Tannins and tannic compounds such as tannic acid present in hazelnut skins of nuts were effective for inhibiting the formation of Aβ in early stages.[213] Bakoyiannis *et al.*[214] reviewed the effect of flavonoid and flavonoid-rich food supplements on memory and cognitive functions in humans. The supplementation with catechins found in green tea and cocoa, and quercetin from blueberry was effective for enhancing memory and different cognitive functions and protecting against cognitive impairment caused by ageing.

Polo-Hernández *et al.*[215] reported that oleic acid (ω9), which is the main MUFA found in nuts, increases the axonal and dendritic growth by promoting neurotrophic factors in neurons.[215] Moreover, oleic acid supplementation showed to decrease the Aβ plaques formation in mice.[216] Additionally, the essential PUFAs are important for the development of the human brain. The maternal deficiency of the essential PUFAs can lead to impairment in the brain tissue development, altering the neurotransmitter metabolism and learning in animals.[217] It has been shown that cultured brain cells need the PUFAs linoleic (C18:2, ω6) and α-linolenic (C18:3, ω3) acids for differentiation and their normal functioning.[217]

Some vitamins play a crucial role in the molecular regulation of brain metabolism. Folate deficiency may cause malformations in the neuronal tube during embryogenesis.[218] Supplementation with folic acid in diets of experimental mice with AD showed neuroprotective effects *via* DNA methylation.[219] Moreover, vitamin B_1 is involved in the cholinergic system function which is involved in the synthesis and regulation of the neurotransmitter acetylcholine.[220] In addition, studies with mice showed that a deficiency of thiamine induces amyloidogenesis. This condition, which is characterized by the aggregation of Aβ peptides and is related to AD, could be reversed with supplementation of thiamine.[221] On the other hand, nicotinamide which results from the niacin (vitamin B_3) metabolism was added to the feed of Aβ-injected mice, and the gene expression of the amyloid precursors was reduced.[222]

It has been reported that vitamin E can prevent the formation of Aβ plaques associated with ROS and inflammatory processes and avoid the hyperphosphorylation of tau protein in neurons.[223] Vitamin E was also associated with memory retention in mice and related to the regulation of the cholinergic system.[224] Choline, thiamine, and vitamin E are involved in the regulation of acetylcholine metabolism. Choline is essential in the synthesis of acetylcholine which is involved in the human neurotransmission mechanisms. Batool *et al.*[225] showed an increase in the acetylcholine content in different parts of the brain of healthy rats which had a diet supplemented with almond extract for four weeks. In rats with induced amnesia, the memory deficit was also improved.[225] These effects on neurotransmission and memory may be associated with the high contents of choline, thiamine, and vitamin E in nuts.

Sphingolipids are compounds involved in various intracellular signaling pathways related to metabolic mechanisms in the mammalian brain.[226] Cerebrosides are sphingolipids that can be found in some nuts such as almonds and hazelnuts.[50] Che *et al.*[227] showed that the supplementation with extracts rich in cerebrosides improved learning, memory deficits, and oxidative stress and decreased the accumulation of Aβ plaques in demented mice.[227]

4.5.3 Anticancer Effect

Cancer, which is one of the major public health problems worldwide, has a significant global impact in the social and economic spheres. The cost of cancer treatments has a great burden on patients, their families, and public health services.[228] It has been estimated that 19.3 million new cases of cancer occurred worldwide in 2020 leading to about 10 million deaths.[229] 22.2 million new cases are expected each year by 2030.[230] Cancer is a multifactorial disease involving a complex sequential process with alterations in the genes that regulate normal functions of the cells. It leads to mutation and a progressive increase in cellular proliferation. An uncontrolled production of cellular growth factors and survival ability without recognizing apoptosis signals is observed at the beginning. The angiogenesis and tissue invasion increase and, as a consequence, the metastasis, which is the ability of the cancer cells to migrate to other tissue, is observed. This involves changes in some cellular functions such as molecular and cell adhesions.[231]

Sex, family history, age, diet, lifestyle, gut microbiota profiles or comorbidities, and environmental factors play a key role in cancer pathogenesis.[232] Epidemiological studies have consistently shown that some dietary patterns have been established as risk factors for multiple types of cancer. Diets rich in vegetables such as fruits, legumes, seeds, nuts, and whole grains are strongly associated with reduced risk of developing chronic diseases including cancer.[233] Systematic review and meta-analysis performed by Wu *et al.*[234] found significant correlations between nut consumption and decreasing risk of colorectal cancer, endometrial cancer, and pancreatic cancer.[234] Hamsters fed with a hazelnut skin extract showed a reduction in secondary bile acids in faeces, an important marker for colon cancer.[192] In addition, extracts of cashew nut shells obtained with petroleum ether showed apoptotic and anti-proliferative effects on cancer HeLa cells.[235] Pepple *et al.*[236] reported the chemopreventive effect of ethanolic cashew shell extract against skin damage and cancer initiation induced by UV radiation.[138] Methanolic extracts of the seed, green husk, and leaf of walnuts showed anti-proliferative properties against kidney and colon cancer cells.[237] Pecan nutshell aqueous extract also presented an anti-tumoral effect against breast cancer cell line.[53] The authors suggested that pecan nutshell extracts may induce the activation of proteins involved in the signaling for cell death by apoptosis and cellular cycle regulation.[54] De la Rosa *et al.*[201] reported the antiproliferative capacity against bladder cancer cell line for extracts of kernels and shells of pecan nuts.[201] It has been reported that the consumption of nuts could reduce the risk of cancer by the synergistic effect of several bioactive compounds in their composition. Abraham *et al.*[238] summarized the anticancer properties of vitamin E and its isoforms (α, β, γ, and δ-tocopherols and tocotrienols).[23] Ju *et al.*[239] reported that the anti-inflammatory effect of vitamin E metabolites is based on their inhibitory capacity of COX-2 and 5-lipooxygenase (LOX-5) enzymes. Eicosanoids resulting from the activity of COX-2 and LOX-5 such as hydroxyeicosatetraenoic acid and hydroxyheptadecatrienoic acid seem to be involved in colon cancer pathogenesis as well as in colon inflammation and carcinogenesis.[239] Vitamin E also showed anti-proliferative effects by inhibition of cell cancer cycle progression and induction of apoptosis.[240] Anti-platelet and anti-aggregation effects were also shown by tocopherols. This effect is very important since platelets protect the circulating tumor cells in the bloodstream by forming platelet clots around them and promote the release of various angiogenic factors.[238] Tocotrienols also show anticancer properties by downregulating secretion of pro-angiogenic factors and vascular endothelial growth factors.[241] In addition, they can inhibit the 3-hydroxy-3-methylglutaryl-coenzyme A reductase (HMG CoA reductase). It has been reported that this enzyme and its metabolites are involved in breast cancer cell proliferation.[242] The flavonoid quercetin found in several nuts has been reported as a suppressor of angiogenesis with anti-proliferative and anticancer activities by induction of apoptosis in cancer cells. The suppression of the proliferation of gastric, breast, colorectal, oral, liver, prostate, thyroid, pancreatic, and lung cancer and leukemia has been reported for quercetin.[243] Moreover, tumoral marker enzymes such as aryl hydrocarbon hydroxylase and 5′-nucleotidase in the serum and lungs of mice treated with benzo-pyrene were improved with quercetin administration. In addition, an increase in antioxidant enzymes was observed.[244] Folate, another important component in the composition of nuts, is an important cofactor involved in several metabolic pathways such as DNA synthesis and normal cell growth and proliferation. Epidemiological and clinical studies showed that the supplementation of folate might decrease the risk of colorectal cancer. The mechanism for this activity was related to DNA methylation which is essential for DNA integrity and stability. Alterations in DNA methylation are common in a variety of tumors as well as in their development.[245,246] The phenolic compounds in nuts and their coproducts have been reported as protective compounds against cancer. These molecules can scavenge free radicals protecting the DNA from harmful effects of oxidative stress caused by endogenous and/or exogenous ROS and NRS.[231] Phenolic compounds such as anthocyanins, kaempferol, quercetin, esters of coumaric acid, and ellagic acid showed properties against tumor growth of human oral (KB, CAL-27), breast (MCF-7), colon (HT-29, HCT-116), and prostate

(LNCaP, DU-145) cancer cell lines.[231] The induction of apoptosis was the anticancer mechanism reported for phenolic compounds such as quercetin, (−)-epigallocatechin-3-gallate (EGCG), apigenin, ellagic acid, and resveratrol. Resveratrol, EGCG, apigenin, quercetin, kaempferol, and rutin are potent inhibitors of the nuclear factor kappa-light-chain-enhancer of activated B cells (NF-κB) which is a protein complex involved in the transcription of DNA, cytokine production cell proliferation, and cell cycle.[231] In addition, catechins, proanthocyanidins, and resveratrol showed anti-angiogenesis capacity by regulation of the vascular endothelial growth factor (VEGF).[247] Quercetin showed to inhibit angiogenesis by the interaction with the epidermal growth factor receptor (EGFR).[248] Szekeres *et al.*[249] reported that resveratrol can inhibit cell migration/invasion and metastasis of breast cancer.[249] Catechins showed anti-metastasis capacity inhibiting the adhesion of lung carcinoma cell lines and endothelial cell lines.[250] Caffeic acid showed inhibition of prostate cancer cells *in vitro*.[251]

Nuts and their coproducts are rich in fibers which have been associated with a lower risk of breast and colorectal cancer. Protective mechanisms of dietary fiber are related to its fermentation by colonic bacteria which results in the formation of short-chain fatty acids (SCFA). These fatty acids are present in the mucosa of the large intestine and they can control cellular proliferation and differentiation. Fiber and SCFA also promote lower intestinal pH inhibiting the conversion of primary bile acids to secondary bile acids. The secondary bile acids are linked to promote cell proliferation increasing the possibility of cell mutations and the risk for replication of abnormal cells. In addition, the fibers can also reduce the time of intestinal transit reducing the permanence of potential carcinogen agents in contact with the large intestine mucosa. Moreover, fiber can prevent breast cancer improving insulin resistance since insulin and insulin-like growth factors seem to be related to the development of such type of cancer.[252,253]

4.5.4 Other Effects

The prebiotic properties of nuts are related to the improvement of the gut microbiome. The high density of nutrients in nuts and their coproducts can stimulate the growth of *Bifidobacteria* and *Eubacterium rectale.*[23] The extracts of almond skin rich in polyphenols showed probiotic potential. The extracts increased the populations of beneficial colonic intestinal flora such as *Bifidobacteria*, *C. coccoides*, and *E. rectale*. A decrease in the number of the pathogen *C. hystolyticum* group was also observed.[42] The authors suggested that the ingestion of almond and almond skin may induce positive alterations in the intestinal microbiota profile in healthy humans. This can increase the activity of β-galactosidase and decrease the activity of β-glucuronidase faecal enzymes.[42] In addition to the probiotic properties, extracts of almond skins showed antimicrobial capacity against gram-positive bacteria (*Staphylococcus aureus* and *Listeria monocytogenes*).[42] Pecan nutshell extracts obtained with hexane showed antimicrobial activity against

Mycobacterium tuberculosis. Aqueous and ethanolic extracts obtained from leaves of pecan nuts showed antimicrobial activity against gram-positive and gram-negative bacteria and yeast. Liquid smoke obtained from roasted and non-roasted pecan nutshells presented antimicrobial activity against food-borne pathogens such as *Escherichia, Salmonella, Listeria monocytogenes,* and *Staphylococcus.*[254] The extracts showed higher antimicrobial activity than pure phenolic compound standards. The authors indicated that the synergic effect of different phenolic compounds increases the antimicrobial activity. The mechanisms proposed include the oxidation of cell membranes, metal chelation, and extracellular enzyme inhibition.[254] The cashew nutshell aqueous extract prevented gastric damage induced by ethanol in mice.[42]

Extract from pecan nutshell enriched with quercetin, caryatin-3′methyl ether, and caryatin-3′methyl ether-7-*O*-β-D-glucoside showed antidiabetic effect (aldose reductase inhibition and decrease in the plasma insulin level) in rats. The results also showed an improvement in the antioxidant status and glycated hemoglobin. An aqueous pecan nutshell extract and methanolic leaves extract administered for 28 days in diabetic-induced Wistar rats showed a reduction in urea, glucose, and glycated hemoglobin in blood, and an increase in the antioxidant activity.[254] Antidiabetic activity was higher for the extract obtained from leaves when compared to the extracts obtained from nutshells. The authors indicated that the contents of phenolics such as rutin and quercitrin were higher in leaves than in the shells. The authors suggest that this antidiabetic effect corresponds to the capacity of phenolic compounds to inhibit some digestive enzymes such as α-lipase. In fact, the extract rich in proanthocyanidin showed the capacity to inhibit lipase, amylase, and trypsin digestive enzymes in an *in vitro* human digestive system.[254] Other studies also found that pecan coproducts and extracts rich in procyanidins from almond skin showed hepatoprotective effect in rats.[42,254] Furthermore, pecan and walnut extracts ameliorated the health impairments against cigarette smoke-induced oxidative damage and pulmonary edema in animal models.[42,254] Dermatological benefits were also reported for pistachio skin extracts against skin erythema and UV-B skin damage. Moreover, the usage of pistachio skin extracts accelerated wound healing in rats.[42] The toxicity of some nut coproducts was studied. Doses of almond skin up to 2.5 $g\,kg^{-1}$, 300 $mg\,kg^{-1}$ of anacardic acids from cashew, and 200 $mg\,kg^{-1}$ of pecan shell did not show toxic effects in animal models.[42,254]

4.6 Potential of Underexplored Nuts

Portuguese chestnut (*Castanea sativa)*, sacha inchi (*Plukenetia volubilis* L.), and native Brazilian nuts such as baru (*Dipteryx alata*), chichá (*Sterculia striata*), gurguéia (*Dipteryx lacunifera*), monguba (*Pachira aquatica*), pracaxi (*Pentaclethara macroloba*), and sapucaia (*Lecythis pisonis*) are some examples of underexplored and little-known nuts.

The European chestnut, or Portuguese chestnut, belongs to the Fagaceae family, which includes several ecologically important species with high economic potential.[255] *C. crenata*, *C. molissima*, *C. seguinii*, *C. davidii,* and *C. henryi* have been reported in Asia; *C. dentata*, *C. pumila*, *C. floridana*, *C. ashei*, *C. alnifolia,* and *C. paucispina* have been reported in North America; and *C. sativa* is predominant in Europe.[256]

Sacha inchi of the family Euphorbiaceae is also known as sacha peanut, mountain peanut, Inca nut, or Inca peanut.[257,258] It is native to the tropical rainforest of the Amazon region including Peru and North Brazil.[259] The cultivation of sacha inchi has been reported in Southeast Asia, because of its great economic potential.[260,261]

Brazil is known for having a great diversity of plants distributed in the biomes Amazon, Cerrado, Atlantic Forest, Caatinga, Pampas, and Pantanal.[262] Nut trees are essential for maintaining the balance between flora and fauna in tropical forests.[263] Baru, chichá, gurguéia, monguba, pracaxi, and sapucaia are Brazilian nuts native from Cerrado e Amazon biomes that are still not well known and underexplored. The production of these nuts is based on extractivism, and they are sold at local fairs.[35,264,265]

Table 4.8 shows the nutritional composition of Portuguese chestnut, sacha inchi, baru, chichá, gurguéia, monguba, pracaxi, and sapucaia.

Portuguese chestnuts have a high content of starch (up to 70% of carbohydrates), low levels of proteins (2–4%), and lipids (2–5%). The main components in the TAG profile were OLL, LLP, OLO, and LOP.[274] On the other hand, sacha inchi is rich in lipids (35–60%) and proteins (25–30%).[258,275,276] Sacha inchi is a source of polyunsaturated fatty acids, mainly linoleic (12–14%) and linolenic (12–16%) acids.[258]

Sapucaia nut is native to the Amazon region and has a high lipid content (58.7–60.8 g 100 g^{-1}).[273] Baru nut is native to Cerrado and has a higher protein content (26–30 g 100 g^{-1}) than peanuts (21–25 g 100 g^{-1}). In addition, it is rich in essential amino acids.[277] Gurguéia is native to Cerrado and has a high content of oleic acid (64.7%).[278] Monguba is native to the Amazon region and is rich in palmitic acid (60.9%).[270] This raw material can be used to structure fats and replace fats such as milk and butter cocoa. Chichá nut from Cerrado has a low lipid content (24.5–28.6 g 100 g^{-1}) compared to cashew and Brazil nuts and it has been reported as a light nut.[267] Pracaxi from the Amazon region is a source of behenic acid (16%) and has emollient properties.[271]

The shell of the Portuguese chestnut can be used as lignocellulosic biomass in the production of biofuels.[279] It has been reported that chestnut shells have between 2.7 and 5.2% (w/w) of polyphenols with antioxidant activity. In addition, this raw material, which is rich in sugars (36% w/w), can be used in the production of biofuels.[280,281] The polysaccharides of chestnut shells have been used for increasing the strength of starch films;[282] as bio adsorbent for heavy metals; phenol substitutes for wood adhesive formulations; and chromium alternatives for leather tanning.[283]

The sacha inchi oil press cake is rich in proteins (32.5–62.0%), carbohydrates (5.0–26.8%), and lipids (4.8–25.5%). The main minerals in this raw material are phosphorus (531.5–1312.5 mg 100 g^{-1}), potassium (1393.5 mg 100 g^{-1}), and calcium (301.7–761.6 mg 100 g^{-1}).[284] The sacha

Table 4.8 Nutritional composition of underexplored nuts (wet base).

Nuts	Moisture ($g\ 100\ g^{-1}$)	Ash ($g\ 100\ g^{-1}$)	Proteins ($g\ 100\ g^{-1}$)	Lipids ($g\ 100\ g^{-1}$)	Total fiber ($g\ 100\ g^{-1}$)	Carbohydrate ($g\ 100\ g^{-1}$)	Energy value ($kcal\ g^{-1}$)
Portuguese chestnut[272]	46.3–53.3	1.5–2.2	4.9–7.4	1.7–3.1	—	34.0–45.6	170.9–239.9
Sacha inchi[258]	3.3–8.3	2.7–6.5	24.2–27.0	3.4–54.3	—	13.4–30.9	181.0–720.3
Baru[264,266]	3.5–7.2	3.2	26.0–30.0	39.0–42.0	9.2–14.5	9.1–12.2	419–546
Chichá[267]	8.2–11.4	3.0–3.2	18.5–22.5	24.5–28.6	4.6–5.8	40.5–45.8	456–530
Gurguéia[268,269]	5.3–8.3	2.3–2.7	13.3–15.4	41.9–45.6	4.7–6.1	34.3–36.2	552–568
Monguba[270]	5.3–6.0	3.5–4.2	12.0–13.0	43.4–53.9	12.3	29.7–35.0	557–677
Pracaxi[271]	4.0	1.9	15.5	53.4	—	25.2	644
Sapucaia[273]	4.1–4.2	2.9–3.5	15.5–60.8	58.7–60.8	16.5	4.9–13.8	616–665

inchi shell has been studied as a substrate for biofuel production.[285,286] Sanchez-Reinoso *et al.*[261] obtained a phenolic content for sacha inchi shell using MAE of 41.9 mg gallic acid g^{-1} shells. The antioxidant activity using FRAP and TEAC values were 374.4 and 384.5 μmol Trolox g^{-1} shell, respectively. The authors reported 3,4-dihydroxy-benzaldehyde, hydroxy-4-chromone, and their derivatives as the main phenolic compounds in the extracts.

The partially defatted baru flour obtained after the oil extraction by pressing on a continuous screw had protein, lipid, fiber, and carbohydrate contents of 29.46, 11.84, 38.80, and 11.57 g 100 g^{-1}, respectively. The authors also reported total phenolic compounds, flavonoids, and condensed tannins of 121.34, 85.41, and 64.39 mg 100 g^{-1}, respectively.[82]

Chichá oil press cold cake has in its composition mainly carbohydrates (41.6 g 100 g^{-1}), proteins (19.1 g 100 g^{-1}), and lipids (21.4 g 100 g^{-1}). The main minerals reported for this raw material are potassium (10 384 μg g^{-1}) followed by magnesium (2620 μg g^{-1}). Ellagic acid (4.61 mg 100 g^{-1}) and protocatechuic acid (2.35 mg 100 g^{-1}) were the main phenolic compounds reported in chichá nutshell extracts obtained with water.[267]

The monguba cake, extracted by SFE (60 °C/20 MPa) and sequential extraction with PLE using ethanol 99.5%, showed total phenolic compounds of 350.3 mg GAE 100 g^{-1}. The activity antioxidant determined for the extracts by DPPH (61.58%), FRAP (28.4 mM Fe^{2+} 100 g^{-1}), and ILP (28.4%) were considered high.[171] A lignocellulosic material developed with monguba peel was subjected to an alkaline treatment with 2-aminopropane (C3H9N 10% m/v) and sodium hydroxide (NaOH 10% m/v), and it presented a desired crystallinity.[287]

Sapucaia cake obtained after the extraction of the oil by cold pressing showed a composition rich in lipids (31.3–37.0 g 100 g^{-1}), fiber (27.9–31.2 g 100 g^{-1}), and proteins (26.5–29.7 g/100 g). The main minerals reported were magnesium (1872–3555 μg g^{-1}) followed by calcium (1583–2473 μg g^{-1}).[188] The phenolic compounds identified in sapucaia nutshell extracts were catechin (2.6–4.0 mg g^{-1}), epicatechin (0.6–1.3 mg g^{-1}), myricetin (0.03–0.05 mg g^{-1}), ellagic acid (0.2–0.4 mg g^{-1}), and ferulic acid (0.03–0.04 mg g^{-1}).[273]

4.7 Conclusions

Nuts and the coproducts obtained after their processing are pleasant and versatile foods rich in nutrients and bioactive compounds with great economic potential. This composition is related to the prevention of cardiovascular and neurodegenerative diseases, diabetes, and cancer. Information on the bioavailability of bioactive compounds in nuts and their impact on human health is still needed. There are many possible uses for nuts and the coproducts resulting from their processing. Specialty oils can be obtained after the pressing of nuts and the defatted wholemeal flour is rich in nutrients and bioactive compounds. These flours can enrich bakery and pastry products, sweets, sauces, chocolates, and ice creams. In addition, they can be used for producing plant-based milk and yogurts. The bioactive compounds present in the by-products of nuts such as shells and skins, which represent between 40 and 50% of the raw material and are usually discarded, can be extracted using green technologies. The extracts can be used in the food, cosmetics, and pharmaceutical industries. The challenge of converting the by-products of nuts and high volumes of feedstock into useful commercial coproducts should be encouraged. In this way, the zero-waste goal can be achieved by the processing industries of nuts in the near future. The relatively unknown unconventional nuts, consumed only in native regions, continue to be a little-explored topic. Therefore, more studies should be carried out to stimulate the large-scale commercialization of unconventional nuts. The full use and exploitation of these raw materials can improve the population's diet and contribute to the sustainable socio-economic development of many regions worldwide.

Abbreviations

EGCG	(−)-Epigallocatechin-3-gallate
DPPH	1,1-Diphenyl-2-picrylhydrazyl
AD	Alzheimer's disease
ISAPP	Association for Probiotics and Prebiotics
CVDs	Cardiovascular diseases
DESs	Deep eutectic solvents
EAE	Enzyme-assisted extraction
GRAS	Generally recognized as safe
HDS and HAS	High number of hydrogen donor and acceptor sites
EPR	Hydroxyl radical scavenging activities using electron paramagnetic resonance
ROS and RNS	Inflammatory process generates reactive oxygen and nitrogen species
ILs	Ionic liquids
LDL	Low-density lipoprotein
MAE	Microwave-assisted extraction
MUFAS	Monounsaturated fatty acids
NADES	Natural deep eutectic solvents
NCCD	Non-communicable chronic diseases
ORAC	Oxygen radical scavenging capacity
PUFAS	Polyunsaturated fatty acids
PLE	Pressurized liquid extraction
SFAs	Saturated fatty acids
SCFA	Short-chain fatty acids
SubFE	Subcritical fluid extraction
SFE	Supercritical fluid extraction
TEAC	Trolox equivalent antioxidant capacity
TNF-α	Tumor necrosis factor-alpha
UAE	Ultrasound-assisted extraction
VEGF	Vascular endothelial growth factor

Acknowledgements

The authors are grateful to CAPES for the PVS/CAPES-PrInt fellowship of Jane Mara Block (process number 88887.571379/2020-00) and Coordination for the Improvement of Higher Education Personnel (CAPES) for the doctoral scholarship of Gabriela Polmann (process 88882.344944/2019-01).

References

1. G. E. Fraser, J. Sabaté, W. L. Beeson and T. H. Strahan, *Arch. Intern. Med.*, 1992, **152**, 1416.
2. J. Sabate and G. E. Fraser, *Curr. Opin. Lipidol.*, 1994, **5**, 11.
3. S. Rohrmann and D. Faeth, *BMC Med.*, 2013, **11**, 165.
4. J. C. King, J. Blumberg, L. Ingwersen, M. Jenab and K. L. Tucker, *J. Nutr.*, 2008, **138**, 1736S.
5. E. Ros, *Nutrients*, 2010, **2**, 652.
6. D. J. McClements, *Future Foods*, 2019, 167.
7. I. J. Ferreira, E. M. Alexandre, J. A. Saraiva and M. Pintado, *Innovative Food Sci. Emerging Technol.*, 2022, 102931.
8. D. Asha and T. Sumathi, *Int. J. Pharm. Sci. Res.*, 2016, **8**, 7.
9. S. Acevedo-Juárez, D. Guajardo-Flores, E. Heredia-Olea and M. Antunes-Ricardo, *Int. J. Food Sci. Technol.*, 2022, **57**, 2226.
10. V. Vargova, R. B. Singh, J. Fedacko, D. Pella, D. Pella, A. Wilczynska, V. Mojto, O. A. Al-bawareed and V. I. Torshin, *Functional Foods and Nutraceuticals in Metabolic and Non-Communicable*, 2022, p. 383.
11. A. M. Tindall, K. S. Petersen, R. Lamendella, G. C. Shearer, L. E. Murray-Kolb, D. N. Proctor and P. M. Kris-Etherton, *Curr. Dev. Nutr.*, 2018, **3**, 11.
12. WHO – World Health Organization, www.who.int/news-room/fact-sheets/detail/healthy-diet (accessed 22.07.22).
13. INC – Nuts & Dried Fruits Statistical Yearbook 2020/2021. Available in: https://www.nutfruit.org/files/tech/1621253983_INC_Statistical_Yearbook_2020-_202 (accessed 26.07.22).
14. I. Mateos-Aparicio, *Food Industry Processing By-Products In Foods*, 2019, p. 239.
15. T. B. Ribeiro, G. B. Voss, M. C. Coelho and M. E. Pintado, Global Trends, Opportunities, and Sustainability Challenges, *Future Foods*, 2022, p. 569.
16. A. A. Adedeji, *J. ASABE*, 2022, **65**, 471.
17. INC – *International Nut and Dried Fruit Council. Zero Waste in the Tree Nut Industry: Adding Value to Coproducts*. Available in: nutfruit.org (accessed on 27.07.22).
18. S. Gomes, P. V. Finotelli, V. F. Sardela, H. M. G. Pereira, R. E. Santelli, A. S. Freire and A. G. Torres, *LWT*, 2019, **116**, 108495.
19. L. G. Maciel, F. L. Ribeiro, G. L. Teixeira, L. Molognoni, J. Nascimento dos Santos, I. Larroza Nunes and J. Mara Block, *Food Res. Int.*, 2020, **127**, 108718.

20. M. D. Salvador, R. M. Ojeda-Amador and G. Fregapane, *Oxidative Stress and Dietary Antioxidants*, 2020, p. 309.
21. M. Venkatachalam and S. K. Sathe, *J. Agric. Food Chem.*, 2006, **54**, 4705.
22. J. Naozuka, E. C. Vieira, A. N. Nascimento and P. V. Oliveira, *Food Chem.*, 2011, **124**, 1667.
23. G. Polmann, V. Badia, R. Danielski, S. R. S. Ferreira and J. M. Block, *Food Rev. Int.*, 2022, 1.
24. C. S. Queirós, S. Cardoso, A. Lourenço, J. Ferreira, I. Miranda, M. J. J. V. Lourenço and H. Pereira, *Biomass Convers. Biorefin.*, 2020, **10**, 175.
25. P. A. Costa, C. A. da Ballus, J. Teixeira-Filho and H. T. Godoy, *Food Res. Int.*, 2010, **43**, 1603–1606.
26. G. D. Fernandes, R. B. Gómez-Coca, M. D. C. Pérez-Camino, W. Moreda and D. Barrera-Arellano, Chemical characterization of major and minor compounds of nut oils: almond, hazelnut, and pecan nut, *J. Chem.*, 2017, **2017**, 26.
27. P. Herbello-Hermelo, J. P. Lamas, M. Lores, R. Domínguez-González, P. Bermejo-Barrera and A. Moreda-Piñeiro, *Food Chem.*, 2018, **254**, 20.
28. G. Polmann, V. Badia, M. Frena, G. L. Teixeira, E. Rigo, J. M. Block, M. M. Camino and M. M. Feltes, *LWT*, 2019, 108283.
29. E. Ryan, K. Galvin, T. P. O'Connor, A. R. Maguire and N. M. O'Brien, *Int. J. Food Sci. Nutr.*, 2006, **57**, 219.
30. I. Ikeda and M. Sugano, *Curr. Opin. Lipidol.*, 1998, **9**, 527.
31. W. Zhao, G. Xu, Z. Yu, J. Li and J. Liu, *Comput. Biol. Med.*, 2021, **138**, 104937.
32. J. Wang, T. Wu, L. Fang, C. Liu, X. Liu, H. Li and W. Min, *Food Funct.*, 2020, **11**, 8112–8121.
33. R. Jahanbani, S. M. Ghaffari, M. Salami, K. Vahdati, H. Sepehri, N. N. Sarvestani and A. A. Moosavi-Movahedi, *Plant Foods Human Nutr.*, 2016, **71**, 402–409.
34. M. D. Bisinotto, D. C. da Silva, L. de Carvalho Fino, F. M. Simabuco, R. M. N. Bezerra, A. E. C. Antunes and M. T. B. Pacheco, *Food Res. Int.*, 2021, **139**, 109906.
35. B. R. Cardoso, G. B. S. Duarte, B. Z. Reis and S. M. F. Cozzolino, *Food Res. Int.*, 2017, **100**, 9.
36. C. Alasalvar and B. W. Bolling, *Br. J. Nutr.*, 2015, **113**, S68–S78.
37. N. Gorji, R. Moeini and Z. Memariani, *Pharmacol. Res.*, 2018, **129**, 115–127.
38. G. G. Kuhnle, C. Dell'Aquila, S M. Aspinall, S. A. Runswick, A. A. Mulligan and S. A. Bingham, *J. Agric. Food Chem.*, 2008, **56**, 7311.
39. L. Xie and B. W. Bolling, *Food Chem.*, 2014, **148**, 300.
40. R. Bodoira and D. Maestri, *J. Agric. Food Chem.*, 2020, **68**, 927–942.
41. M. Woźniak, A. Waśkiewicz and I. Ratajczak, *Molecules*, 2022, **27**, 4326.
42. S. K. Chang, C. Alasalvar, B. W. Bolling and F. Shahidi, *J. Funct. Foods*, 2016, **26**, 88–122.

43. A. Chandrasekara and F. Shahidi, *J. Agric. Food Chem.*, 2012, **58**, 6706.
44. C. M. Liyana-Pathirana and F. Shahidi, *J. Agric. Food Chem.*, 2006, **54**, 1256–1264.
45. M. Nardini and A. Ghiselli, *Food Chem.*, 2004, **84**, 137.
46. B. A. Acosta-Estrada, J. A. Gutiérrez-Uribe and S. O. Serna-Saldívar, *Food Chem.*, 2014, **152**, 46–55.
47. C. Alasalvar and B. W. Bolling, *Br. J. Nutr.*, 2015, **113**, S68.
48. P. E. Milbury, C. Y. Chen, G. G. Dolnikowski and J. B. Blumberg, *J. Agric. Food Chem.*, 2006, **54**, 5027.
49. J. A. John and F. Shahidi, *J. Funct. Foods*, 2010, **2**, 196.
50. B. W. Bolling, C. Y. O. Chen, D. L. McKay and J. B. Blumberg, *Nutr. Res. Rev.*, 2011, **24**, 244–275.
51. D. Del Rio, L. Calani, M. Dall'Asta and F. Brighenti, *J. Agric. Food Chem.*, 2011, **59**, 9935.
52. C. Alamprese, S. Ratti and M. Rossi, *J. Food Eng.*, 2009, **95**, 272–279.
53. J. Hilbig, V. R. Alves, C. M. O. Müller, G. A. Micke, L. Vitali, R. C. Pedrosa and J. M. Block, *Food Res. Int.*, 2018, **106**, 549.
54. J. Hilbig, P. Policarpi, V. M. A. S. Grinevicius, N. S. R. S. Mota, I. M. Toaldo, M. T. B. Luiz and J. M. Block, *J. Ethnopharmacol.*, 2018, **211**, 256.
55. A. Rabadán, M. Álvarez-Ortí, R. Gómez, A. Alvarruiz and J. E. Pardo, *LWT*, 2017, **83**, 79.
56. A. Rabadán, M. Álvarez-Ortí, J. E. Pardo and A. Alvarruiz, *Food Chem.*, 2018, **259**, 31.
57. B. S. Kamel and Y. Kakuda, *Technological Advances in Improved and Alternative Sources of Lipids*, Springer, Hong Kong, 1994.
58. A. Thanonkaew, S. Wongyai, D. J. McClements and E. A. Decker, *LWT*, 2012, **48**, 231.
59. O. V. Santos, N. C. F. Corrêa, F. A. S. M. Soares, L. A. Gioielli, C. E. F. Costa and S. C. S. Lannes, *Food Res. Int.*, 2012, **47**, 253–258.
60. E. Sena-Moreno, J. E. Pardo, A. Pardo-Giménez, R. Gómez and M. Alvarez-Ortí, *Int. J. Food Prop.*, 2016, **19**, 2750.
61. Y. Mazaheri, M. Torbati, S. Azadmard-Damirchi and G. P. Savage, *J. Food Process. Preserv.*, 2019, **43**, 1.
62. M. M. Özcan, F. Al Juhaimi, K. Ghafoor, E. E. Babiker and M. M. Özcan, *J. Food Sci. Technol.*, 2020, **57**, 955.
63. I. A. Mohamed Ahmed, F. Y. Al-Juhaimi, M. M. Özcan, M. A. Osman, M. A. Gassem and H. A. A. Salih, *J. Oleo Sci.*, 2019, **68**, 167.
64. J. M. de Carvalho, R. W. Figueiredo, P. H. M. de Sousa, F. M. T. de Luna and G. A. Maia, *Int. J. Food Sci. Technol.*, 2018, **53**, 930.
65. T. Oro, H. M. A. Bolini, D. B. Arellano and J. M. Block, *J. Am. Oil Chem. Soc.*, 2009, **86**, 971.
66. J. Scapinello, J. D. Magro, J. M. Block, M. Di Luccio, M. V. Tres and J. V. Oliveira, *J. Food Sci. Technol.*, 2017, **54**, 3366.
67. A. P. L. Ribeiro, F. F. Haddad, T. S. Tavares, K. T. Magalhães, C. J. Pimenta and C. A. Nunes, *Emirates J. Food Agric.*, 2020, **32**, 295.

68. A. I. Kartika, P. Y. Pontalier and L. Rigal, *Ind. Crops Prod.*, 2010, **32**, 297–304.
69. O. Z. Sampaio Neto, E. A. C. Batista and A. J. A. Meirelles, *J. Cleaner Prod.*, 2018, **180**, 866.
70. J. Yang, *LWT*, 2009, **42**, 1573.
71. A. Cilla, A. Alegría, B. De Ancos, C. Sánchez-Moreno, M. P. Cano, L. Plaza and R. Barberá, *J. Agric. Food Chem.*, 2012, **60**, 7282.
72. E. Yildiz and D. Gocmen, *J. Food Sci. Technol.*, 2021, **58**, 940.
73. A. G. O. Sartori, S. M. de Alencar, D. H. M. Bastos, M. A. B. R. d'Arce and L. H. Skibsted, *Eur. Food Res. Technol.*, 2020, **244**, 1657.
74. B. Ling, B. Zhang, R. Li and S. Wang, *J. Am. Oil Chem. Soc.*, 2016, **93**, 689.
75. M. L. Martínez, M. P. Fabani, M. V. Baroni, R. N. M. Huaman, M. Ighani, D. M. Maestri and G. E. Feresin, *J. Food Sci. Technol.*, 2016, **53**, 2260.
76. M. Paciulli, M. Rinaldi, A. Cavazza, T. Ganino, M. Rodolfi, B. Chiancone and E. Chiavaro, *LWT*, 2018, **98**, 451.
77. L. Marchetti, A. N. Califano and S. C. Andrés, *LWT*, 2018, **95**, 85.
78. M. O. Adegunwa, B. I. Kayode, R. M. O. Kayode, S. A. Akeem, A. A. Adebowale and H. A. Bakare, *J. Food Meas. Charact.*, 2020, **14**, 1998.
79. J. C. M. Barreira, M. A. Nunes, B. V. da Silva, F. B. Pimentel, A. S. G. Costa, M. Alvarez-Ortí and M. B. P. P. Oliveira, *Food Sci. Human Wellness*, 2019, **8**, 292.
80. L. H. Channaiah, M. Michael, J. C. Acuff, K. Lopez, R. K. Phebus, H. Thippareddi and G. Milliken, *Int. J. Food Microbiol.*, 2019, **294**, 27.
81. C. Jia, Y. S. Kim, W. Huang and G. Huang, *Food Res. Int.*, 2008, **41**, 930.
82. L. L. O. Pineli, M. V. de Carvalho, L. A. de Aguiar, G. T. de Oliveira, S. Celestino, M. C. Sô, R. B. A. Botelho and M. D. Chiarello, *LWT*, 2015, **60**, 50.
83. J. Xu, Y. Zhang, W. Wang and Y. Li, *Trends Food Sci. Technol.*, 2020, **103**, 200–213.
84. M. Arslan, A. Rakha, Z. Xiaobo and M. A. Mahmood, *Trends Food Sci. Technol.*, 2019, **83**, 194–202.
85. T. De Pilli, K. Jouppila, J. Ikonen, J. Kansikas, A. Derossi and C. Severini, *J. Food Eng.*, 2008, **87**, 495.
86. M. O. Aremu, A. Olonisakin, D. A. Bako and P. C. Madu, *Pak. J. Nutr.*, 2006, **5**, 328.
87. D. Turan, E. Capanoglu and F. Altay, *LWT*, 2015, **63**, 758.
88. K. H. M. Watters and P. J. Cherry, *J. Food Sci.*, 1977, **42**, 1444–1447.
89. J. Santos, M. Alvarez-Ortí, E. Sena-Moreno, A. Rabadán, J. E. Pardo and P. P. Oliveira, *J. Sci. Food Agric.*, 2018, **98**, 1813.
90. M. P. Garcia-Mendoza, F. A. Espinosa-Pardo, R. Savoire, C. Etchegoyen, C. Harscoat-Schiavo and P. Subra-Paternault, *Ind. Crops Prod.*, 2021, **167**, 113546.
91. F. G. Fonseca and A. P. Soares Dias, *Renewable Energy*, 2021, **180**, 1380.

92. S. Malayil, A. N. Surendran, K. Kate and J. Satyavolu, *SSRN Electron. J.*, 2022, **19**, 101150.
93. N. Dejang, O. Somprasit and S. Chindaruksa, A Preparation of Activated Carbon from Macadamia Shell by Microwave Irradiation Activation, *Energy Procedia*, 2015, **79**, 727.
94. E. Surek and A. O. Buyukkileci, *Carbohydr. Polym.*, 2017, **174**, 565.
95. T. Cheng, J. Li, X. Ma, L. Zhou, H. Wu and L. Yang, *Environ. Pollut.*, 2022, **295**, 118714.
96. E. Sánchez, R. Zabaleta, M. P. Fabani, R. Rodriguez and G. Mazza, *J. Environ. Manage.*, 2022, **318**, 115604.
97. G. Mandalari, A. Tomaino, G. T. Rich, R. Lo Curto, T. Arcoraci, M. Martorana and M. S. J. Wickham, *Food Chem.*, 2010, **122**, 1083.
98. Z. Liu, X. Lin, G. Huang, W. Zhang, P. Rao and L. Ni, *Anaerobe*, 2014, **26**, 1.
99. P. Sharma, V. K. Gaur, R. Sirohi, C. Larroche, S. H. Kim and A. Pandey, *Ind. Crops Prod.*, 2020, **152**, 112550.
100. A. C. P. do Prado, H. S. da Silva, S. M. da Silveira, P. L. M. Barreto, C. Vieira, M. Maraschin and J. M. Block, *Ind. Crops Prod.*, 2014, **52**, 552.
101. S. Sethi, S. K. Tyagi and R. K. Anurag, *J. Food Sci. Technol.*, 2016, **53**, 3408.
102. A. R. A. Silva, M. M. N. Silva and B. D. Ribeiro, *Food Res. Int.*, 2020, **131**, 108972.
103. S. I. aan den Toorn, E. Worrell and M. A. van den Broek, *J. Ind. Ecol.*, 2020, **24**, 601.
104. E. F. Aydar, S. Tutuncu and B. Ozcelik, *J. Funct. Foods*, 2020, **70**, 103975.
105. O. E. Mäkinen, V. Wanhalinna, E. Zannini and E. K. Arendt, *Crit. Rev. Food Sci. Nutr.*, 2016, **56**, 339.
106. I. Felberg, R. Antoniassi, R. Deliza, S. C. Freitas, R. C. Modesta and R. C. Della, *Cienc. Tecnol. Aliment.*, 2009, **29**, 609.
107. Y. Alozie and A. E. Yetunde, *World J. Dairy Food Sci.*, 2015, **10**, 117.
108. A. Cristina Popovici, A. Alexei Baerle and P. Tatarov. *Biochem. Aspects Walnut Dairy Free Milk*, 2017, vol. 56, pp. 28–33.
109. K. E. Scholz-Ahrens, F. Ahrens and C. A. Barth, *Eur. J. Nutr.*, 2020, **59**(1), 19–34.
110. I. Fructuoso, B. Romão, H. Han, A. Raposo, A. Ariza-Montes, L. Araya-Castillo and R. P. Zandonadi, *Nutrients*, 2021, **13**, 9.
111. A. Shakerardekani, R. Karim, H. M. Ghazali and N. L. Chin, *Int. J. Mol. Sci.*, 2013, **14**, 4223.
112. S. Chalupa-Krebzdak, C. J. Long and B. M. Bohrer, *Int. Dairy J.*, 2018, **87**, 84–92.
113. E. L. K. Martínez-Padilla, H. B. Frandsen, M. S. Joehnke, E. Vargas-Bello-Pérez and I. L. Petersen, *Foods*, 2020, **9**, 1.
114. G. R. Gibson, R. Hutkins, M. E. Sanders, S. L. Prescott, R. A. Reimer, S. J. Salminen and G. Reid, *Nat. Rev. Gastroenterol. Hepatol.*, 2017, **14**, 491.

115. K. S. Swanson, G. R. Gibson, R. Hutkins, R. A. Reimer, G. Reid, K. Verbeke and M. E. Sanders, *Nat. Rev. Gastroenterol. Hepatol.*, 2020, **17**, 687.
116. R. M. Lamuel-Raventos and M. P. S. Onge, *Crit. Rev. Food Sci. Nutr.*, 2017, **57**, 3154.
117. N. Bernat, M. Cháfer, A. Chiralt and C. González-Martínez, *Food Sci. Technol. Int.*, 2015, **21**, 440.
118. Y. E. Tunçil, *Food Chem.*, 2020, **316**, 126338.
119. F. Hesam, B. G. Tarzi, M. Honarvar and M. Jahadi, *J. Food Meas. Charact.*, 2021, **15**, 33.
120. D. Nabarlatz, X. Farriol and D. Montané, *Ind. Eng. Chem. Res.*, 2005, **44**, 7746.
121. M. C. Rebouças, M. C. P. Rodrigues and M. R. A. Afonso, *J. Food Sci.*, 2014, **79**, 51393.
122. M. Rebouças, M. Rodrigues, S. de Freitas and B. Ferreira, *Beverages*, 2016, **2**, 23.
123. P. Sfakianakis and C. Tzia, *Foods*, 2014, **3**, 176.
124. C. T. Udayarajan, K. Mohan and P. Nisha, *Trends Food Sci. Technol.*, 2022, **124**, 208.
125. A. Shakerardekani, R. Karim, H. Ghazali and N. L. Chin, *J. Am. Oil Chem. Soc.*, 2015, **92**, 1015.
126. J. R. Lima, D. S. Garruti and L. M. Bruno, *LWT*, 2012, **45**, 180.
127. R. Mônaco, T. Giancone, S. Cavella and P. Mais, Properties of Hazelnut Spreads, *J. Texture Stud.*, 2008, **39**, 460.
128. B. V. Pallavi, R. Chetana, R. Ravi and S. Y. Reddy, *J. Food Sci. Technol.*, 2015, **52**, 1663.
129. A. Padmashree, G. K. Sharma, K. A. Srihari and A. S. Bawa, *J. Food Sci. Technol.*, 2012, **49**, 335.
130. C. Cunha, C. R. Cruz, C. L. de, C. V. Duarte, D. Souza, J. M. L. de and V. S. N. Silva, Cereal Bars with Brazil Nut and Green Banana Flour. Research Bulletin Number 9 – Embrapa, 1–59, 2011. Retrieved from http://ainfo.cnptia.embrapa.br/digital/bitstream/CNPAT-2010/4779/1/Bp-023.pdf.
131. E. Eke-Ejiofor and C. Okoye, *J. Agric. Food Chem.*, 2018, **64**, 5334.
132. R. Bodoira, M. Martínez, A. Velez, M. C. Cittadini, P. Ribotta and D. Maestri, *J. Sci. Food Agric.*, 2022, **102**, 239.
133. B. Schuur, T. Brouwer, D. Smink and L. M. Sprakel, Green solvents for sustainable separation processes, *Curr. Opin. Green Sustainable Chem.*, 2019, **18**, 57.
134. S. Erşan, O. Güçlü Üstündağ, R. Carle and R. M. Schweiggert, *J. Agric. Food Chem.*, 2016, **64**, 5.
135. M. K. Albuquerque Mendes, C. B. dos Santos Oliveira, M. D. A. Veras, B. Q. Araujo, C. Dantas, M. H. Chaves and E. C. Vieira, *Talanta*, 2019, **205**, 120100.
136. M. Bener, F. B. Şen, A. N. Önem, B. Bekdeşer, S. E. Çelik, M. Lalikoglu and R. Apak, *Food Chem.*, 2022, **385**, 132633.

137. V. Hessel, N. N. Tran, M. R. Asrami, Q. D. Tran, N. V. D. Long, M. Escribà-Gelonch and K. Sundmacher, *Green Chem.*, 2022, **2**, 1.
138. A. P. Sánchez-Camargo, M. Bueno, F. Parada-Alfonso, A. Cifuentes and E. Ibáñez, *TrAC, Trends Anal. Chem.*, 2019, **118**, 227.
139. F. Chemat, M. A. Vian and G. Cravotto, *Int. J. Mol. Sci.*, 2012, **13**, 8615.
140. L. Ruiz-Aceituno, S. Rodríguez-Sánchez, J. Sanz, M. L. Sanz and L. Ramos, *Food Chem.*, 2014, **153**, 450.
141. A. Dailey and Q. V. Vuong, *Processes*, 2015, **4**, 2.
142. D. Labuckas, D. Maestri and A. Lamarque, *Nat. Prod. Commun.*, 2016, 11.
143. R. Bodoira, A. Velez, L. Rovetto, P. Ribotta, D. Maestri and M. Martínez, *J. Food Sci.*, 2019, **84**, 963.
144. S. T. Wong, M. C. Tan and C. H. Geow, *J. Food Process. Preserv.*, 2019, **43**, e14138.
145. M. A. Salem, N. M. Aborehab, A. A Al-Karmalawy, A. R. Fernie, S. Alseekh and S. M. Ezzat, *Antioxidants*, 2022, **11**, 462.
146. Y. H. Choi and R. Verpoorte, *Curr. Opin. Food Sci.*, 2019, **26**, 87.
147. F. Chemat, M. Abert Vian, H. K. Ravi, B. Khadhraoui, S. Hilali, S. Perino and A. S. Fabiano Tixier, *Molecules*, 2019, **24**, 3007.
148. I. Fierascu, R. C. Fierascu and E. Sieniawska, Advanced techniques for recovery of active compounds from food by-products, in *Valorization of Agri-Food Wastes and By-Products*, Academic Press, 2021, pp. 693–710.
149. A. A. T. Hijo, C. Alves, F. O. Farias, V. S. Peixoto, A. J. Meirelles, G. H. Santos and G. J. Maximo, *Food Res. Int.*, 2022, **157**, 111194.
150. H. Zhang, J. M. Vicent-Luna, S. Tao, S. Calero, R. J. Jiménez Riobóo, H. Ferrer and M. C. Gutiérrez, *ACS Sustainable Chem. Eng.*, 2022, **10**, 1232.
151. X. Fu, T. Belwal, Y. He, Y. Xu, L. Li and Z. Luo, *Food Chem.*, 2022, **370**, 131042.
152. J. F. Osorio-Tobón, *J. Food Sci. Technol.*, 2020, **57**, 4299.
153. A. Dailey and Q. V. Vuong, *Cogent Food Agric.*, 2015, **1**, 1115646.
154. A. Valdés, L. Vidal, A. Beltran, A. Canals and M. C. Garrigós, *J. Agric. Food Chem.*, 2015, **63**, 5395.
155. T. S. Ballard, P. Mallikarjunan, K. Zhou and S. O'Keefe, *Food Chem.*, 2010, **120**, 1185.
156. H. Han, S. Wang, M. Rakita, Y. Wang, Q. Han and Q. Xu, *Food Nutr. Sci.*, 2018, **9**, 1034.
157. A. A. Salvador, R. Podestá, J. M. Block and S. R. S. Ferreira, *J. Supercrit. Fluids*, 2016, **116**, 215.
158. J. dos Santos Alves, T. C. Confortin, I. Todero, A. S. Rodrigues, S. R. Ribeiro, C. P. Boeira and C. S. da Rosa, *J. Supercrit. Fluids*, 2019, **153**, 104598.
159. M. Balvardi, J. A. Mendiola, P. Castro-Gómez, J. Fontecha, K. Rezaei and E. Ibáñez, *J. Am. Oil Chem. Soc.*, 2015, **92**, 1503.
160. Y. E. Rossi, L. P. Bohl, N. L. V. Braber, M. B. Ballatore, F. M. Escobar, R. Bodoira and M. A. Montenegro, *J. Funct. Foods*, 2020, **67**, 103862.

161. H. P. Cornelio-Santiago, R. B. Bodini, A. L. Oliveira and A. L. Oliveira, *J. Am. Oil Chem. Soc.*, 2020, **98**, 3.
162. O. R. Alara, N. H. Abdurahman and C. I. Ukaegbu, *Curr. Res. Food Sci.*, 2021, **4**, 200.
163. J. F. Osório-Tobón, *J. Food Sci. Technol.*, 2020, **57**, 4299.
164. M. I. L. Neves, M. M. Strieder, R. Vardanega, E. K. Silva and M. A. A. Meireles, *RSC Adv.*, 2020, **10**, 112.
165. F. Chemat, N. Rombaut, A. G. Sicaire, A. Meullemiestre, A. S. Fabiano-Tixier and M. Abert-Vian, *Ultrason. Sonochem.*, 2017, **34**, 540.
166. M. Vinatoru, T. J. Mason and I. Calinescu, *TrAC Trends Anal. Chem.*, 2017, **97**, 159–178.
167. M. G. Pereira, F. Hamerski, E. F. Andrade, A. D. P. Scheer and M. L. Corazza, *J. Supercrit. Fluids*, 2017, **128**, 338.
168. C. Wen, J. Zhang, H. Zhang, C. S. Dzah, M. Zandile, Y. Duan and X. Luo, *Ultrason. Sonochem.*, 2018, **48**, 538.
169. G. Ballistreri, E. Arena and B. Fallico, *Molecules*, 2009, **14**, 4358.
170. S. Jin, M. Gao, W. Kong, B. Yang, H. Kuang, B. Yang and H. Li, *3 Biotech*, 2020, **10**, 1.
171. M. M. Özcan, I. A. Mohamed Ahmed, N. Uslu, F. Al-Juhaimi, K. Ghafoor, E. E. Babiker and H. A. Alqah, *J. Food Meas. Charact.*, 2021, **15**, 2481.
172. X. Ma, X. Y. Zhou, Q. Q. Qiang and Z. Q. Zhang, *J. Sep. Sci.*, 2014, **37**, 1834.
173. Q. W. Zhang, L. G. Lin and W. C. Ye, *Chin. Med.*, 2018, **13**, 1–26.
174. G. L. Teixeira, L. G. Maciel, S. Mazzutti, R. C. T. Barbi, R. H. Ribani, S. R. S. Ferreira and J. M. Block, *J. Cleaner Prod.*, 2021, **306**, 127223.
175. C. G. Pereira and M. A. A. Meireles, *Food Bioprocess Technol.*, 2010, **3**, 340.
176. K. Y. Khaw, M. O. Parat, P. N. Shaw and J. R. Falconer, *Molecules*, 2017, **22**, 1186.
177. A. M. Farías-Campomanes, M. A. Rostagno, J. J. Coaquira-Quispe and M. A. A. Meireles, *Bioresour. Bioprocess.*, 2015, **2**, 1–13.
178. B. W. Bolling, J. B. Blumberg and C. Y. Oliver Chen, *Food Chem.*, 2010, **123**, 1040.
179. V. Deon, C. Del Bo', F. Guaraldi, F. Abello, S. Belviso, M. Porrini, P. Riso and O. Guardamagna, *Clin. Nutr.*, 2018, **37**, 1193.
180. G. A. Roth, G. A. Mensah, C. O. Johnson, G. Addolorato, E. Ammirati, L. M. Baddour, N. C. Barengo, N. C. Beaton, A. Benjamin, E. J. Benziger, C. P. Bonny, A. Brauer, M. Brodmann, M. Cahill, T. J. Carapetis, J. R. Catapano, A. L. Chugh, S. Cooper, L. T. Coresh and J. V. Fuster, *J. Am. Coll. Cardiol.*, 2020, **76**, 2982.
181. T. Senoner and W. Dichtl, *Nutrients*, 2019, **11**, 2090.
182. S. K. Biswas, *Oxid. Med. Cell. Longev.*, 2016, 5698931.
183. R. Estruch, E. Ros, J. Salas-Salvadó, M. Covas, D. Corella, F. Arós, E. Gómez-Gracia, V. Ruiz-Gutiérrez, M. Fiol, J. Lapetra, R. M. Lamuela-Raventos, L. Serra-Majem, X. Intó Basora, J. Muñoz, M. A. Sorlí, J. V. Martínez, J. A. Fitó, M. Gea and A. M. A. Martínez-González, *N. Engl. J. Med.*, 2018, **378**, e34.

184. F. Mach, C. Baigent, A. L. Catapano, K. C. Koskinas, M. Casula, L. Badimon, M. J. Chapman, G. G. De Backer, V. Delgado, B. A. Ference, I. M. Graham, A. Halliday, U. Landmesser, B. Mihaylova, T. R. Pedersen, T. R. Riccardi, G. Richter, D. J. Sabatine, M. S. Taskinen and M. R. R. S. Patel, *ESC/EAS Eur. Heart J.*, 2020, **41**, 111.
185. S. Parthasarathy, A. Raghavamenon, M. O. Garelnabi and N. Santanam, *Methods Mol. Biol.*, 2010, **610**, 403.
186. S. S. K. Wijeratne, M. M. Abou-Zaid and F. Shahidi, *J. Agric. Food Chem.*, 2006, **54**, 312–318.
187. O. Chen, P. E. Milbury and J. B. Blumberg, *Antioxidants*, 2019, **8**, 95.
188. F. Demoliner, P. de Britto Policarpi, J. C. Ramos, V. L. A. F. Bascuñan, R. A. Ferrari, I. Jachmanián, A. de Francisco de Casas, L. F. L. Vasconcelos and J. M. Block, *Food Res. Int.*, 2018, **108**, 27.
189. U. Tinggi, *Environ. Health Prev. Med.*, 2008, **13**, 102.
190. M. Sakamoto, A. Yasutake, A. Kakita, M. Ryufuku, H. M. Chan, M. Yamamoto, S. Oumi, S. Kobayashi and C. Watanabe, *Environ. Sci. Technol.*, 2013, **47**, 2862.
191. M. Roman, P. Jitaru and C. Barbante, *Metallomics*, 2014, **6**, 25.
192. A. Caimari, F. Puiggròs, M. Suárez, A. Crescenti, S. Laos, J. A. Ruiz, V. Alonso, J. Moragas, J. M. Del Bas and L. Arola, *Food Chem.*, 2015, **167**, 138.
193. C. E. Cabral and M. R. S. T. Klein, *Arq. Bras. Cardiol.*, 2017, **109**, 475.
194. H. Gylling and P. Simonen, *Nutrients*, 2015, **7**, 7965.
195. Z. Qiang, S. O. Lee, Z. Ye, X. Wu and S. Hendrich, *Phytother. Res.*, 2012, **26**, 1048.
196. J. A. Nettleton, J. A. Lovegrove, R. P. Mensink and U. Schwab, *Ann. Nutr. Metab.*, 2016, **68**, 249.
197. S. T. Auti and Y. A. Kulkarni, *Int. J. Vitam. Nutr. Res.*, 2017, **87**, 99.
198. X. Chen, J. Drew, W. Berney and W. Lei, Neuroprotective Natural Products for Alzheimer's Disease, *Cells*, 2021, **10**, 1309.
199. E. D. AlFadly, P. A. Elzahhar, A. Tramarin, S. Elkazaz, H. Shaltout, M. M. Abu-Serie, J. Janockova, O. Soukup, D. A. Ghareeb, A. F. El-Yazbi, R. W. Rafeh, N. M. Z. Bakkar, F. Kobeissy, I. Iriepa, I. Moraleda, M. N. S. Saudi, M. Bartolini and A. S. F. Belal, *Eur. J. Med. Chem.*, 2019, **167**, 161.
200. Z. Bahaeddin, A. Yans, F. Khodagholi, H. Hajimehdipoor and S. Sahranavard, *Nutr. Neurosci.*, 2017, **20**, 317–326.
201. L. A. De la Rosa, A. A. Vazquez-Flores, E. Alvarez-Parrilla, J. Rodrigo-García, O. N. Medina-Campos, A. Ávila-Nava, S. González-Reyes and J. Pedraza-Chaverri, *J. Funct. Foods*, 2014, **7**, 219.
202. P. Sharma, P. K. Verma, S. Sood, N. K. Pankaj, S. Agarwal and R. Raina, *Toxicol. Rep.*, 2021, **8**, 223.
203. J. H. Kim, Q. Wang, J. M. Choi, S. Lee and E. J. Cho, *Nutr. Res. Pract.*, 2015, **9**, 480.
204. M. J. Kim, A. R. Seong, J. Y. Yoo, C. H. Jin, Y. H. Lee, Y. J. Kim, J. Lee, W. J. Jun and H. G. Yoon, *Mol. Nutr. Food Res.*, 2011, **55**, 1798.

205. T. N. Ahmed, W. Setzer, S. Fazel Nabavi, I. Erdogan Orhan, N. Braidy, E. Sobarzo-Sanchez and S. Mohammad Nabavi, *Curr. Pharm. Des.*, 2016, **22**, 1350.
206. S. E. Chastain, *Therapeutic potential of catechins and derivatives for the prevention of Alzheimer's disease*, 2016.
207. H. J. Lim, S. B. Shim, S. W. Jee, S. H. Lee, C. J. Lim, J. T. Hong, Y. Y. Sheen and D. Y. Hwang, *J. Nutr. Biochem.*, 2013, **24**, 1302.
208. Y. Shimmyo, T. Kihara, A. Akaike, T. Niidome and H. Sugimoto, *J. Neurosci. Res.*, 2008, **86**, 368.
209. M. Y. Yamakawa, K. Uchino, Y. Watanabe, T. Adachi, M. Nakanishi, H. Ichino, K. Hongo, T. Mizobata, S. Kobayashi, K. Nakashima and Y. Kawata, *Nutr. Neurosci.*, 2016, **19**, 32.
210. M. A. Ansari, H. M. Abdul, G. Joshi, W. O. Opii and D. A. Butterfield, *J. Nutr. Biochem.*, 2009, **20**, 269.
211. J. K. Kim, S. J. Choi, H. Y. Cho, H. J. Hwang, Y. J. Kim, S. T. Lim, C. J. Kim and H. K. Kim, *Biosci., Biotechnol., Biochem.*, 2010, **74**, 397–401.
212. X. Zhao, C. Liu, M. Xu, X. Li, K. Bi and Y. Jia, *PLoS One*, 2016, **11**, e0152772.
213. K. Ono, K. Hasegawa, H. Naiki and M. Yamada, *Biochim. Biophys. Acta, Mol. Basis Dis.*, 2004, **1690**, 193.
214. I. Bakoyiannis, A. Daskalopoulou, V. Pergialiotis and D. Perrea, *Biomed. Pharmacother.*, 2019, **109**, 1488.
215. E. Polo-Hernández, V. Tello, A. A. Arroyo, M. Domínguez-Prieto, F. De Castro, A. Tabernero and J. M. Medina, *Brain Res.*, 2014, **1570**, 13.
216. K. Youn, E. Y. Yun, J. Lee, J. Y. Kim, J. S. Hwang, W. S. Jeong and M. Jun, Molecular Docking Studies, 2014.
217. J. M. Bourre, *J. Nutr., Health Aging*, 2004, **8**, 163.
218. M. Hinterberger and P. Fischer, *J. Neural Transm.*, 2013, **120**, 211.
219. W. Li, H. Liu, M. Yu, X. Zhang, Y. Zhang, H. Liu, J. X. Wilson and G. Huang, *Mol. Neurobiol.*, 2016, **53**, 6548–6556.
220. N. Zhao, C. Zhong, Y. Wang, Y. Zhao, N. Gong, G. Zhou, T. Xu and Z. Hong, *Neurobiol. Dis.*, 2008, **29**, 176.
221. Q. Zhang, G. Yang, W. Li, Z. Fan, A. Sun, J. Luo and Z. J. Ke, *Neurobiol. Aging*, 2011, **32**, 42.
222. E. J. Kim and S. J. Yang, *Clin. Nutr. Res.*, 2017, **6**, 130.
223. V. Boccardi, M. Baroni, F. Mangialasche and P. Mecocci, *Alzheimer's Dementia: Translational Res. Clin. Interventions*, 2016, **2**, 182.
224. A. Eidi, M. Eidi, G. Mahmoodi and S. Oryan, *Eur. Neuropsychopharmacol.*, 2006, **16**, 101.
225. T. Batool, E. A. Makky, M. Jalal and M. M. Yusoff, *Appl. Biochem. Biotechnol.*, 2016, **178**, 900.
226. S. Sonnino and A. Prinetti, *J. Neurochem.*, 2016, **137**, 485.
227. H. Che, L. Du, P. Cong, S. Tao, N. Ding, F. Wu, C. Xue, J. Xu and Y. Wang, Cerebrosides from Sea Cucumber Protect Against Oxidative Stress in SAMP8 Mice and PC12 Cells, *J. Med. Food*, 2017, **20**(4), 392–402.

228. A. S. Choudhari, P. C. Mandave, M. Deshpande, P. Ranjekar and P. Prakash, *Front. Pharmacol.*, 2020, **10**, 1614.
229. H. Sung, J. Ferlay, R. L. Siegel, M. Laversanne, I. Soerjomataram, A. Jemal and F. Bray, *CA: Cancer J. Clin.*, 2021, **71**, 209.
230. F. Bray, A. Jemal, N. Grey, J. Ferlay and D. Forman, *Lancet Oncol.*, 2012, **13**, 790.
231. A. Basli, N. Belkacem, I. Amrani, *Phenolic Compounds – Biological Activity*, 2017.
232. AICR & WCRF, Diet, Nutrition, Physical Activity and Cancer: a Global Perspective A summary of the Third Expert Report, 2020.
233. S. E. Steck and E. A. Murphy, *Nat. Rev. Cancer*, 2019, 125.
234. L. Wu, Z. Wang, J. Zhu, A. L. Murad, L. J. Prokop and M. H. Murad, *Nutr. Rev.*, 2015, **73**, 409.
235. S. M. Ashraf and K. Rathinasamy, Antibacterial and anticancer activity of the purified cashew nut shell liquid: implications in cancer chemotherapy and wound healing, *Natural Prod. Res.*, 2018, **32**, 2856.
236. N. M. Pepple, W. U. Ekoriko, F. M. Idih and V. O. Chidozie, *J. Med. Plants Res.*, 2020, **14**, 488.
237. M. Carvalho, P. J. Ferreira, V. S. Mendes, R. Silva, J. A. Pereira, C. Jerónimo and B. M. Silva, *Int. J. Published Br. Industrial Biol. Res. Assoc.*, 2010, **48**, 441.
238. A. Abraham, A. J. Kattoor, T. Saldeen and J. L. Mehta, *Crit. Rev. Food Sci. Nutr.*, 2019, **59**, 2831.
239. J. Ju, S. C. Picinich, Z. Yang, Y. Zhao, N. Suh, A. N. Kong and C. S. Yang, *Carcinogenesis*, 2010, **31**, 533.
240. A. K. Smolarek, J. Y. So, P. E. Thomas, H. J. Lee, S. Paul, A. Dombrowski, C. X. Wang, C. L. L. Saw, T. O. Khor, A. N. T. Kong, K. Reuhl, M. J. Lee, C. S. Yang and N. Suh, *Mol. Carcinog.*, 2013, **52**, 514.
241. B. B. Aggarwal, C. Sundaram, S. Prasad and R. Kannappan, *Biochem. Pharmacol.*, 2010, **80**, 1613.
242. S. Wada, *Forum Nutr.*, 2009, **61**, 204.
243. A. Vafadar, Z. Shabaninejad, A. Movahedpour, F. Fallahi, M. Taghavipour, Y. Ghasemi, M. Akbari, A. Shafiee, S. Hajighadimi, S. Moradizarmehri, E. Razi, A. Savardashtaki and H. Mirzaei, *Cell Biosci.*, 2020, **10**(1), 32.
244. S. Kamaraj, R. Vinodhkumar, P. Anandakumar, S. Jagan, G. Ramakrishnan and T. Devaki, *Biol. Pharm. Bull.*, 2007, **30**, 2268.
245. P. M. Das and R. Singal, *Off. J. Am. Soc. Clin. Oncol.*, 2004, **22**, 4632–4642.
246. C. T. Kuo, C. Chang, W. Lee and W. Sen, *Sci. Rep.*, 2015, **5**, 1.
247. I. A. Siddiqui, A. Malik, V. M. Adhami, M. Asim, B. B. Hafeez, S. Sarfaraz and H. Mukhtar, *Oncogene*, 2008, **27**, 2055.
248. H. Huynh, T. Nguyen, E. Chan and E. Tran, *Int. J. Oncol.*, 2003, **23**, 821.
249. T. Szekeres, P. Saiko, M. Fritzer-Szekeres, B. Djavan and W. Jäger, *Ann. N. Y. Acad. Sci.*, 2011, **1215**, 89.
250. F. J. Zheng, L. Shi, J. Yang, X. H. Deng, Y. Q. Wu, X. Q. Yan and N. Huang, *Asian Pac. J. Cancer Prev.*, 2012, **13**, 3751.

251. E. P. Lansky, G. Harrison, P. Froom and W. G. Jiang, *Invest. New Drugs*, 2005, **23**, 121–122.
252. M. S. Farvid, N. D. Spence, M. D. Holmes and J. B. Barnett, *Cancer*, 2020, **126**, 3061.
253. N. Murphy, T. Norat, P. Ferrari, M. Jenab, B. Bueno-de-Mesquita, G. Skeie, C. C. Dahm, K. Overvad, A. Olsen, A. Tjønneland, F. Clavel-Chapelon, A. Boutron-Ruault, M. C. Racine, A. Kaaks, R. Teucher, B. Boeing, H. Bergmann, M. M. Trichopoulou, D. Trichopoulos and E. Riboli, *PLoS One*, 2012, **7**(6), 39361.
254. E. Alvarez-Parrilla, R. Urrea-López and L. A. Rosa, *J. Food Bioactives*, 2018, **1**, 56.
255. P. S. Manos, Z. K. Zhou and C. H. Cannon, *Int. J. Plant Sci.*, 2001, **162**, 1361.
256. G. Bounous, *Acta Horticulturae*, 2005, **693**, 33.
257. L. A. Follegatti-Romero, C. R. Piantino, R. Grimaldi and F. A. Cabral, *J. Supercrit. Fluids*, 2009, **49**, 323.
258. S. Wang, F. Zhu and Y. Kakuda, *Food Chem.*, 2018, **265**, 316.
259. R. Chirinos, R. Pedreschi, G. Domínguez and D. Campos, *Food Chem.*, 2015, **173**, 1203.
260. L. F. Gutiérrez, Y. Quiñones-Segura, Z. Sanchez-Reinoso, D. L. Díaz and J. I. Abril, *Food Chem.*, 2017, **237**, 581.
261. Z. Sanchez-Reinoso, W. I. Mora-Adames, C. A. Fuenmayor, A. E. Darghan-Contreras, C. Gardana and L. F. Gutiérrez, *Chem. Eng. Process.*, 2020, **153**, 107922.
262. O. Freitas-Silva and A. Venâncio, *Food Res. Int.*, 2011, **44**, 1434.
263. L. H. O. Wadt, K. A. Kainer, C. L. Staudhammer and R. O. P. Serrano, *Biol. Conserv.*, 2008, **141**, 332.
264. A. M. Kluczkovski and M. Martins, in *Encyclopedia of Food and Health*, 1st edn, 2015.
265. F. S. Martins, L. L. Borges, J. R. Paula and E. C. Conceição, *Brazilian J. Pharmacognosy*, 2013, **23**, 521.
266. A. G. de O. Sousa, D. C. Fernandes, A. M. Alves, J. B. Freitas and M. M. V. Naves, *Food Res. Int.*, 2011, **44**, 2319.
267. P. B. Policarpi, L. Turcatto, F. Demoliner, R. A. Ferrari, V. L. A. Bascuñan, J. C. Ramos and J. M. Block, *Food Res. Int.*, 2018 **106**, 736.
268. V. Queiroga-Neto, P. S. Bora, Z. N. Diniz, J. M. D. O. Cavalheiro and P. A. S. Souza, *CYTA – J. Food*, 2009, **7**, 23.
269. F. S. C. Ribeiro, V. A. B. Souza and A. C. A. Lopes, *Rev. Cienc. Agron.*, 2012, **43**, 301.
270. A. P. Rodrigues, G. A. Pereira, P. H. F. Tomé, H. S. Arruda, M. N. Eberlin and G. L. Pastore, *Food Res. Int.*, 2019, **121**, 880.
271. G. L. Teixeira, L. G. Maciel, S. Mazzutti, C. B. Gonçalves, S. R. S. Ferreira and J. M. Block, *Biocatal. Agric. Biotechnol.*, 2020, **24**, 101521.
272. O. Borges, B. Gonçalves, J. L. S. de Carvalho, P. Correia and A. P. Silva, *Food Chem.*, 2008, **106**, 976.

273. F. Demoliner, L. T. de Carvalho, G. R. de Liz, E. S. Prudêncio, J. C. Ramos, V. L. A. F. Bascuñan and J. M. Block, *Int. J. Food Sci. Technol.*, 2020, **55**, 3031.
274. J. C. M. Barreira, S. Casal, I. C. F. R. Ferreira, M. B. P. P. Oliveira and J. A. Pereira, *J. Agric. Food Chem.*, 2009, **57**, 2836.
275. Z. Q. Cai, *Ind. Crops Prod.*, 2011, **34**, 1235.
276. R. Chirinos, G. Zuloeta, R. Pedreschi, E. Mignolet, Y. Larondelle and D. Campos, *Food Chem.*, 2013, **141**, 1732.
277. A. G. De Oliveira Sousa, D. C. Fernandes, A. M. Alves, J. B. de Freitas and M. M. V. Naves, *Food Res. Int.*, 2011, **44**, 2319.
278. F. D. Da Silva-Araújo, C. V. R. De Moura and M. H. Chaves, *Quim. Nova*, 2010, **33**, 1671.
279. D. Pinto, M. L. Cádiz-Gurrea, A. Vallverdú-Queralt, C. Delerue-Matos and F. Rodrigues, *Food Res. Int.*, 2021, **144**, 110364.
280. A. Morana, G. Squillaci, S. M. Paixão, L. Alves, F. La Cara and P. Moura, *Energies*, 2017, **10**(10), 1–14.
281. F. M. Vella, B. Laratta, F. La Cara and A. Morana, *Nat. Prod. Res.*, 2018, **32**, 1022.
282. L. Amaral, F. Rodrigues, A. Silva, P. Costa, C. Delerue-matos and E. F. Vieira, *Food Chem.*, 2022, **396**, 133609.
283. G. Vázquez, O. Mosquera, M. S. Freire, G. Antorrena and J. González-álvarez, *Chem. Eng. J.*, 2012, **184**, 147.
284. E. G. Torres Sánchez, B. Hernández-Ledesma and L. F. Gutiérrez, *Food Rev. Int.*, 2021, **1**, 148–159.
285. B. Kumar, K. Smita, E. Sánchez, C. Stael and L. Cumbal, *Ecol. Eng.*, 2016, **93**, 152.
286. Y. J. Rueda-Ordóñez, M. A. Mariño-Bohórquez and D. A. Rueda-Ordóñez, *Bioresour. Technol. Rep.*, 2021, **15**, 100807.
287. M. S. Carvalho and C. F. Virgens, *Microchem. J.*, 2018, **143**, 410.

CHAPTER 5

Chemical and Biological Valorization of Tomato Waste

M. GARCIA-MARTI AND J. SIMAL-GANDARA*

Universidade de Vigo, Nutrition and Bromatology Group, Analytical Chemistry and Food Science Department, Faculty of Science, E32004 Ourense, Spain
*Email: jsimal@uvigo.es

5.1 Introduction

The amount of food by-products along the food cycle from agricultural practices to consumers has become a major concern for environmental sustainability due to its adverse consequences on natural resources, food safety, the environment, and human health.[1] The Food and Agriculture Organization (FAO) estimated that about 1.3 billion tons, which represents about 30% of all food manufactured for human consumption in the world, is wasted globally. Moreover, food demand is expected to rise by seventy percent to sustain a human population predicted to reach 9.8 billion by 2050,[2] so it is an urgent call to look for solutions about the valorization of these waste products.

Tomatoes (*Solanum Lycopersicum* L.) are the second most produced and consumed vegetable crop. Nearly 180 million tons are produced annually according to FAOSTAT. Tomato is a native plant to South America and was introduced in Europe by colonizers. Over the years, the aim of tomato domestication and plant selective breeding was to increase the size of the fruit and to achieve more productive varieties with more desirable characteristics for the consumer.[3] Nowadays, China is by far the largest producer of tomatoes, followed by India. In Europe, Italy is the main producer followed by

Green Chemistry Series No. 78
Agri-food Waste Valorisation
Edited by Pankaj Chowdhary and Abhay Raj

Published by the Royal Society of Chemistry, www.rsc.org

Spain and Portugal.[4] Tomato is consumed raw and processed because of its nutrients and benefits to human health. A part of the total production is eaten as fresh fruit, but another part is intended for industrial processes in order to produce tomato derivatives such as paste, juice or sauces. Because of high consumption, tomato fruit and tomato processed products can generate major losses of food throughout their life cycle.

Losses and food by-products are generated during all the phases of the manufacturing process, taking into account agricultural practices and harvesting, industrial processing as classification and grading, storage, marketing, processing and distribution or transport to packinghouses or markets.[5] Although most studies do not distinguish between food loss and food waste, it is essential to clarify the differences between these two types of by-products. On the one hand, food losses refer to a decrease in the total volume of food that is edible for human consumption, and that typically occurs during the distribution chain as production, harvest and processing phases. On the other hand, food waste involves food losses occurring at the retail, consumption and disposal stages and depends on the behaviour of retail stores and consumers.[6]

Tomato losses not only are about the wasting of fresh fruit or manufactured products, but also involve the wasting of critical resources such as water, fertilizers, lands, chemical energy and labour too. Horticultural practices produce the major waste into the landscape, and it is approximated that 25% of the irrigation water during the production season is lost. Moreover, food waste caused about 8% of global greenhouse gas emissions.[7] That causes a detrimental impact on the environment, contributing to the pollution as well as to the degradation and depletion of natural resources, which is extremely damaging for food safety and sustainable development.[8] These general losses, in spite of their different causes and different solutions, can be evaluated quantitatively and qualitatively. Quantitatively it is easy to define since it refers to the total volume of losses; however, qualitatively it is difficult to assess and quantify because it is related to a decrease in the nutritional value, its caloric intake and the consumer acceptability although the product continues to be nutritionally beneficial for health.[9]

Tomato by-products integrate unused or unconsumed parts including leaves and roots during the harvest season, tomato fresh fruit with undesirable characteristics for consumers, skin, pulp, seeds and pomace in industrial processing.[10] Losses of fresh tomatoes usually occur due to consumer or retailer behaviour because they are discarded due to size, shape, colour or ripening stage although the fruit is fit for consumption.[11] During tomato industrial processes, a significant quantity of tomato pomace, which represents about 5–30% of the main product, is discarded as food waste. Both wastes are commonly used as livestock feed or disposed of in landfills, but in the best of cases, fresh tomatoes can also be used in the industry for tomato derivatives.[12]

Recent projects are focused on added-value products of tomato waste produced during food chain industrial production, processing and manufacturing to recover bioactive and nutraceutical compounds and industrial enzymes, produce biofuel or be a biological resource for agricultural practices.[13] Some of those by-products can be re-utilized in the human diet such as sugars, organic acids, fibre, proteins, oil antioxidants and vitamins. Moreover, there are many products including food, cosmetics and pharmaceutical formulations where waste materials can be a great natural source of compounds in order to develop high added-value products.[14]

Taking into account the worldwide presence of tomato losses and the amount of both production and industry stages, increasing efforts are focused on the possibility of reusing and upgrading agro-industrial waste in order to decrease the effect on the environment and give tomato by-products a second life.[15] Few years ago, food by-products were considered neither a cost nor a benefit, but this point of view has become obsolete due to several reasons involving environmental concerns and practices to minimize the impact of waste.[16] The current trend in the recovery of waste consists of treating it as a potential source of compounds because it has been demonstrated that wastes can be a natural source and minimize resources or decrease the high costs of waste disposal. Tomato waste can contain a high quantity of reusable substances of great value that can be useful in isolation or included within more complex processes. These by-products can be transformed into other commercial matters either as raw materials for secondary processes or being part of different novel products. For that, an adequate technology in order to separate and recycle which is absolutely cost efficient is required.[17]

From a chemical and biological point of view, many studies aim at the re-utilization of tomato waste from fresh fruits and industry to isolate high value-added compounds such as minerals, vitamins, dietary fibres, proteins, polyphenols and carotenoids.[18] In addition, agricultural and food processing waste has become an interesting feedstock for chemical substances fermentation processes in the biorefinery industry.[19] On the other hand, tomato plant residues are a great source of organic waste because they involve an important source of carbon and chemical elements that can also be incorporated into the composting process as a feasible way of transforming plant residues into organic amendment[20] (Figure 5.1).

Current predictions regarding the rise in world population and environmental solutions for food losses and wastes make an urgent finding of alternatives for agricultural and industrial by-products to give a second life to these wastes.

5.2 Tomato Waste as a Source of Biochemical Compounds

The main generated waste during tomato industrial processes is tomato pomace. It is nearly 33% seeds, 27% skin and 40% pulp.[21] The total volume of tomato pomace can have variations depending on the processing states

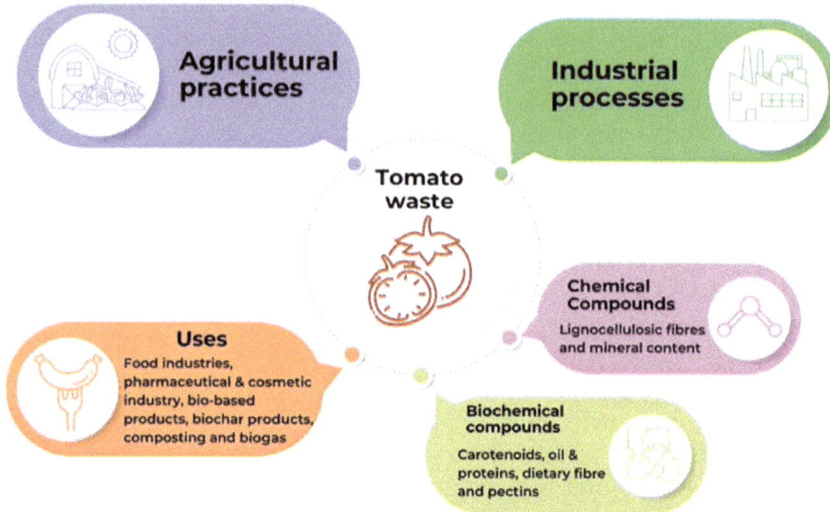

Figure 5.1 Chemical and biological valorization of tomato waste.

and tomato varieties. In general, it has been described that it accounts for up to 5–10% of the whole tomato, which represents a high volume of organic matter to be reused in other processes.[22] When tomato pomace is dried during different phases of manufacturing, it has an average composition of 50% neutral fibre, 25% total sugar, 19% protein, 6% lipids, 8% pectins and 4% mineral content on dry weight. Moreover, tomato pomace is a great fount of carotenoids or phenolic compounds that have been recognized as antioxidants due to strong free radical scavenging activities.[23]

In a comparative study of fresh tomatoes and tomato industrial processing by-products, different bioactive compounds were analysed. The results showed that molecules such as carotenes, terpenes and polyphenols were detected both in raw tomatoes and in waste products, with carotenes being the major bioactive compound present in tomato and industrial waste. It is concluded that, except for vitamin C, phytochemical compounds were able to withstand industrial processing.[24]

5.2.1 Carotenoids

Carotenoids are phytochemicals naturally present in tomatoes and other fruits. Carotenoids have antioxidant properties, and they interact with other antioxidants to prevent cell damage.[25] These are water-insoluble pigments synthesized by many vegetal and microorganisms, and these are responsible for yellow, orange or red tonalities of fruits. Most of them are tetra-terpenes, with a 40-carbon skeleton conjugated and polyene chain. Carotenoids with more than 40–50 carbon skeleton are named higher carotenoids and carotenoids with fewer than 40 carbons are called apocarotenoids. The latter are present in nature due to the degradation of C40 carotenoids such as lycopene or β-carotene. Carotenoids can be selected in two different groups: on one hand carotenes, such as α-carotene, β-carotene and lycopene, and xanthophylls such as lutein. Carotenes are cycled or linear hydrocarbons, while xanthophylls contain oxygen atoms and can be present as other complex molecules.[26] These bioactive compounds have an important health benefit in the human diet as they act as a disease preventer,[27] but the main compound with a high added value in tomato is lycopene.

5.2.1.1 Lycopene

Lycopene is an acyclic carotenoid with 11 conjugated double bonds and it gives a deep red colour to tomatoes at different ripe stages. It is the main carotenoid found in tomato by-products and plays a fundamental role in human health because of its antioxidant and anti-inflammatory characteristics. In addition, lycopene favours immune response, is photo-protective and has hypoglycaemic properties in organisms[28–30] so there is an increasing interest in using tomato pomace as a source for lycopene extraction.

The lycopene content in tomatoes can vary depending on the variety of fruits, geographic location or growth or vegetal development, crop management and environmental conditions.[31] There is a strong interest in lycopene extraction because it is considered a natural pigment suitable for the market and a functional healthy-promotion ingredient that can be used in nutraceutical, cosmetic and food formulations. The utilization of tomato waste could supply extra income for the tomato industry and reduce by-product disposal problems and expenses.[32] The composition of the tomato residue matrix influences the bioavailability of lycopene. Lycopene is found mostly in the skin of tomatoes, attached to the membrane of the chromoplasts, which contain polysaccharides including cellulose, hemicelluloses and pectins.[33]

Numerous advances have been successfully applied for extracting lycopene from tomato by-products in order to recover and reuse tomato waste. Traditionally, organic solvents have been used for it, but given the current concern about food safety and more clean techniques utilization, new sustainable extraction techniques are being investigated.[5] Lycopene can easily isomerize (*trans–cis*), as it is a compound sensitive to external conditions such as light, heat and oxygen. All-*trans* lycopene is the main specie in tomato because it is the most thermodynamically stable form. That is an important point because it has been shown that lycopene-based food products can be very changeable systems, as during the processing and storage of processed foods, the *cis*-isomers increase because they are more unstable than the *trans*-isomers.[34] That must be taken into account for lycopene incorporation into nutraceuticals or functional foods in order to avoid the *trans–cis* isomerisation. Furthermore, reusing and recycling components with eco-friendly technologies is not very advanced yet. When tomatoes are grown exclusively to extract lycopene, they are not economically viable for many industries. Therefore, it is necessary to invest in developing new strategies for the effective management of lycopene extraction in tomato industrial wastes.

5.2.1.2 β-Carotene

β-Carotene is the major consumed compound plant precursor of vitamin A, and it gives the orange and green bright colour to fruits such as carrots, oranges, green-leafy vegetables and fruits such as apricots and tomatoes.[35] Although β-carotene is in lower concentrations than lycopene in tomatoes, it has been extensively studied, since, after lycopene, it is the compound most frequently found in industrial tomato waste.[14]

β-Carotene has immunological functions, and antioxidant capacity and its consumption can reduce the risk of diseases such as diabetes and obesity.[36] However, vitamin A precursor is the most clearly proven function of that compound, which is deficient in nutritionally vulnerable populations. Xerophthalmia is a disease caused by vitamin A deficiency, a leading cause of preventable blindness. On the other hand, lutein is another different carotenoid present in ripe tomatoes and it is also related to eyes and vision problems.[37] At that point, tomato pomace from industrial waste represents an affordable source of bioactive compounds in order to reduce vitamin A and lutein deficiency and improve human diseases related to those carotenoids.[38]

5.2.2 Oil and Proteins

As noted above, tomato pomace waste involves pulp, peels and seeds. Although these products have been recognized as great sources of bioactive compounds with nutraceutical characteristics, the content of seeds is quite different from other parts because it is formed from about 32% proteins, 27% total fat and 18% fibre depending on the tomato variety.[39] Tomato seeds have been described as a high natural fount of antioxidants due to their concentration in phytochemicals such as gallic acid, *trans*-cinnamic acid and quercetin. Seeds also contain non-saponifiable compounds such as α-tocopherol γ-tocopherol and demethylsterol.[40] Besides that, the most remarkable aspect of the composition of the seeds is their applicability to obtain oil and proteins.

Tomato seed oil shows a high potential to be manufactured as a marketable edible oil. During tomato waste processing, seed generated about 20–35% of oil, composed of unsaturated fatty acid, and it has a high quantity of linolenic and oleic acids. Although bioactive compounds such as carotenoids, phenolic compounds and fatty acids can be adversely affected by heat treatments applied during industrial processes, tomato seed oil can be compared with other commercial edible oils including olive oil or soybean oil with high nutritional quality and antioxidant activity.[41] Due to their high thermal stability and antioxidant characteristic, tomato seed oil has been used in food preservation and as an ingredient in bakery products and food preservation.[42] However, it is necessary to focus on future research about acceptance by consumers and the health benefits of tomato oil.

On the other hand, proteins are also highly present in tomato seeds. Tomato protein extracted from tomato seed can be compared to other plant proteins: tomato seeds have a much higher protein content than grape seeds or peanuts, but not as much as apple seeds. Taking into account the amino acid content in tomato seeds protein, glutamic acid is the major present, followed by aspartic acid.[43] Moreover, tomato seed has a high concentration of lysine, than its counterparts from various cereals.[44]

Thanks to these seed components, especially the high level of protein, seeds have a great potential for being used as a nutrient supplement in animal feeding.[45] Up to now, tomato wastes derived from the food industry have been used directly as a feed with farm animals, rodents, hens and fishes, but lately it is usually dehydrated or ensiled in order to extend its shelf life and improve its nutritional profile and storage properties.

Environmental conditions and parameters involved in seed protein extraction may vary the nutritional value and functionality of the proteins For example, temperature, time, pH or extraction method can affect protein functional properties if denaturation or dissociation occurs.[46] In that sense, it would be convenient to develop optimal extraction methods to support a specific application to tomato protein isolates in commercial products.

Proteins are also an important nutrient for human health. Protein deficiency persists globally and increases the risks for diseases associated with low protein intake, especially in developed countries.[47] Furthermore, there is a growing concern about reducing the amount of animal or animal-byproducts of protein consumed in the diet. From that nutritional perspective, the supply of protein has not satisfied its high request, so there is a fundamental challenge to exploit different natural protein sources such as tomato seeds.

5.2.3 Dietary Fibre and Pectins

The interest in connection between fibre consumption and its beneficial effects on the digestive tract have been increased due to the growing amount of proof in human nutrition. Generally, dietary fibre does not have a well-defined chemical profile, although it is known that a mix of substances including cellulose, hemicellulose, lignin and pectic substances can provide functional characteristics added in the food industrial process.[48] Consuming an appropriate amount of soluble and insoluble fibres reduces the risks of different diseases.[49] In this sense, dietary fibres are conceived as a supplement or functional compound in healthy foods due to our understanding of their health benefits.

The chemical characterization of tomato pomace shows an average composition of 60% neutral detergent fibre, especially present in tomato peel, since it represents more than half of its dry weight content. This suggests that the quality of food products can be improved by adding fibre extracted from tomato waste as dietary fibre.[50]

Pectin is a hetero-polysaccharide present in the majority of plant cells and has been regarded as an additive, gelling agent, thickener and emulsifier in the food industry. Traditionally pectin has been isolated from citrus peel and apple pomace at the commercial scale; however, the global demand for pectin has increased in the last years, so these traditional sources are insufficient to meet the market demand.[51]

Currently, tomato by-products are being studied as an alternative to the extraction of this component, but large-scale production techniques are insufficient to isolate this compound from tomato pomace. Although pectin extracted from tomato pomace fulfils all the standard requirements in the food industry, it is important to promote more research investigation focusing on sustainability, profitability and efficiency of pectin extraction from tomato waste.[52]

5.3 Tomato Waste as a Source of Chemical Compounds

As mentioned above, industrial tomato processing generates a high volume of waste called tomato pomace. The biochemical valorization of tomato pomace and its uses in the food industry has been previously described; however, other tomato discarded products can be isolated and added to pharmaceutical and cosmetic industry products. On the other hand, producing compost or biochar from tomato pomace is another alternative to tomato horticultural practice residues including stalks and other parts such as leaves and roots.[53]

Tomato cultivation, harvesting and industrial processing generate a number of wastes besides tomato pomace. Although it does not have marketable value, the reuse of tomato waste can generate another source of income for farmers and industrial tomato companies. Chemical characterization of tomato wastes includes different tomato wastes produced such as polymers and mineral elements. In addition to the use of tomato pomace as a source of biochemical compounds, many research projects aim for tomato by-product conversion into renewable useful products such as biodiesel, biopolymers and as a source of chemical compounds in different processes.

5.3.1 Lignocellulosic Fibres

Lignocellulosic raw and enzymatic treated materials originating from agri-waste products can be utilized as a natural source of polymers including hemicellulose (20–40%), cellulose (30–60%) and lignin (10–25%) which present very complex structures.[54] However, this content can be different between tomato varieties and processing extraction methods. The main fraction of the fibres is composed of cellulose, a polysaccharide consisting of a linear chain of β-(1–4) linked D-glucose units, the fibrils of which are strongly bonded by a hemicellulose layer, partially soluble in water due to its hydroxyl and acetyl groups. Moreover, lignin is an aromatic polymer highly complex with low water sorption including in lignocellulosic pulp.[55]

Tomato production and processing industry wastes have high value to develop bio-based material products. To achieve that, it has been described that the treatment of lignocellulosic raw tomato agri-waste with enzymes is a sustainable process focused on the recovery of polymeric matrices and good quality fibres which leads to realizing more economical value-added bio-based products.

5.3.2 Mineral Content

The physical and chemical properties of tomato waste have been recently studied to verify if it is possible to use this by-product as a thermochemical energy source. Tomato harvest waste is an interesting source of carbonaceous materials such as carbon, hydrogen, oxygen and volatile compounds. At that point, the high potential for energy recovery of these residues by different techniques such as pyrolysis, gasification or combustion is suggested. Besides that, the contents of mineral elements including nitrogen, potassium, chloride, calcium, iron and manganese are present in lower quantities than hydrocarbon ratios, which could limit the direct use of tomato waste as a biofuel.[56] The distribution of these minerals varies across different tissues of tomato plants. Elements such as Ca, Cu, Mn, Zn and Se are associated with antioxidant enzymes and are present mainly in tomato peel. The content of Na is higher in fruits than in other parts such as leaves or stems; in contrast, elements such as Ca, Mg, K or P are present both in fruits and tomato leaves. However, the mineral content of tomatoes, especially fruits, can be different during the ripening stages or vary depending on tomato species or cultivars.[57]

Volatile organic compounds are substances responsible for tomato aroma and involve different chemical groups including aldehydes, alcohols, ketones, esters, sulphur compounds, phenols and terpenes.[58] Among the most desirable organoleptic characteristics for consumers, tomato aroma and flavour are among the most important. However, consumers also take into account the physical characteristics of the fruit such as its size, texture and ripeness. Besides their importance in quality and marketable characteristics, plants emit that compound as chemical signs interacting and communicating for recognition, attraction and defence from abiotic and biotic stress purposes, both above the environment and the rhizosphere microenvironment. The role of root exudates belowground is mediating the plant–microbe interactions and attracting beneficial bacteria to plant roots.[59]

5.4 Tomato Waste Uses

As has been commented before, tomato wastes contain high levels of bioactive and chemical compounds that can be considered as a natural source for reusing in different products. Biochemical compounds contribute to human health added to food, pharmaceutical and cosmetic products. Fibre can be used in industrial processes and the composition of different bio-products. Moreover, seeds, skin and peels are usually reused in different products with non-food applications such as biochar, composting and biogas production.

Management of tomato food by-products is contemplated as a novel environmental solution with social and economic benefits. However, the potentiality of the food waste chain to become a valuable alternative natural source for different industrial products needs to be deeply developed in order to empower the production of added-value products that can replace usual products and traditional practices of waste management.

5.4.1 Food Industry

Traditionally, tomato waste has been used as a nutritive supplement due to its high potential as a source of biochemical compounds. In the last few years, the food industry has focused on the development of functional ingredients and food with beneficial properties on health. This component addition was always added in tomato peels and seeds powder form, but it has been concluded that the use of tomato powder has both advantages and disadvantages in the foods in which it is incorporated. The favourable part resides in increasing the nutrient contents such as lycopene, phenols, dietary fibre or proteins. On the other hand, this incorporation sometimes causes damage to textural and sensory characteristics of food.[39]

5.4.1.1 Flour-based Products

Commonly, wheat flour-based products including bread, pasta or bakery products serve as a great energy source in many countries, and these industrial products have a great potential to be mixed with tomato waste components. Many studies have concluded that the addition of tomato wastes to bread raised their nutritional values and increased their shelf life, but although these products have a good consistency, the texture is softer. The addition of tomato pomace powder as a replacement for wheat flour increases the protein, dietary fibre and lycopene concentration; however, colour and taste may suffer alterations which are accepted up to a certain extent by consumers.[60,61] These alterations could be attributed to a decrease in gluten levels in tomato waste that replaced the original ingredients.[62]

Pasta is another popular wheat flour-based food due to its easy cooking, feasibility and accessible price, but it has a modest nutritional value. So many industrial company foods are focused on the development of functional pasta using tomato powder from industrial processing.[63] The addition of tomato peel flour to these foods increases the dietary fibre, carotenoid contents and antioxidant activities, but also causes damage to the elasticity, odour and firmness of pasta.[64,65]

Soda crackers and cookies have been evaluated by consumers comparing different proportions between usual ingredients and their replacement for tomato waste powder. Consumers noted that it has a little bitterness in crackers, but the addition of tomato seed flour produces a higher moisture content and a softer texture and improves their nutritional value.[66]

To sum up, the incorporation of tomato wastes in flour-based foods is possible, but their consequences can suffer variations depending on the tomato pomace composition, proportions or food selected for replacement. Moreover, it is possible that consumers accept these different sensorial effects if the benefits of these functional foods are communicated appropriately.

5.4.1.2 Meat Products

Naturally-derived colourants are preferred by consumers because they are linked with healthy and good quality food. The incorporation of dried tomato peels into the food improves the perception of red colour and sensory properties in products such as dry fermented sausages, beef hamburgers, ham or frankfurters.[67–69] Moreover, tomato peel powder in meat products leads to a significant improvement in texture characteristics including springiness, hardness and cohesiveness of beef because of an increase in the fibre content.[70] Another advantage of adding tomato powder to these food products is that it reduces lipid oxidation and microbial activity during the ripening and storage of these products extending their shelf life. Besides that, the addition of tomato powder to these products improves the intake of lycopene deriving in a healthier food.

5.4.1.3 Tomato Pastes and Derivatives

As has already been mentioned throughout this document, tomato waste is mainly derived from the industrial processing of tomatoes. Consecutively, the recycling of tomato pomace as an ingredient of tomato paste derivatives such as sauces, extracts, ketchup, puree juice and paste is strongly recommended in order to reduce tomato losses in the course of industrial processes.

The main quality parameters in these tomato products are texture and colour and a decrease in this attribute quality is easily perceived by the final consumers. At this point, it has been concluded that the flavour and appearance of the resulting tomato pomace are not compromised or altered when it is mixed with lyophilized tomato powder coming from tomato wastes.[71] Furthermore, this enriched tomato product has a higher micronutrient content including lycopene, lutein and β-carotene.[72]

Nonetheless, taking in mind the type of tomato derivative foods, the application of tomato wastes and their compounds can produce undesirable characteristics on these products coming from the insoluble particulates of tomato pomace and peel portion especially. Thus, in order to reduce the unfavourable properties affecting the sensory quality of the products, tomato residues are often treated and modified prior to their application.[73]

5.4.1.4 Sweet, Oil and Dairy Products

The growing concern about nutrition issues and their relationship with a healthy lifestyle has led to the development of products with a lower glycaemic index. An example of these products is tomato jam, where sucrose is replaced with fructose and sativoside and contains more dietary fibre and less carbohydrate content than other commercial jams.

Tomato extract is a natural antioxidant used in the manufacture of dairy products as an additive. It has been added as an ingredient in traditional Tunisian butter in order to extend its shelf life.[74] On another note, the enrichment of olive and sunflower oils with dried tomato wastes releases carotenoids and lycopene used to prevent oxidation reactions instead of solvents or chemicals.[75] Besides that, the extraction of tomato peel added in functional ice creams improves the quality and sensorial properties such as flavour, melting, colour and texture.[76]

To summarize, it has been demonstrated that tomato wastes can be used in different foods with enriched results in the majority of these cases. However, further research is necessary to avoid undesirable properties caused due to these different mixtures and additions. Table 5.1 summarizes the uses of tomato waste in the food industry.

5.4.2 Pharmaceutical and Cosmetic Industry

In addition to their use in the food industry, carotenoids are the main bioactive compounds extracted from tomato waste used in pharmaceutical and cosmetic industries due to their high antioxidant properties. Carotenoids have a high market value, and these are specially formulated and distributed as water-soluble and stable emulsions. Traditionally, these

Table 5.1 Uses of tomato waste in the food industry.

Source	Food product	Factors	Function
Tomato powder	Flour-based product, pasta, bread, soda crackers and cookies[63,66]	Increases the protein content, dietary fibre, lycopene and carotenoids	Replacement of wheat[60,61]
Tomato peel powder	Meat products, sausages, beef, hamburgers and frankfurters[67–69]	Increases the fibre content and lycopene	Perception of red colour and reduction of microbial activity
Tomato pomace	Sauces, extracts, ketchu, puree and juice	Increases the lycopene content, lutein and carotenoids[72]	Improves the nutritional value[71]
Tomato seeds	Oil, butter, sweets and ice creams	Increases lycopene and carotenoid contents	Extending the shelf life, antioxidant effects and improving sensorial properties[74,75,78]

carotenoids have been chemically synthesized, but nowadays consumers demand extracted carotenoids from natural sources such as tomato processing waste.

Lycopene is fundamentally recognized as a potent antioxidant. It traps reactive oxygen species, increasing the overall antioxidant potential and reducing the oxidative damage in cells. The intake of lycopene from tomato products is inversely proportional to the risk of prostate, stomach, colon, lung and skin so it has a protective effect inhibiting cancer cells. On the other hand, correct levels of lycopene in the blood lead to protection against cardiovascular heart disease and age-related macular degeneration.[77]

Because of its high content of linolenic acid, tomato seed oil can be used in medical applications due to its properties against eczema, ageing and UV damage of the skin. Tomato seed oil is an important ingredient for cosmetics, and it is used in cleansers, anti-ageing creams, sunscreens and lotions along with lycopene, which decrease inflammation processes and improve skin protection against UV sun exposure damage.[78]

5.4.3 Bio-based Products

Tomato waste management and its conversion into high-value products is one of the major challenges in the food processing industry. A lot of strategies try to stimulate the use of tomato waste from the processing industry and cultivation as a natural fount of bio-based materials that can be utilized for many purposes.

In horticulture, gardeners and farmers use pots and cell trays, which are usually produced from petroleum derivate materials and cause environmental problems when they are discarded in landfills or combusted with emission of toxic gases both into the atmosphere and into the soil. In that sense, tomato seed and peel fibre wastes are good materials to be converted into biodegradable pots that are environmentally friendly and easily available.[79]

On the other hand, fibres from tomato wastes have been used as a raw material for particleboard due to pressure on wood resources. It has been proved that applying tomato stalks as a base for that product has a high potential, but it is necessary for future research into broader applications in the furniture industry.[80] Additionally, lignocellulose from tomato agri-waste is being used in the paperboard preparation with successful results. Fibres can be utilized in thermoforming paperboards with physical and mechanical properties very similar to those of commercial cellulose ones.

5.4.4 Biochar Production

The need for food waste minimization under the circular economy concept makes it urgent to use fruit and vegetable by-products as new resources. The main part of tomato waste is commonly used in the food industry, but its seed, skin and peel contain valuable products including pectins, hemicellulose and cellulose that can be isolated and used by the pharmaceutical and cosmetic industry. When that kind of industry cannot utilize tomato wastes as a new source of compound, conversion to other added-value materials can be achieved through the production of compost or biochar. The chemical composition and fibres of tomato pomace and crop residues can be brighter feedstock ingredients for the production of biochar with promising adsorption properties.

Biochar is a product rich in carbon produced from materials with a high organic content such as vegetable and fruit crop waste. The biomass–biochar converting procedure has different steps including drying, grinding, pyrolysis and separation, which can affect the properties of final biochar along the basic characteristic of the feedstock materials and their organic content.[81] The advantages of biochar production are the abundant existence of feedstock and its low production cost; however, its disadvantages are pore blocking, flammability and hygroscopicity.[82]

Biochar from tomato waste has different purposes, but its production and reuse are still scarce. Tomato residues are lignocellulosic compounds, so they have a higher concentration of carbon residues and a lower nitrogen content compared to other biochar sources such as animal waste or urban wastewater.[83] It has been described that tomato processing by-products can be used as organic enhancers to improve the fertility and agricultural and crop productivity both in hydroponics and soil and, in addition, the use of these by-products reduces the physicochemical degradation of the soil.[84,85] Biochar has a high carbon content and high adsorption capacity. These particular qualities improve the structural stability, water retention and fertility of amended soils, among physicochemical characteristics such as granulometry, density, porosity, pH, conductivity and cation exchange capacity.[82]

The improvement of soil physicochemical and biological properties directly influences seed germination, plant growth and the subsequent crop yield and production quality. Therefore, depending on the climatology conditions and crop management, the positive effects of biochar application become more evident in the long term, providing proven benefits over cropping seasons.

5.4.5 Composting and Vermicomposting

One of the main goals of the concept of circular economy is recycling organic residues as tomato waste to decrease the number of organic by-products that end in landfills. In that sense, composting and vermicomposting are considered prevention activities for land applications that contribute to reducing problematical issues related to the management and disposal of by-products in horticultural and industry processes.[86]

According to the literature, composting is an aerobic process where organic matter is converted into compost through both mesophilic and thermophilic microorganisms. The early phase of composting is thermophilic when compost temperatures reach 45–70 °C. Later, during the mesophilic phase, recalcitrant organic compounds are degraded slowly until the disappearance of phytotoxicity. Composting has many other phases in order to maintain aerobiosis such as aeration and watering.[87]

Vermicomposting is a similar process to composting, but in this case, the fragmentation of a substrate and the biochemical degradation are carried out by both microorganisms and earthworms.[88] One of the differences between composting and vermicomposting resides in temperature during the thermophilic phase, as vermicomposting does not usually heat up above ambient temperature. However, one of the advantages of that kind of organic material reuse is that it can be combined with the initial phase of pre-composting at high temperatures followed by a final stage of vermicomposting.[89] Overall, both compost and vermicompost products can be used as a substrate for horticultural practices with beneficial effects on plant growth and quality.

It has been described that tomato plant vermicomposting has a major mass-reducer potential than tomato waste compost, as well as higher humus content and lower phytotoxicity. On the other hand, vermicomposting has more nutrient and microbial activity than this kind of compost.[90] It has been proved that tomato waste vermicomposting facilitates the transformation of horticultural by-products into other useful products with successful results; nonetheless, depending on the natural composition of these residues, the worm activity can be inhibited, which decreases the efficacy of the process.[91] Tomato wastes must be combined with other organic wastes as animal manure to enhance worm activity and the consequentially vermicomposting quality. Sustainable strategies like that must be taken into account for development as well as implementation in the future in order to minimize the impact of agriculture and food processing on the environment.

5.4.6 Biogas Production

Biogas is obtained from organic materials in the absence of oxygen through a biological degradation called anaerobic digestion. The result of the process is a nutrient-rich solid part and a mixture of gases known as biogas consisting of different compounds. The main gas obtained from anaerobic digestion is methane (50–75%) followed by carbon dioxide (25–45%) and water vapour (2–7%) depending on the nature of the organic material. There are other gases present in a minor proportion compared to 1% of the total as oxygen, nitrogen, ammonia or hydrogen sulphide.[92,93]

Biogas from tomato wastes can be used in different processes such as heat and electricity production in a combined heat power. Other uses of biogas are the feed-in of upgraded biogas to biomethane to the natural gas grid and as a fuel in the chemical industry.[94] However, as occurs in vermicomposting issue, the use of tomato waste for biogas production has major potential if it is combined with other organic feedstock such as animal manure or other horticultural crop wastes.[92] On the other hand, some studies confirm biodiesel can be technically processed from the oily fraction of tomato wastes, but this research must be deeply developed in order for these residues to be chosen as a commercial fuel.[95]

5.5 Conclusion

As has been commented throughout this chapter, food waste is recognized as a global issue due to its environmental, economic and social impacts with global consequences and a remarkable amount of natural resource losses. The cultivation, harvesting and industrial processing of tomatoes generate significant quantities of waste that are being traditionally disposed of; however, management of the tomato plant residues and tomato processing wastes can supply the demand for sustainable practices and waste reuse for this by-product.

The remarkable properties of tomato wastes can be possible through the chemical and biological valorization of these by-products. The extraction of biocomponents from industrial tomato residues is one of the greatest solutions that the industry has actually carried out as has been described, and molecules such as carotenoids, especially lycopene, and oils and proteins extracted from different parts of the fruit are incorporated as functional ingredients, colourants or additives in different foods such as meat, pasta or dairy products. On the other hand, products of chemical nature such as lignocellulose from crop residues or the mineral content of plants have different applications as a natural source in industrial processes.

Although there is a growing interest in finding new proceedings for the reuse of tomato wastes, known processes must be deeply studied too in order to reduce the environmental impact of these practices and develop high-value-added products with a sustainable approach. This will contribute to a long-term improvement of the problems derived from food losses.

References

1. L. Xue, *et al.*, Missing food, missing data? A critical review of global food losses and food waste data, *Environ. Sci. Technol.*, 2017, **51**(12), 6618–6633.
2. V. Lineham *et al.*, *Food Demand to 2050: Opportunities for Australian Agriculture, Algebraic Description of Agrifood Model*, Abares, 2012.
3. H. Razifard, *et al.*, Genomic evidence for complex domestication history of the cultivated tomato in Latin America, *Mol. Biol. Evol.*, 2020, **37**(4), 1118.
4. Tomatoes Statistics, European Commision official website, [updated 2022 June 29; cited 2022 August 1]. Available from: https://agriculture.ec.europa.eu.
5. V. N. Madia, *et al.*, Recent advances in recovery of lycopene from tomato waste: A potent antioxidant with endless benefits, *Molecules*, 2021, **26**(15), 4495.
6. K. Hueso-Kortekaas, J. C. Romero and R. González-Felipe, Energy-Environmental Impact Assessment of Greenhouse Grown Tomato: A Case Study in Almeria (Spain), *World*, 2021, **2**(3), 425–441.

7. World Resources Institute, Creating a Sustainable Food Future: A Menu of Solutions, [updated 2019 July 19; cited 2022 August 1]. Available from: https://www.wri.org/research/creating-sustainable-food-future.
8. M. M. de Oliveira, *et al.*, Food loss and waste in the context of the circular economy: A systematic review, *J. Cleaner Prod.*, 2021, **294**, 126284.
9. A. C. Fărcaş, *et al.*, Tomato waste as a source of biologically active compounds, *Bull. UASVM Food Sci. Technol.*, 2019, **73**(2), 55–60.
10. T. Løvdal, *et al.*, Valorization of tomato surplus and waste fractions: A case study using Norway, Belgium, Poland, and Turkey as examples, *Foods*, 2019, **8**(7), 229.
11. V. Thole, *et al.*, Analysis of Tomato Post-Harvest Properties: Fruit Colour, Shelf Life, and Fungal Susceptibility, *Curr. Protoc. Plant Biol.*, 2020, **5**(2), e20108.
12. R. K. Saini, M. H. So and K. Young-Soo, An updated review on use of tomato pomace and crustacean processing waste to recover commercially vital carotenoids, *Food Res. Int.*, 2018, **108**, 516–529.
13. N. Martins and I. C. Ferreira, Wastes and by-products: Upcoming sources of carotenoids for biotechnological purposes and health-related applications, *Trends Food Sci. Technol.*, 2017, **62**, 33–48.
14. N. Kalogeropoulos, *et al.*, Bioactive phytochemicals in industrial tomatoes and their processing byproducts, *LWT – Food Sci. Technol.*, 2012, **49**, 213–216.
15. K. Kumar, *et al.*, Food waste: A potential bioresource for extraction of nutraceuticals and bioactive compounds, *Bioresources Bioprocessing*, 2017, **4**(1), 1–14.
16. A. Baiano, Recovery of biomolecules from food wastes – A review, *Molecules*, 2014, **19**(9), 14821–14842.
17. I. Esparza, *et al.*, Fruit and vegetable waste management: Conventional and emerging approaches, *J. Environ. Manage.*, 2020, **265**, 110510.
18. V. M. Madia, *et al.*, Recent advances in recovery of lycopene from tomato waste: A potent antioxidant with endless benefits, *Molecules*, 2021, **26**(15), 4495.
19. S. S. Hassan, G. A. Williams and A. K. Jaiswal, Lignocellulosic biorefineries in Europe: current state and prospects, *Trends Biotechnol.*, 2019, **37**(3), 231–234.
20. I. Tabrika, *et al.*, Composting of tomato plant residues: Improvement of composting process and compost quality by integration of sheep manure, *Org. Agric.*, 2020, **10**(2), 229–242.
21. M. M. Poojary and P. Passamonti, Extraction of lycopene from tomato processing waste: kinetics and modelling, *Food Chem.*, 2015, **173**, 943–950.
22. P. A. Silva, *et al.*, Characterization of tomato processing by-product for use as a potential functional food ingredient: nutritional composition, antioxidant activity and bioactive compounds, *Int. J. Food Sci. Nutr.*, 2019, **70**(2), 150–160.

23. K. Szabo, A. F. Cătoi and D. C. Vodnar, Bioactive compounds extracted from tomato processing by-products as a source of valuable nutrients, *Plant Foods Human Nutr.*, 2018, **73**(4), 268–277.
24. S. Magazù, F. Migliardo and M. T. F. Telling, Structural and dynamical properties of water in sugar mixtures, *Food Chem.*, 2008, **106**(4), 1460–1466.
25. W. Stahl and H. Sies, Antioxidant activity of carotenoids, *Mol. Aspects Med.*, 2003, **24**(6), 345–351.
26. T. Maoka, Carotenoids as natural functional pigments, *J. Nat. Med.*, 2020, **74**(1), 1–16.
27. W. C. Willett, The WHI joins MRFIT: a revealing look beneath the covers, *Am. J. Clin. Nutr.*, 2010, **91**(4), 829–830.
28. M. Imran, *et al.*, Lycopene as a natural antioxidant used to prevent human health disorders, *Antioxidants*, 2020, **9**(8), 706.
29. K. K. Das, *et al.*, Hypoglycemic activity of curcumin synthetic analogues in alloxan-induced diabetic rats, *J. Enzyme Inhib. Med. Chem.*, 2016, **31**(1), 99–105.
30. J. H. Cha, *et al.*, Anti-inflammatory effect of lycopene in SW480 human colorectal cancer cells, *Nutr. Res. Pract.*, 2017, **11**(2), 90–96.
31. R. C. Ranveer, *et al.*, Extraction, purification and characterization of lycopene from tomato (Cv Vijeta) processing industry waste, *EJNFS*, 2020, **12**, 83–90.
32. M. S. Lenucci, *et al.*, Possible use of the carbohydrates present in tomato pomace and in byproducts of the supercritical carbon dioxide lycopene extraction process as biomass for bioethanol production, *J. Agric. Food Chem.*, 2013, **61**(15), 3683–3692.
33. J. Jeffery, A. Holzenburg and S. King, Physical barriers to carotenoid bioaccessibility. Ultrastructure survey of chromoplast and cell wall morphology in nine carotenoid-containing fruits and vegetables, *J. Sci. Food Agric.*, 2012, **92**(13), 2594–2602.
34. K. N. Honest, H. W. Zhang and L. Zhang, Lycopene: Isomerization effects on bioavailability and bioactivity properties, *Food Rev. Int.*, 2011, **27**(3), 248–258.
35. D. Weber and G. Tilman, The contribution of β-carotene to vitamin A supply of humans, *Mol. Nutr. Food Res.*, 2012, **56**(2), 251–258.
36. A. G. Maria, R. Graziano and D. O. Nicolantonio, Carotenoids: potential allies of cardiovascular health?, *Food Nutr. Res.*, 2015, **59**(1), 26762.
37. P. Chaudhary, *et al.*, Bioactivities of phytochemicals present in tomato, *J. Food Sci. Technol.*, 2018, **55**(8), 2833–2849.
38. J. C. Sherwin, *et al.*, Epidemiology of vitamin A deficiency and xerophthalmia in at-risk populations, *Trans. R. Soc. Trop. Med. Hyg.*, 2012, **106**(4), 205–214.
39. Z. Lu, *et al.*, Sustainable valorisation of tomato pomace: A comprehensive review, *Trends Food Sci. Technol.*, 2019, **86**, 172–187.
40. S. Azabou, *et al.*, Phytochemical profile and antioxidant properties of tomato by-products as affected by extraction solvents and potential application in refined olive oils, *Food Biosci.*, 2020, **36**, 100664.

41. D. Shao, *et al.*, Study of optimal extraction conditions for achieving high yield and antioxidant activity of tomato seed oil, *J. Food Sci.*, 2012, **77**(8), E202–E208.
42. D. Shao, *et al.*, Thermal and storage characteristics of tomato seed oil, *LWT - Food Sci. Technol.*, 2015, **63**(1), 191–197.
43. S. J. Latlief and D. Knorr, Tomato seed protein concentrates: effects of methods of recovery upon yield and compositional characteristics, *J. Food Sci.*, 1983, **48**(6), 1583–1586.
44. M. Mechmeche, *et al.*, Optimization of extraction parameters of protein isolate from tomato seed using response surface methodology, *Food Anal. Methods*, 2017, **10**(3), 809–819.
45. K. Gebeyew, *et al.*, The effect of feeding dried tomato pomace and concentrate feed on body weight change, carcass parameter and economic feasibility on hararghe highland sheep, Eastern Ethiopia, *Vet. Sci. Technol.*, 2015, **6**(2), 1.
46. S. M. Meshkani, *et al.*, Optimization of protein extraction and evaluation of functional properties of tomato waste and seeds from tomato paste plants, *Biosci. Biotechnol. Res. Asia*, 2016, **13**(4), 2387–2401.
47. R. R. Wolfe, *et al.*, Optimizing protein intake in adults: interpretation and application of the recommended dietary allowance compared with the acceptable macronutrient distribution range, *Adv. Nutr.*, 2017, **8**(2), 266–275.
48. F. Yangilar, The application of dietary fibre in food industry: structural features, effects on health and definition, obtaining and analysis of dietary fibre: a review, *J. Food Nutr. Res.*, 2013, **1**(3), 13–23.
49. T. M. Barber, *et al.*, The health benefits of dietary fibre, *Nutrients*, 2020, **12**(10), 3209.
50. M. del Valle, M. Cámara and M. E. Torija, Chemical characterization of tomato pomace, *J. Sci. Food Agric.*, 2006, **86**(8), 1232–1236.
51. A. S. Sengar, *et al.*, Comparison of different ultrasound assisted extraction techniques for pectin from tomato processing waste, *Ultrason. Sonochem.*, 2020, **61**, 104812.
52. M. M. Alancay, *et al.*, Extraction and physicochemical characterization of pectin from tomato processing waste, *J. Food Meas. Charact.*, 2017, **11**(4), 2119–2130.
53. Z. H. Kheiralla, *et al.*, Upgrading of tomato (Solanum lycopersicum) agroindustrial wastes, *J. Microb. Biochem. Technol.*, 2018, **10**, 46–48.
54. H. Yang, *et al.*, Characteristics of hemicellulose, cellulose and lignin pyrolysis, *Fuel*, 2007, **86**(12–13), 1781–1788.
55. C. Covino, *et al.*, Lignocellulosic fibres from enzyme-treated tomato plants: characterisation and application in paperboard manufacturing, *Int. J. Biol. Macromol.*, 2020, **161**, 787–796.
56. B. Khiari, M. Moussaoui and M. Jeguirim, Tomato-processing by-product combustion: Thermal and kinetic analyses, *Materials*, 2019, **12**(4), 553.
57. S. Vats, *et al.*, Unexplored nutritive potential of tomato to combat global malnutrition, *Crit. Rev. Food Sci. Nutr.*, 2022, **62**(4), 1003–1034.

58. Z. Xiao, *et al.*, Optimization of headspace solid-phase micro-extraction and its application in analysis of volatile compounds in cherry tomato by gas chromatography, *Food Anal. Methods*, 2017, **10**(3), 596–609.
59. S. Gulati, *et al.*, A non-invasive soil-based setup to study tomato root volatiles released by healthy and infected roots, *Sci. Rep.*, 2020, **10**(1), 1–11.
60. M. Majzoobi, *et al.*, Effect of tomato pomace powder on the physicochemical properties of flat bread (Barbari bread), *J. Food Process. Preservation*, 2011, **35**(2), 247–256.
61. V. Nour, M. E. Ionica and I. Trandafir, Bread enriched in lycopene and other bioactive compounds by addition of dry tomato waste, *J. Food Sci. Technol.*, 2015, **52**(12), 8260–8267.
62. F. Isik and C. Topkaya, Effects of tomato pomace supplementation on chemical and nutritional properties of crackers, *Italian J. Food Sci.*, 2016, **28**(3), 525.
63. M. Li, *et al.*, Natural additives in wheat-based pasta and noodle products: opportunities for enhanced nutritional and functional properties, *Compr. Rev. Food Sci. Food Saf.*, 2014, **13**(4), 347–357.
64. L. Padalino, *et al.*, Functional pasta with tomato by-product as a source of antioxidant compounds and dietary fibre, *Czech J. Food Sci.*, 2017, **35**(1), 48–56.
65. A. Pasqualone, *et al.*, Functional, textural and sensory properties of dry pasta supplemented with lyophilized tomato matrix or with durum wheat bran extracts produced by supercritical carbon dioxide or ultrasound, *Food Chem.*, 2016, **213**, 545–553.
66. M. A. Bhat and A. Hafiza, Physico-chemical characteristics of cookies prepared with tomato pomace powder, *J. Food Process. Technol.*, 2016, **7**(1), 543.
67. E. Eyiler and A. Oztan, Production of frankfurters with tomato powder as a natural additive, *LWT – Food Sci. Technol.*, 2011, **44**(1), 307–311.
68. M. M. Calvo, M. L. García and M. D. Selgas, Dry fermented sausages enriched with lycopene from tomato peel, *Meat Sci.*, 2008, **80**(2), 167–172.
69. S. Yadav, *et al.*, Development of dietary fibre enriched chicken sausages by incorporating corn bran, dried apple pomace and dried tomato pomace, *Nutr. Food Sci.*, 2016, **46**(1), 16–29.
70. M. L. García, M. M. Calvo and M. D. Selgas, Beef hamburgers enriched in lycopene using dry tomato peel as an ingredient, *Meat Sci.*, 2009, **83**(1), 45–49.
71. A. Torbica, *et al.*, Nutritional, rheological, and sensory evaluation of tomato ketchup with increased content of natural fibres made from fresh tomato pomace, *Food Bioproducts Process.*, 2016, **98**, 299–309.
72. L. Previtera, *et al.*, Chemical and organoleptic characteristics of tomato purée enriched with lyophilized tomato pomace, *J. Sci. Food Agric.*, 2016, **96**(6), 1953–1958.
73. X. Hua, *et al.*, Effects of high-speed homogenization and high-pressure homogenization on structure of tomato residue fibres, *Food Chem.*, 2017, **232**, 443–449.

74. Y. Abid, *et al.*, Storage stability of traditional Tunisian butter enriched with antioxidant extract from tomato processing by-products, *Food Chem.*, 2017, **233**, 476–482.
75. A. Benakmoum, *et al.*, Valorisation of low quality edible oil with tomato peel waste, *Food Chem.*, 2008, **110**(3), 684–690.
76. E. M. Rizk, A. T. El-Kady and A. R. El-Bialy, Charactrization of carotenoids (lyco-red) extracted from tomato peels and its uses as natural colorants and antioxidants of ice cream, *Ann. Agric. Sci.*, 2014, **59**(1), 53–61.
77. R. M. Stoica, *et al.*, Tomato by-products as a source of natural antioxidants for pharmaceutical and food industries – A mini-review, *Sci. Bull. Ser. F. Biotechnol.*, 2018, **22**, 200–204.
78. W. Stahl, *et al.*, Lycopene-rich products and dietary photoprotection, *Photochem. Photobiol. Sci.*, 2006, **5**(2), 238–242.
79. E. Schettini, *et al.*, Recycled wastes of tomato and hemp fibres for biodegradable pots: Physico-chemical characterization and field performance, *Resour., Conserv. Recycl.*, 2013, **70**, 9–19.
80. I. Taha, M. S. Elkafafy and H. E. Mously, Potential of utilizing tomato stalk as raw material for particleboards, *Ain Shams Eng. J.*, 2018, **9**(4), 1457–1464.
81. M. Belhachemi, *et al.*, Characterization of biomass-derived chars, *Char and Carbon Materials Derived from Biomass*, Elsevier, 2019, pp. 69–108.
82. L. Zhu, D. Shen and K. H. Luo, A critical review on VOCs adsorption by different porous materials: Species, mechanisms and modification methods, *J. Hazard. Mater.*, 2020, **389**, 122102.
83. S. J. Dunlop, *et al.*, Closing the loop: Use of biochar produced from tomato crop green waste as a substrate for soilless, hydroponic tomato production, *HortScience*, 2015, **50**(10), 1572–1581.
84. S. Marinos *et al.*, Biochar production from the pyrolysis of tomato processing residues, *Tomato Processing by-Products*, Academic Press, 2022, pp. 171–200.
85. A. Parmar, P. K. Nema and T. Agarwal, Biochar production from agro-food industry residues: a sustainable approach for soil and environmental management, *Curr. Sci.*, 2014, **107**(10), 1673–1682.
86. A. A. Zorpas, Strategy development in the framework of waste management, *Sci. Total Environ.*, 2020, **716**, 137088.
87. A. A. Zorpas, Recycle and reuse of natural zeolites from composting process: a 7-year project, *Desalin. Water Treat.*, 2014, **52**(37–39), 6847–6857.
88. F. Fornes, *et al.*, Composting versus vermicomposting: a comparative study of organic matter evolution through straight and combined processes, *Bioresour. Technol.*, 2012, **118**, 296–305.
89. M. J. Fernández-Gómez, E. Romero and R. Nogales, Feasibility of vermicomposting for vegetable greenhouse waste recycling, *Bioresource Technol.*, 2010, **101**(24), 9654–9660.
90. R. Nogales, J. Domínguez and S. Mato, *Vermicompostaje*, ed. J. Moreno and R. Moral, 2008, pp. 187–208.

91. C. A. Edwards and N. Q. Arancon, The Use of Earthworms in Organic Waste Management and Vermiculture, *Biology and Ecology of Earthworms*, Springer, New York, NY, 2022, pp. 467–527.
92. Y. Li, *et al.*, Reactor performance and economic evaluation of anaerobic co-digestion of dairy manure with corn stover and tomato residues under liquid, hemi-solid, and solid state conditions, *Bioresour. Technol.*, 2018, **270**, 103–112.
93. Y. Li, *et al.*, Reactor performance and energy analysis of solid state anaerobic co-digestion of dairy manure with corn stover and tomato residues, *Waste Manage.*, 2018, **73**, 130–139.
94. R. Sarada and K. Nand, Start-up anaerobic digestion of tomato-processing wastes for methane generation, *Biol. Wastes*, 1989, **30**(3), 231–237.
95. H. Sivasubramanian, *et al.*, Investigation of biodiesel obtained from tomato seed as a potential fuel alternative in a CI engine, *Biofuels*, 2020, **11**(1), 57–65.

CHAPTER 6

Agro-wastes for Cost-effective Production of Industrially Important Microbial Enzymes

CHARU GUPTA

Amity Institute of Herbal Research & Studies, Amity University-Uttar Pradesh, Sector-125, Noida-201313, India
Email: cgupta@amity.edu

6.1 Introduction

In the age of international mechanization, enzymes are being used broadly in numerous segments comprising foodstuff handling. Due to the high expense of enzymes, many ingenuities have been undertaken by the study and growth segment for the advancement of new methods or enhancement of the prevailing methods for making budget-friendly enzymes. With the development in the arena of biotechnology, several bioprocesses are being used for the exploitation of diverse agro-industrial remains for the manufacture of innumerable enzymes. Humans have utilized enzymes since the customary periods for treating many food products such as beer, wine, cheese, yogurt, kefir, and other fermented brewery products, but due to their numerous uses, the call in many food segments has augmented greatly. The usual cradles of the enzymes include florae, faunae, and microbes. In distinction to the vegetable and visceral sources, microbial sources are favoured owing to commercial as well as other technical aids such as greater harvests acquired within the shortest fermentation time. Furthermore, microorganisms can be more easily subjected to genetic handling in order to increase the efficiency or catalytic attribute of the enzyme.[1] Diverse microbial bases

Green Chemistry Series No. 78
Agri-food Waste Valorisation
Edited by Pankaj Chowdhary and Abhay Raj

Published by the Royal Society of Chemistry, www.rsc.org

such as bacteria, yeasts, and fungi have been applied for the enzyme manufacture by employing diverse substrates such as starch, glycerol, glucose, and other inorganic salts. A main part of the manufacturing price of the enzymes is primarily owed to the charge of the fermentation media and procedures. Hence, to curtail the price of manufacture and to achieve the industrial burdens and tasks, a range of microbes and inexpensive agro-industrial substrates have been tried to ease the commercial making of enzymes.[2,3]

Likewise, managing unused agro-industrial products is very imperative as it stances a severe risk to the surroundings. A huge quantity of agro-industrial leftovers, such as rice straw, wheat straw, rice husk, bagasse, industrial effluents, sludge, *etc.*, are formed yearly, which generate problems related to the discarding of these solid and liquid excesses. The management of by-products is an augmented approach to achieving low carbon dioxide emissions.[4] The agronomic remains left after harvesting are either burned on-site (which produces air pollution and disturbs soil class) or deposited in the form of piles in the field. Hence, managing the agricultural unused matter is becoming one of the grave problems today. Yet, solid waste can be consumed as a probable source for the manufacture of a range of value-added technologically useful products including enzymes, chemicals, biofuels, and animal feed owing to their nourishing potential.[5] Enzymes alike cellulase and xylanase have industrialized significance and are broadly consumed in many productions including fabric, mush and paper, washing and cleanser, wine and brewery, foodstuff processing, and agrarian industries and bio-ethanol production.[6] The price of these enzymes is a great matter tackled by these industries, and pains are going on to decrease the price by strain enhancement, better fermentation, and retrieval scheme and exploitation of effortlessly obtainable low-cost substrates.[7] These agri-industrial litters can be recycled as an inexpensive substrate for the manufacture of numerous industrially significant enzymes by many businesses.

6.2 Enzymes and Substrates

Enzymes have been a vital share of humanoid life since the olden periods for the manufacture of several foodstuffs, and hence, they are largely being explored in many food trade segments. Amongst the numerous normal sources, microbes are the ideal sources for enzyme manufacture owing to a number of benefits. Due to the high cost of making rising from the fermentation medium and process settings, the trend has moved towards the usage of agro-industrial litters for making cost-effective enzymes. The employment of agro-industrial trashes for microbial fermentation not only increases the method's finances, but also can lessen the ecological matters associated with their clearance. Though diverse residues have been exploited and tried for their possibility as substrates for microbial fermentation of enzymes, still the dairy and cereal industry wastes/by-products have notable prospective for enzyme making. Furthermore, among the diverse types of fermentation (solid state and submerged) tested for enzyme production, maximum work has been carried out under solid-state conditions by batch fermentation. Although several microbial sources have been used for enzyme making, novel and effective microbial sources are desired for the ideal use of agro-industrial wastes to attain higher harvests of the enzymes.[8]

6.3 Agro-industrial Substrates: Types

An agro-industrial substrate usually comprises the wastes created during the industrialized treatment of agronomic or visceral goods or those obtained due to agrarian actions in the form of grass, twig, branch, greeneries, husk, shell, rind, lint, kernels, flesh, pulses or cereals (rice, wheat, corn, sorghum, and barley), bagasse from sugarcane or sweet sorghum milling, spent coffee grounds, brewer's spent grains, and many others. These trashes are mostly composed of sweeties, filaments, proteins, and minerals. The principal ingredients of such agro-industrial litters comprise cellulose, hemicelluloses, and lignin (lignocellulosic materials).[9] Distinct litters exhibit an array of comparative amounts of these chief components. This variability involves an assured level of specificity/preference for their use.

Low-cost agro-industrial bases such as wheat bran, soybean meal, corn steep liquor, sugarcane bagasse, whey, *etc.* have been used as carbohydrate as well as nitrogen sources in lieu of artificial ones.[10] Diverse kinds of treatments (physical, chemical, and enzymatic) can be carried out on these by-products in order to make them effortlessly used up by microbes.[11] Herein the different types of agri-horticultural trashes produced from the diverse foodstuff trades are discussed.

6.3.1 Fruitlet and Vegetal Productions

The fruitlet and vegetal production division has extended enormously due to the fast-increasing response of canned and pre-processed foodstuffs. The rising processing of fruitlets and vegetals has led to the creation of huge quantities of litters each in the form of greeneries or chaff during the treatment, seeds after processing, or mush and pomace.[12] The litters produced from these productions are varied in nature, dependent upon the procedures used and the goods developed.[13] The peel and kernels of fruitlets such as mango, papaya, pineapple, and taro are acquired as the by-products during their treatment.[14] Other main by-products are acquired from the citrus rind commerce. Around 50–60% of the citrus rind leftover is created from citrus fruits, which acts as a likely substrate for any fermentative process.[15] Equally, apple pomace is also a favourable substrate, which is the by-product left over after cider and juice making.[16] Amongst the vegetables, the trashes lingering after the treatment of tomato into juice and puree are mainly in the form of rinds, kernels, and tomato pomace. Tomato treatment by-products are one of the inexpensive and nutritious bases for bacterial development. These types of litters are probable substrates for microbial cultivation for directed foods as these are rich in antioxidants, antimicrobial compounds, phytochemicals, and vitamins to sustenance these operations.[17]

6.3.2 Cereals Manufacturing

Owing to the growing fitness responsiveness amongst individuals, the cereal treating division has enlarged enormously, and thus, the litters/by-products created from these productions have also enlarged equivalently. The trashes produced from this manufacturing are, normally, the outcome of desiccated or damp refining of the cereals. Furthermore, the kind of leftover depends upon the cereal used and on the treatment procedures. The ancient and informal process of arraying the by-products was their usage as visceral feedstuff, but the current drifts are focused on using them for the generation of advanced value or noteworthy goods. Both cellulose and hemicellulose create nearly 50% of the vegetal matter dry biomass. These are tarnished into their individual monomers (simple soluble sugars) by enzymatic hydrolysis. These hydrolytic yields act as a low-cost and renewable energy source for bacterial fermentation.[11] The main by-products of cereal processing are bran and germ obtained during both the wet and dry milling processes. Through the treatment of wheat, rough and fine middling are also obtained as the by-products. The foremost by-products obtained from the maize processing industry after germ and bran are the gluten meal obtained during the wet milling. Separately from these, an additional by-product obtained during the wet milling of corn is corn steep liquor, which is used as a nutrient basis in feed and fermentation methods. It is a composite broth containing carbohydrates, proteins, organic acids, minerals, *etc.* During the processing of rice, rice hulls are obtained and they have a high quantity of silica. De-hulled rice is further processed, which gives rice bran, rice polishing, and broken rice as by-products. Rice polishing is a notable source of thiamin and crude fat and a poor source of crude fiber. Rice bran, another by-product, has a high oil content and is a good source of vitamin B complex, amino acids, and proteins.[18] De-oiled cake left behind oil removal from the rice bran also constitutes a vast accumulated bulk of biomass, and it undergoes justifiable biotransformation into value-added products.

6.3.3 Sugar Production

The chief fresh ingredients used for sugar manufacturing are sugar beet and sugar cane. Grinding of the cane stems for sweetie manufacture results in the production of many by-products such as wastewater, syrup, bagasse, *etc.*[19] Sugar mills produce about 1000 L of waste water per ton of cane crushed. Additional excess generated from the sugar industry is the sugarcane bagasse, which is a stringy deposit left after the crushing and extraction of fluids from the cane stalk. It contains a high quantity of fiber and moisture with fewer amounts of soluble solids, preferably sugars. Molasses is the dense sticky black fluid obtained during the manufacture of raw and refined sugar from sugarcane or sugar beet.[20]

6.3.4 Dairy Production

The dairy industry has been one of the foremost businesses that are liable for the production of great titres of discharge. About 0.2–10 L of litters are produced per litre of treated milk. Cheese whey, a by-product of the cheese processing industry left behind after the milk clotting and casein exclusion, creates a main ecological difficulty due to its high biological oxygen. The foremost constituents of whey include lactose 4.5–5% (w/v), proteins 0.6–0.8% (w/v), and lipids 0.4–0.5% (w/v). The high percent of lactose present in the whey increases the BOD value, thus generating difficulties in its removal and thus producing ecological problems.[21] However, lactose and protein present in liquid whey may also serve as a nutritious basis for the development of microorganisms. Moreover, liquid whey may be processed to yield lactose and/or proteins and nutrients for their additional varied practice. Besides, a huge volume of surplus liquid (approx. 2.5 L) is produced from dairy manufacturing during the washing and treatment methods. The discarded liquid comprises casein and inorganic salts, together from cleansers and antiseptics, which increase the BOD and COD contents, and thus lead to ecological difficulties.[22]

6.3.5 Distilling Manufacturing

Litters such as spent grain and spent hops are the greatest collective kinds of trashes created from distilling manufacturing in great quantities. The spent grain is the vital by-product of distilling manufacturing and holds a great quantity of pentosan, lignin, protein, and other nutritive components.[23] The transformation of such agro-industrial by-products into value-added goods has lately become a highly-active field of study; thus, many biotechnological methods and tactics are being applied to convert these litters into value-added goods as well as for making enzymes[24] (Figure 6.1).

6.4 Agro-industrial Remains Used for the Manufacture of Microbial Enzymes

A culture medium is the basic and foremost requirement for the growth and multiplication of any type of microorganisms. A majority of culture media used in research studies are very costly and therefore there is a need to hunt for cheap and freely accessible raw materials that can sustain the development and multiplication of microorganisms. In this direction, the use of agro-industrial waste would be an attractive and cost-effective way of reducing the manufacturing or production expenses.[25] The Food and Agriculture Organization (FAO), in 2013, testified that 250 million tons of

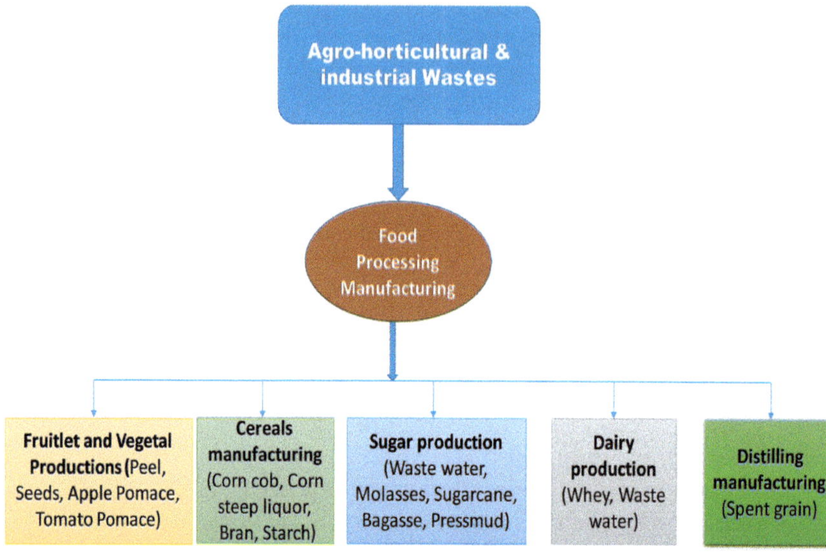

Figure 6.1 Production of agro-industrial by-products from diverse foodstuff treating productions.

agro-industrial trashes are created worldwide per year, during treatment of diverse plant crops, primarily cereals, starchy roots, fruits, and other vegetables.[26] Ordinary untreated ingredients and by-products created by agro-manufacturing industries can be extensively recycled as a culture medium owing to their little price, after the constituents of the medium can signify from 38 to 73% of the complete manufacturing cost[25] (Panesar *et al.*, 2015). Agro-horticultural and related industrial remains are unprocessed and very less used, although being rich in nutrient components, such as sugars, proteins, fibers, minerals, and vitamins. The exploitation of this leftover not only removes the discarding harms, but also the surrounding contamination including noxiousness to marine life, contamination of shallow and ground waters, transformed soil quality, phyto-toxicity, odorous and colored natural waters, and harmful impression on humanoid and visceral health.[27,28] The ecological apprehension is linked to the content of these wastes, as a maximum of them comprise phenolic complexes that have noxious prospective; besides this, these wilds show a high value of biological oxygen demand (BOD), chemical oxygen demand (COD), and other suspended solids that can be reflected as toxins to the environment.[29] Besides, plant cell walls present in the agro-industrial remains containing chiefly lignocellulose, a resistant constituent, comprising cellulose, hemicellulose, and lignin, are also an atmosphere contaminant. Retrieval of high value-added constituents from the discarded and their reuse as nutrition seasonings or therapeutics are additional fascinating facets to valorise these agro-industrial remains.

Concerning the trashes, pre-treatment is vital for encouraging the degradation of these remains, mostly produced by cellulose, hemicellulose, and lignin. This degradation will upsurge the accessibility of the nutrients from the substrate to the microbes. Several pre-treatment procedures such as physical, chemical, biological (enzymatic), and joined procedures are obtainable. In some of the studies, the agro-industrial wastes were pre-treated. Some instances of physical pre-treatment deal with paper mill sludge and sugarcane bagasse that are pulverized and strained to attain even size of particles,[30] in addition to onion peels and mung bean husk,[31] petiole oil palm fronds,[32] and fruits and vegetable wastes.[33] Chemical treatments of the leftover were carried out in some readings. Bengal gram crust was pre-treated with hydrochloric acid, encouraging acid hydrolysis to increase the accessibility of the substrate.[34] Sugarcane bagasse was pre-treated under alkaline conditions with sodium hydroxide and afterwards hydrolysed with a viable cellulase complex.[35] In this instance, there was a mixture of biological management. Similarly, unused bakery products were also hydrolysed with enzymes formed by *Aspergillus awamori* and *Aspergillus oryzae* prior to their use.[36] In the bulk of the readings, the agro-industrial wastes are recycled as carbon, nitrogen, and micronutrient sources by microbes.

Brazil is one of the major agronomic product manufacturers, as it is reflected as the world's major manufacturer of sugar, coffee, orange juice, and soybeans.[37]

The different types of wastes generated are maltose sauce, cassava wastewater, sugarcane (juice, molasses, and hydrolysed bagasse), solid coffee waste, crude glycerol, rice parboiling water, slaughterhouse wastewater, corn steep liquor (or corn maceration water), grape waste, cheese whey, soybean meal, feather meal, soy protein, and rice crust. The average composition of the agro-industrial waste in terms of carbon and/or nitrogen source is mentioned in Table 6.1.

Agro-industrial remains decline the prices of the fermentation method. Intensification of the enzyme making while reducing the manufacture expenses has been the objective of the present methods applied to yield microbial enzymes at big scales. Culture medium optimization is mainly important to acquire the maximum available of the making. Optimizing the medium comprises monitoring the settings of the fermentation, such as the temperature, pH, airing, stirring, and media constituents. Process optimization methods have used statistical experimental designs and response surface analysis with limited use of artificial intelligence (AI) systems such as genetic algorithms.[38]

6.5 Bacterial Enzymes for Foodstuff Industries

The bacterial making of enzymes can be accomplished together by submerged (SMF) and solid-state fermentation (SSF). Solid-state fermentation is a more favoured process of manufacture in the trade region. However, the preference has started to move towards SSF owing to the engagement of diverse agro-industrial litters, as it costs less in terms of carbon and nitrogen sources; thus lowering the cost for enzyme making.[39] Diverse entities are

Table 6.1 Average composition of agro-industrial waste.

Agro-industrial residue	Carbon composition	Nitrogen composition
Bakery waste	Starch (37.42 g)	Protein (14.72 g) and other nitrogen sources (2.58 g)
Carob pulp syrup	Sugar (75 $g L^{-1}$ reducing sugar)	Nil
Carrot peel	Carbohydrates (2.98 g)	Proteins (9.7 g)
Cassava wastewater	Carbohydrates (58.11 g)	Total nitrogen (1.94 g)
Cheese whey	Lactose (77%)	Proteins (13%)
Coffee pulp and husk	Carbohydrates (17%)	Proteins (24%)
Corn cob	Cellulose (32%) and hemicellulose (39%)	Nil
Corn steep liquor	Sugar (0.1–11% glucose)	Total nitrogen (2.7–4.5%)
Feather meal	Nil	Proteins (84%)
Grape waste	Carbohydrates (21%)	Proteins (10%)
High test molasses	Sugar (74–79%)	Low amounts (0.1%)
Liquid pineapple waste	Sugars (reducing sugar, glucose, sucrose, and fructose)	Nil
Maltose syrup	Maltose (70 $g L^{-1}$) and glucose (2.23 $g L^{-1}$)	Nil
Mesquite pods	Sucrose, glucose, and fructose	Nil
Mung bean husk	Sugar (328 g)	Proteins (262 g)
Olive pomace	Fatty acids (59–64%)	Proteins (2.43 to 3.8%) and other nitrogen sources (0.39–0.62%)
Onion peels	Sugar (851 $mg g^{-1}$)	Proteins (165 $mg g^{-1}$)
Orange peel	Cellulose (71 $g kg^{-1}$) and hemicellulose (128 $g kg^{-1}$)	Crude protein (57.2 $g kg^{-1}$)
Paper mill sludge	Cellulose (62%), hemicellulose (15%), and lignin (20%)	Nil
Peanut seed oil	Fatty acids (oleic or linoleic acid 80%)	Nil
Petiole oil palm fronds	Cellulose (44%), hemicellulose (27%), and lignin (10%)	Nil
Rice bran	Nil	Protein (114 $g kg^{-1}$)
Rice husk	Cellulose (31%), hemicellulose, and lignin (22% each)	Nil
Soybean meal	Carbohydrates (35%)	Proteins (50%)
Soybean protein	Carbohydrates (12%)	Proteins (79%)
Sugarcane bagasse	Cellulose (32–34%), hemicellulose (19–24%), and lignin	Nil
Sugarcane juice	Sugar (19 g reducing sugar)	Nil
Sweet whey	Lactose (4.5%)	Proteins (0.8%)
Wastes of potato chips	Starch (61.49%)	Low amounts of protein

accountable for the making of enzymes that may differ in properties including hydrolysing, oxidizing, or reducing enzymes.

The over-all outline for the industrial enzyme making is presented in Figure 6.2.

Numerous bacterial species can be used for the manufacture of commercially important enzymes. Some of the industrially significant enzymes are discussed below.

6.5.1 Pectinases

Pectinases (E.C. 3.2.1.15) are a set of enzymes that break the pectin existing in plant cells by several appliances and are dispersed into those that breakdown glycosidic bonds alongside the polymer backbone-polygalacturonase, pectin lyase and pectate lyase, and those that disrupt the methoxy groups' pectin esterase.[40] Pectic enzymes are, normally, utilized by the fruitlet treating trade in the method of abstraction of the fruitlet saps to support better abstraction and to decrease the viscosity of the sap preparation. Moreover, this enzyme is broadly utilized in the wine, coffee, and tea-treating industries.[8] Bacteriological fermentation of pectinase had been described to be done by consuming several agro-industrial remains such as wheat fiber, soybean, apple pomace, cocoa, sugarcane bagasse, lemon, and orange rind. Pectinase has been manufactured using sugarcane bagasse, orange bagasse, and wheat bran using the thermophilic fungus *Thermoascus aurantiacus*.[41] The research showed greater yields of extracellular polygalacturonase (PG) and pectic lyase (PL) enzyme manufacture.

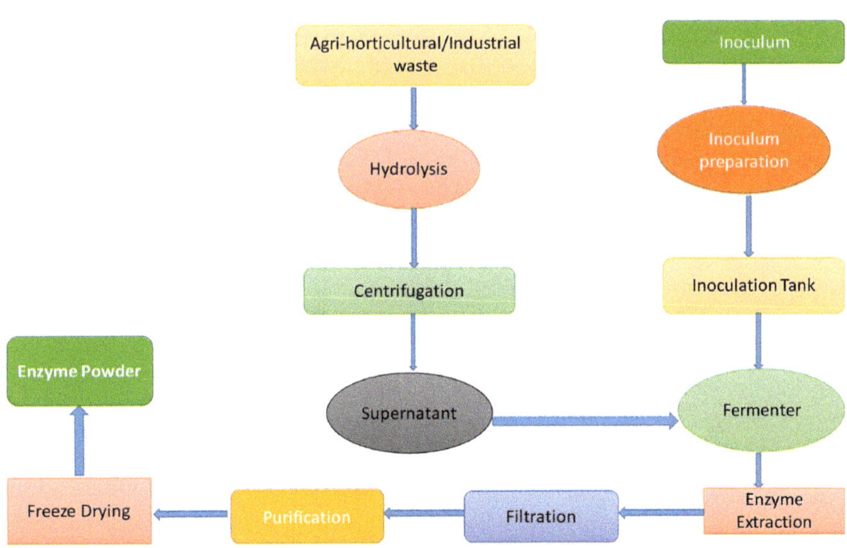

Figure 6.2 Diagrammatic representation of the manufacture of an industrial enzyme.

The maximum PG activity (43 U g^{-1}) was attained on wheat bran or orange bagasse as a substrate between the 2nd and 4th days of fermentation, while addition of sugarcane bagasse both to orange bagasse and wheat bran substrate did not have a noteworthy influence on the harvest. As matched to the polygalacturonase making, maximum polygalacturonase output (40 180 U g^{-1}) has been reported to be found between the 8th and 10th days of fermentation from a mixture of 10% sugarcane and 90% orange bagasse. Orange bagasse, when used as a substrate for manufacturing pectinase, acts as an inducer for pectinase production as it incorporates vast amounts of solvable carbohydrates (fructose, glucose, sucrose, and pectin) and unsolvable cellulose.[42] Citrus commerce by-products are common bases in the manufacture of pectinase. Another substrate, lemon pulp, a leftover after the removal of lemon juice, has been used as a fresh material for pectinase manufacture by means of two fungal strains *Aspergillus niger* and *Trichoderma viride* during submerged fermentation. Amongst the two strains, the highest pectinase production of 9.01 U ml^{-1} was gained from *T. viride* as related to 1.27 U ml^{-1} from *A. niger*.[43]

A ratio of 1 : 1 wheat bran and orange bagasse mixture is used for the manufacture of endo and exo-poly-galactouranase, pectin lyase, and pectin esterase from *Penicillium viridicatum* RFC3; the fermentation is done in flasks and polypropylene bags at 70% and 80% moisture content, respectively.[44] The highest action of endo-poly-galactouranase (5.6 U g^{-1} of the substrate) has been found from both 70% and 80% moisture, whereas exo-polygalactouranase activity (71.2 U g^{-1}) was found from 80% moisture level. Similarly, 80% moisture produced maximum pectin lyase (480 U g^{-1}) in comparison with the others. Additionally, the highest production of all the enzymes was attained when the fermentation was carried out in propylene bags at an 80% moisture level, signifying the likelihood for the upgrade lessons with respect to the volume of the substrate to be fermented. Furthermore, pectin esterase and other hydrolytic enzymes such as amylase, xylanase, and endoglucanase were also produced, as described by Silva *et al.* 2005. The relative research was carried out on pectinase production under combined submerged and solid-state fermentations using deseeded sunflower head from *A. niger* DMF27 and DMF45; enhanced with green gram husk indicated higher yields of endo- (19.8 U g^{-1}) and exo-pectinase (45.9 U g^{-1}) from *A. niger* DMF45 in solid state fermentation, as compared to 18.9 U ml^{-1} and 30.3 U ml^{-1} of endo-pectinase and exo-pectinase, respectively from DMF27 by submerged fermentation.[45]

Another cost-effective, orange bagasse was also found to be a probable raw material for the manufacture of the highest titres of pectinase (32 U ml^{-1}) under solid-state fermentation using *Botryosphaeria rhodina* MAMB-05.[46] Likewise, orange bagasse in amalgamation with wheat bran resulted in the making of 120 IU ml^{-1} exo-poly-galactouranase under solid-state fermentation, whereas under submerged fermentation, a harvest of 13 IU ml^{-1} was attained by means of thermophilic *Thermomucor indicae-seudaticae*. However, in contrast to solid substrate fermentation (SSF), the enzyme produced from submerged fermentation is more thermostable with great solidity at acidic pH.[47] The manufacture of a diversity of enzymes by a solitary strain would be useful as the amylase produced would be able to eliminate the starch from the banana, apple, and pear juice during the clarification process. The solid state fermentation for the making of exo-poly-galacturonase from the blend of orange bagasse and molokhia stalks (1 : 3) caused the highest production of 3270 $U g^{-1}$ dry substrate from *Penicillium pinophilum* Hedg 3503 NRRL.[48] Though many substrates have been established for pectinase production, molokhia stalks were used for the first time for enzyme production. The high values attained from the combination of orange bagasse and molokhia stalks show the prospective of these substrates for enzyme manufacture at little price. Additionally, due to the high quantity of pectic substances obviously existing in the orange rind, they were used as possible substrates for enzyme manufacture using *A. niger*, which showed the highest enzyme productivity of 117.1 $\mu m\, ml^{-1} min^{-1}$ at 144 h during submerged fermentation.[49]

Statistical procedures, response surface methodology (RSM), and Plackett–Burman design were utilized for the optimization of polygalactouranse (PGase) production using *Bacillus licheniformis* SHG10 from diverse agro-industrial wastes such as orange rind waste, lemon rind waste, banana rind waste, artichoke rind waste, and pomegranate rind waste and also from synthetic carbon sources.[50] Amongst all the tried sources, the highest PGase production (2.69 μg galacturonic acid per min per mg) was found by using orange rind as a solitary carbon source.

In the research by Biz *et al.*, the microbial fermentation of pectinase was carried out in a packed bed reactor by utilizing 52% citrus flesh and 48% sugarcane bagasse, where the sugarcane bagasse delivered the strength to the bed with great permeability, thus avoiding the difficulties connected with bed contraction and creation of agglomerates within the substrate.[51] Though, moulds are most commonly used for the manufacture of enzymes, yeast strain *Saccharomyces cerevisiae* also leads to the highest enzyme manufacture by means of orange rind, groundnut oil bar, and $MnSO_4$ during 48 h of fermentation period.[52]

6.5.2 Amylases

Amylases (E.C. 3.2.1.1) are hydrolytic enzymes accountable for the whole hydrolysis of starch and have a marked business status. These enzymes have likely usages in the food industry, mainly in the roasting, infusing, and fruitlet industries.[24] Due to the manufacturing demand for these enzymes, determinations have been made to search numerous know-hows to acquire large production of the enzymes at little price. Therefore, to diminish the charge of making, numerous inexpensive, easily accessible agro-industrial litters have been discovered for the microbial manufacture of amylase. For example, coconut oil cake is used as a medium for the manufacture of α-amylase by *A. oryzae* under SSF conditions.[53] Coconut oil cake is found to sustain the development of the organism, thus leading to the making of 1372 U ml^{-1} α-amylase in a 24 h fermentation time. Supplementation of the basal medium with 0.5% starch and 1% peptone improved the enzyme production by making 3388 U ml^{-1} of the enzyme, suggesting the effectiveness of coconut oil bar as a probable substrate for α-amylase manufacture. Diverse agro-industrial remains (millet cereal, crushed wheat, corn cobs, molasses, wheat flakes, barley bran, and rice bran) were tried for their probable use for the enzyme making in SSF.[54] Amongst all the tried remainders, wheat fiber has been eminent to cause in highest enzyme making, displaying an enzyme activity of 261 U g^{-1} trailed by millet cereal.

Heat steady α-amylase, taking the activity of about 464 000 U g^{-1} bacterial bran, was formed by means of wheat fiber as a substrate enhanced with glycerol, L-proline, soybean meal, and vitamin B-complex from the thermophilic bacterial strain *Bacillus* sp. PS-7.[55] Furthermore, Kammoun *et al.* used gruel, a by-product from wheat crushing, as a medium for amylase manufacture using *A. oryzae*, which gave the highest yield of 148 U ml^{-1} during the submerged fermentation.[56] Diverse agro-industrial trashes (wheat bran, rice bran, mustard oil cake, *Imperata cylindrica* grass, banana leaves, and potato peel) were also tried for the microbial manufacture of α-amylase from *Bacillus subtilis* DM03, which showed the highest enzyme manufacture of 532 ± 5 U per gram dry substrate under optimized conditions.[57] While numerous substrates have been reported for amylase production, only little substrates such as wheat fiber and potato rind have the prospective to encourage amylase manufacture from microorganisms. Also, a combination of rice and wheat fiber in the ratio of 1 : 2 provided the highest enzyme manufacture of 549.11 IU ml^{-1} amongst the new agricultural remainders using *Streptomyces* in the submerged fermentation.[58]

Six diverse agro-industrial litters (rice hay, rice fiber, corn flecks, wheat fiber, wheat flecks, and ground wheat germ) were used for the extracellular enzyme making from a new isolate *A. oryzae* IIB-6.[10] Amongst all the remainders, wheat fiber gave the highest harvest of 7800 U ml^{-1} at 80% moisture content in 72 h of fermentation. Likewise, numerous agro-industrial remainders, *viz.*, tuna fish powder discarded, wheat gluten leftover, soybean meal as nitrogen sources, and wheat gruel were used as carbon basis for the making of the enzyme under solid state conditions from *A. oryzae* S2. Though all the trashes used as nitrogen basis had prospective for the enzyme manufacture, the highest output of 22118.34 U g^{-1} dry substrate was found with the soybean meal.[59] Separately from these, feather meal, rape seed bar, and potato skin were employed as fresh substrates for amylase manufacture from *Bacillus subtilis* PF1, which displayed the highest making (16.39 ± 4.95 μg ml^{-1}) in the next 96 h of incubation period.[60] Concurrent making of protease and amylase was carried out using unused brewery products by *Bacillus subtilis* UO-01, which caused the highest level of protease (9.87 EU ml^{-1}) and amylase (9.35 EU ml^{-1}) production respectively, after 15 h of incubation under optimized conditions.[61]

6.5.3 Laccase

Laccases, the common name for p-diphenol dioxygen oxido-reductase (E.C. 1.10.3.2), belongs to the family of polyphenol oxidases and is one of the first calculated enzymes possessing a molecular weight of about 60–90 kDa. These enzymes are broadly disseminated in the environment, in plant life such as in cabbages, pears, apples, and potatoes, in insects, and in microscopic sources such as bacteria and fungi, particularly in white rot.[62] Laccases are accountable for catalysing the oxidation of both phenolic and nonphenolic composites. They are also proficient in mineralizing many artificial colorants.

Laccase making from white rot fungus *Trametes versicolor* was carried out by submerged fermentation by means of lignocellulosic trashes such as grape seed, grape stems, and barley bran.[63] In contrast to the others, barley bran was revealed to deliver the highest laccase activity of 500–600 U l^{-1} from the 27th to 37th days. Moreover, banana left over wastes, pseudo-stems, and greeneries have also been recycled for making extracellular lignolytic enzymes from *Pleurotus ostreatus* and *P. sajor-caju*, which showed the highest manufacture of 1.7106 IU mg^{-1} protein and 1.6669 IU mg^{-1} protein from greeneries, due to their greater surface area that promotes the microscopic growth.[64] The usage of banana trashes for the making of laccase implies their significance as a media to yield several enzymes as matched to the supplementary agro-industrial remainders.

Remains from the groundnut treating trade such as groundnut shells and kernels were also tried for their possibility in laccase making from *Trametes hirsuta* in a solid state.[65] Using groundnut kernel as a medium, the highest laccase activity was observed during the 14th day of fermentation that was roughly a 10 time greater rate gained than that with groundnut shells. Though groundnut shells are much more appropriate for enzyme making due to their great cellulose content, permeability, and coarseness, still the highest enzyme making from groundnut kernels was given to the point that the kernels comprise a huge quantity of vitamins and amino acids that would have stirred the enzyme manufacture.

Oat peelings and litters from the paper industry *viz.* fiber sludge and pooled sludge of fiber and deinking were also examined by Winquist *et al.* as solid backs for laccase creation from white rot Basidiomycetes fungi *Cerrena unicolour* T71.[66] The investigation group testified the highest laccase activity (178 nkat per g DW) found from oat shells enhanced with 20% collective fiber and deinking sludge. In distinction to the other agro-industrial litters, oat peelings showed to be the encouraging substrate as they possessed other nutrients and lignocellulosic material at 2–10%, which may turn into an inducer for laccase enzyme manufacture.

Joint roughage and de-inking slurry carry fewer nutrients and additional heavy metals, and fiber slurry holds kaoline, which bases the substrate to be thick; therefore, causing problems in the growth of mould on the substrate. Likewise, additional white rot fungus *Pleurotus florida* NCIM 1243 developed on hard supports of several rinds (banana, mandarin, and cantaloupe) exhibited the highest activity of 5.4 U g^{-1} from banana skin, 3.1 U g^{-1} from mandarin skin, and 4.0 from cantaloupe rind.[67] The mixture of these rinds in the ratio of 5 : 2 : 3 has been proved to afford better laccase making up to 6.8 U g^{-1}, maybe due to the synergistic action of several constituents existing in these litters. Solid state fermentation of the laccase consuming *Phanerochaete chrysosporium* from apple pomace and brewery wastes resulted in the highest enzyme activity of 789 U ml^{-1} and 841 U ml^{-1}, respectively at 80% moisture, 3 mmol kg^{-1} veratryl alcohol, and 1.5 mmol kg^{-1} copper sulphate.[68]

6.5.4 Lipases

Triacylglycerol-acyl hydrolases or lipases (E.C. 3.1.1.3) are the utmost useful biocatalysts that are proficient in catalysing numerous reactions such as esterification, trans-esterification, and inter esterification of lipids, as well as whole or half hydrolysis of triacylglycerols. These enzymes have abundant prospects in the food industry, particularly for the production of mono- and diglycerides, making of lipids having a great quantity of polyunsaturated fatty acids, flavour growth, and during the ripening of cheese.[69] Hence, to decrease the price of enzyme for its appropriateness in manufacturing uses, exhaustive exertions were made to minimalize the rate of enzyme manufacture by several agro-industrial residues by means of microbial bases. Amongst the several microbial bases such as bacteria, yeast, fungi, Actinomycetes, Archaea, *etc.*, microorganisms of the genera *Candida*, *Rhizopus*, *Rhizomucor*, *Geotrichum*, *Penicillium*, *Aspergillus*, *Bacillus*, *Pseudomonas*, and *Staphylococcus* are the best-favoured sources for the viable manufacture of lipases.[70]

Gutarra *et al.* linked SSF making of lipase from *Penicillium simplicissimum* by means of tray-type and packed-bed bioreactors with babassu cake as the basal medium.[71] Their research showed that lipase manufacture in the packed-bed reactor was about 30% greater than that of tray type-bioreactors. Furthermore, in the tray-type bioreactor, the concentration of nitrogen did not exert any influence on the making, while temperature, moisture content, and carbon source had an intense result on the manufacture. While using a packed-bed reactor, the aeration rate and temperature had deleterious influence on the lipase manufacture.

Lipase has also been prepared by Mala *et al.* using a varied substrate of wheat fiber and gingelly oil cake in the ratio 3 : 1 from *A. niger* MTCC 2594. It showed an enzyme activity of 384.3 ± 4.5 U g^{-1} dry substrate, which was about 36% greater than that acquired from the submerged fermentation.[72] The measure of lipase production on 100 g and 1 kg trays has revealed to improve the enzyme making by 95% and 84%, individually. Additional agro-industrial waste, palm oil mill run-off added with 0.45% peptone and 0.65% Tween 80, was recycled as a substrate for making lipase by submerged fermentation from *Candida cylindracea* ATCC 14830 that directed to the manufacture of 20.26 U ml^{-1} lipase.[3]

The tri-substrate fermentation (TSF) method was first developed by Edwinoliver *et al.* for the manufacture of lipase from *A. niger* MTCC 2594 using wheat bran, coconut oil cake, and wheat rawa, which caused the highest enzyme activity of 628.7 ± 13 U g^{-1} dry substrate. The usage of TSF for the fermentation method ended the procedure economically practicable; hence, this method can be used for additional manufacturing applications.[73]

Relative research was carried out for the production of alkaline lipase from *A. fumigatus* by both solid and submerged fermentations, where the enzyme production in both fermentation methods was described to be alike. However, lipase making from SSF was revealed to be steady up to 15 days in distinction to the enzyme manufactured by submerged fermentation.[74]

Together with the above trashes, several other agro-industrial remainders (castor bean bar, cassava bar, sugarcane bagasse, and corn shell) have been tried for lipase manufacture by utilizing *Trichoderma harzianum* both singly and in mixtures such as castor bean bar with cassava bar, castor bean bar with corn crust, and castor bean bar with sugarcane bagasse, for each in the ratio of 1 : 1, saturated with diverse oils such as olive oil, corn oil, sunflower oil, and soybean oil. Amongst all the remainders, the highest enzyme production of 4 U g^{-1} was attained from castor bean cake in a mixture with sugarcane bagasse saturated with olive oil.[75] Also, the manufacture of lipase from *Fusarium* isolate FCLA-MA41 has been related to the combined submerged and solid-state fermentations using a range of agro-industrial residues.[76] In the submerged fermentations, an output of 3.0 ± 0.25 U ml^{-1} lipase was eminent to be attained in the research using crambe oil augmented with Triton X-100, ammonium sulphate, and yeast extract, whereas lipase manufacture of 5.0 U ml^{-1} was found during the solid fermentation using crambe meal moisturized with phosphate buffer.

Diverse oilseed cakes (sesame oil cake, groundnut oil cake, and coconut oil cake) have been recycled as substrates for lipase manufacture using *Candida rugosa* NCIM 3462 as a microbial basis. The highest enzyme harvest of 22.40 U g^{-1} substrate was gained by consuming a sesame oil bar, and further, by augmenting the diverse method factors by means of a statistical technique.[77]

Besides the fungal species, bacteria have also been studied for the manufacture of lipase by means of diverse agro-industrial remainders. Lipase has been formed by consuming *Pseudomonas* strain BUP6 in a basal medium enhanced with diverse oil cakes, *viz.*, groundnut, cottonseed, gingelly, soybean, and coconut cakes. Amongst numerous remains observed by Faisal *et al.*, groundnut cake sustained the progress and manufacture of lipase.[78] Palm fiber and palm alkaline soap stock remainders gained from the palm oil industry have been recycled for the enzyme manufacture from *A. niger*. In comparison with submerged fermentation, greater harvest was gained from solid fermentation, showing an activity of 15.41 IU ml^{-1}, thereby signifying the possibility of using palm oil industry litters for the industrialized manufacture of enzymes.[79]

6.5.5 Tannase

Tannase, the common name of tannin acyl hydrolase (E.C. 3.1.1.20), catalyses the breakdown of ester and depside links, thus liberating gallic or ellagic acid and glucose. This enzyme has a varied array of trade uses such as clearing up alcoholic drinks and brews, built-up of precooked tea, decreasing the anti-nutritional aspects of tannins, *etc.* Tannase can be prepared by utilizing agro-industrial trashes as substrates both by submerged or by SSF from bacteria, yeasts, and moulds.[80]

Tannase manufacture by consuming cashew apple bagasse as a substrate enhanced with tannic acid, ammonium sulphate, moisture, and sucrose as a supplementary carbon source has been testified to reach a harvest of 4.63 U g^{-1} dry substrate. In total, altered agro-industrial trashes (paddy crust, rice fiber, millet covering, and groundnut covering) have been related to their efficacy during tannase manufacture from *A. oryzae*.[81] In distinction to all the other sources, rice bran has been described to give the highest enzyme manufacture at the temperature array of 35–40 °C at pH 5.5.

Tannase is prepared using diverse substrates such as bahera fruit, myrobalan, wattle powder, and quebracho powder appended with wheat bran. Bahera fruit has been a well-known substrate for giving the highest activity of 528 U ml^{-1}. Besides, adding wheat bran as an extra compound boosted the enzyme making in this system. Bahera fruit is a good source of tannic acid, further other constituents, which makes it fit for tannase making. RSM has been practical to attain supreme concentrations of the enzyme consuming tamarind seed powder as a substrate. Under the optimum conditions of 3.22% tannic acid concentration, pH 4, a temperature of 35.1 °C, and a fermentation time of 96 h, the highest activity of 139.3 IU ml^{-1} was testified.[82]

Coffee pulp, the by-product of the coffee treatment industry, has been recycled as a substrate for tannase production from *Penicillium verrucosum*, which gave the highest output of 28.173 ± 1.4 U ml^{-1}.[83] Furthermore, by optimizing the several procedure settings, there was a 3.93-fold rise in the output, showing the highest activity of 115.995 U ml^{-1}. These outcomes specify the prospective of the coffee mush as the main basis for the production of value-added goods. Besides, SSF of tannase using *Aspergillus terreus* from diverse agro-industrial remains (red gram husk, green gram husk, ground-nut waste, cotton seed waste, wheat bran, rice bran, coffee husk, tamarind seed powder, cashew apple bagasse, corn powder, and coconut powder) has been performed.[84] Amongst all the tried remainders, wheat fiber was found to be the appropriate substrate for tannase manufacture, where the highest tannase activity of 46.8 U mg^{-1} was perceived under the adjusted settings.

6.5.6 Invertase

Invertase (EC 3.2.1.26), also known as β-fructofuranosidase, belongs to the GH32 clan of glycoside hydrolases and is liable for catalysing the transformation of sucrose into D-glucose and D-fructose.[85] Due to these properties, this enzyme has several likely uses in the foodstuff trade, comprising confectionaries, syrups, condensed milk, infant foods, and beverages for the production of artificial honeybun, plasticizing means in cosmetics, and in the analytical fields for building sucrose biosensors.[86] The chief bases of the microbial making of invertase contain *Aspergillus, Bacillus, Saccharomyces*, and *Fusarium.*[87] Amongst all the sources, *Saccharomyces cerevisiae* or *Saccharomyces carlsbergensis* is the best extensively used microorganism owing to their sucrose fermenting capacity as sucrose is the main basis recycled in the fermentation.[88]

The maximum yield of invertase (272.5 U g^{-1}) has been attained consuming red carrot jam processing remainder as related to the additional food processing wastes (lemon pulp, orange pulp, sugarcane bagasse, grape juice residue, apple pomace, and soya bean residue) used during SSF performed by *S. cerevisae* NRRL Y-12632 after 4 days at 90% original humidity content. The prospective of the diverse fruit rinds such as pomegranate, orange, and pineapple has been tried by means of *A. flavus* during the submerged fermentation.[89] Amongst all the tried substrates, the highest yield of 25.8 IU ml^{-1} was attained by using orange peel as a raw material.

Struggles have been made to employ the litters created from the sugar commerce, like press-mud and the surplus created from distillery such as spent yeast for the solid state fermentation of invertase. The highest specific activity of 430 U mg^{-1} was observed after 72 h of fermentation, when a combination of spent yeast and cultured yeast in the ratio of 7 : 3 was used.[90]

Diverse agro-industrial by-products (rinds of orange, pomegranate, sapota, pineapple, lemon, grapes, and sugarcane bagasse) recycled for the enzyme synthesis from *S. cerevisiae* showed an enzyme production of 0.48 ± 0.011 IU ml^{-1} from 4% orange rind. Carrot rinds, by-products obtained after treating carrot and a form of leftover having a great quantity of sucrose and other nutritional factors, have been employed as a raw material for the solid state fermentation of invertase from *A. niger*, which yielded an activity of 7.95 ± 0.1 U ml^{-1} at 90% moisture content after 72 h of fermentation.[91] Likewise, other agro-industrial wastes such as sunflower waste, cotton waste, rice shell, date syrup, and syrup have been tried to acquire the concentrated enzyme making by submerged fermentation by means of *A. niger* IBGE 01.[92] Under the optimum conditions of agitation (150 rev per min), pH (6.0), inoculum size (5×10^6 conidia), and temperature (40 °C), a maximum activity of 8.23 U ml^{-1} was attained using syrup as a substrate after 72 h of fermentation.

6.5.7 Protease

Protease (EC 3.4.21.40) is a hydrolytic enzyme, which catalyses the hydrolysis of the peptide bonds connecting the amino acids in the polypeptide chain. Amongst the four kinds of proteases (serine proteases, aspartic proteases, cysteine proteases, and metallo-proteases), alkaline proteases are broadly used and have a varied array of uses in detergent, leather, pharmaceutical, and chemical trades. These enzymes have an abundant prospective in several segments of the foodstuff commerce such as tenderization of meat, making of cheese, bakery products, and waste managing. Though proteases can be extracted from a diversity of normal bases, still microbes are the favoured bases for enzyme manufacture. Amongst the diverse microscopic types (bacteria, yeasts, and fungi), *Bacillus* is of pronounced select in terms of great harvest of the enzyme.[93]

Agro-industrial remainders like pigeon pea leftovers, pineapple leftovers, orange skin leftovers, rice fiber, wheat fiber, raw potato starch, raw sweet potato starch, and sugarcane bagasse powder have been explored in the solid state fermentation of protease from the thermoalkaliphilic *Bacillus* JB-99, where the highest enzyme manufacture was achieved from pigeon pea remainders (12 430 ± 120 U ml^{-1}), trailed by pineapple discarded and orange rind. *Bacillus circulans* has been used for enzyme making using green gram husk as a solid substrate and optimization of several method factors by Taguchi orthogonal array. The results showed the highest enzyme activity of 32 000–72 000 U g^{-1} of the material under enhanced settings. Microbial making of protease has a distinctive connection with the progression of the microbial strain; thus, the production of the enzyme is associated with the culture conditions used. Green gram husk, a cheap basis, has been examined for enzyme manufacture from *Bacillus*, which exhibited the highest enzyme activity of 9550 U g^{-1} biomass.[94] The making of protease was affected both by the physiological and chemical features of green gram husk and by the growth of the microbe.

Inexpensive agro-industrial litters such as mustard oil bar, wheat roughage, rice fiber, banana greeneries, and kitchen litters like used tea greeneries, potato peelings, and *I. cylindrica* grass were assessed to define their possibility during protease making by means of a thermophilic bacterial strain *Bacillus subtilis* DM 04. In comparison with all the other bases, potato skin has been described by Mukherjee *et al.* as an effective substrate resulting in the highest harvest of the enzyme followed by *I. cylindrica* grass. Furthermore, the grouping of potato skin and *I. cylindrica* grass in the ratio of 1 : 1 (w/w) considerably improved the protease manufacture as related to single substrates; this method is suitable in dropping the making cost as no extra additions of carbon and nitrogen bases are essential during the manufacture.[95] Castor rind, a by-product found during castor oil processing, was evaluated during alkaline protease making by *Bacillus altitudinis* GVC11, which provided an enzyme activity of 419 293 U g^{-1} of husk under the optimum settings of moisture level, incubation period, and particle size.[96] Taguchi orthogonal array was utilized for the optimization of protease making from *Pseudomonas aeruginosa* using wheat fiber as a substrate.[97] Under the optimum conditions of pH, temperature, inoculum size, and agitation speed, the highest enzyme activity of 582.25 ± 9.2 U ml^{-1} was attained.

Additional manufacturing residues such as dal mill leftover, oil mill unused, molasses, fruit leftover, and vegetal trash have been tried for enzyme making from *Fusarium oxysporum*, which discovered the highest making of 8.8 μg ml^{-1} after 7 days of fermentation from the dal mill leftover, oil mill leftover, molasses, and vegetable trash in distinction to the fruit waste.[98] Budget friendly making of protease by means of numerous agronomic remainders (wheat fiber, cotton bar, chicken quill, mustard bar, soybean meal, maize fiber, rice covering, cane bagasse, gram covering, and corncob) from *Bacillus subtilis* K-1 under solid-state fermentation showed the efficacy of cotton seed oil cake during the solid state fermentation.[99] Cotton seed cake maintained highest protease production (728 U ml^{-1}), trailed by gram husk (714 U ml^{-1}), mustard cake (680 U ml^{-1}), and soya bean meal (653 U ml^{-1}).

6.5.8 Naringinase

Naringinase, a de-bittering enzyme complex, includes both α-L-rhamnosidase (E.C.3.2.1.40) and β-D-glucosidase (E.C. 3.2.1.21). α-L-Rhamnosidase is accountable for the breakdown of naringin to prunin plus rhamnose, and β-D glucosidase is mainly accountable for the hydrolysis of prunin to naringenin and glucose, correspondingly. The main manufacturing applications of this enzyme include de-bittering of the citrus fruitlet fluids, elimination of the hesperidin crystals from the orange sap, improvement of flavour, and fragrance in the wine. The enzyme manufacture is carried out mostly by SSF owing to the capability of the procedure for consumption of several low-cost agro-industrial remainders, besides the other advantages.[100] Diverse wastes have been described for naringinase production using SSF.

Naringinase manufacture has been completed using orange and grapefruit skin, which showed the highest enzyme making using grapefruit skin as a substrate.[100] The great harvest of the enzyme from grapefruit peel is mostly due to the occurrence of high narginin amount (572 μmol of narginin per gram of fresh weight), which turns as an inducer for the enzyme manufacture, thus removing the usage of added naringin as a supplement. Diverse agro-industrial remains such as rice fiber, sugar cane bagasse, citrus rind, and press mud in SSF were tried for their possibility during enzyme making using filamentous fungi *A. niger* MTCC 1344.[101] Amongst all these substrates, rice fiber was found to be the best likely substrate during naringinase making, displaying the highest activity of 58.1 ± 1.6 U g^{-1} dry substrate. Likewise, inexpensive and easily accessible agrarian remains (sugarcane bagasse, soy bean hulls, and rice straw) were evaluated for their ability in enzyme making from *A. niger* strain.[102] The highest enzyme making (1.92 U ml^{-1}) was found by using 0.14 g of sugarcane bagasse, 1.25 g of soybean casings, and 3.05 g of rice chaff. Additionally, by improving the moisture level to 75.5% and pH to 4.0, the activity was improved to 3.02 U ml^{-1}. A statistical technique was applied to enhance the parameters to attain an extreme amount of the enzyme using orange peel as a substrate from *A. niger*. Under the optimal parameter conditions, the highest enzyme activity of 13.89 U ml^{-1} was obtained.[103]

6.5.9 β-Galactosidase

β-Galactosidase (β-D-galactoside galactohydrolase, E.C. 3.2.1.23), also identified as lactase, is one of the very significant enzymes used in foodstuff treating, which has together hydrolytic (hydrolysis of lactose to its constituent monosaccharides, glucose and galactose) and transgalactosylation activities.[104] This enzyme has numerous uses in the food, dairy and fermentation productions, and thus offers its variety of applications in dietary, industrial, and conservational areas. A variety of microscopic bases have been utilized for the manufacture of β-galactosidase.

But the most extensively used microbial sources are *Kluyveromyces* sp. and *Aspergillus* sp. Also, a synthetic medium (lactose) and food industry by-products have been tried for producing β-galactosidase. Among the diverse by-products, whey and wheat fiber have shown noteworthy prospective as substrates in β-galactosidase making. However, most of the study has been supported by consuming whey as a substrate for β-galactosidase manufacture.

A variety of yeasts, which are able to consume lactose, can be cultivated on whey and later be used for β-galactosidase making. Different researchers for the production of this enzyme from a whey-based medium have used yeasts such as *Candida pseudotropicalis, Kluyveromyces marxianus, Kluyveromyces fragilis,* and *Kluyveromyces lactis*. In maximum cases, there is a necessity for the addition of whey with diverse nutrients for the optimal making of the enzyme.[105] The optimal procedure conditions for making the enzyme differ significantly for diverse yeast strains. The supplementation of yeast extract, ammonium sulphate, and potassium dihydrogen orthophosphate to the whey can be beneficial for lactase making from *C. pseudotropicalis*. *A. carbonarius* was also used for the preparation of β-galactosidase developed on de-proteinized cheese whey with highest enzyme making.[106]

K. marxianus MTCC 1389 has been verified for β-galactosidase manufacture from whey.[107] Among the four methods tried for the extraction of β-galactosidase, the SDS-chloroform process was found to be preeminent trailed by toluene–acetone, sonication, and homogenization with glass beads. SDS-chloroform is an inexpensive and simple process for enzyme abstraction from *Kluyveromyces* cells, which results in greater enzyme activity in contrast to the other extraction methods.

K. marxianus NCIM 3551 has been recycled for β-galactosidase production and ethanol fermentation from whey.[108] Ideal β-galactosidase manufacture and ethanol fermentation were attained over an incubation period of 20 h at pH 5.0 and at 25 °C with 16 h old culture. It has been shown that addition of nitrogen did not have much effect on β-galactosidase making and ethanol fermentation.

A permeate-based substrate has been utilized for β-galactosidase making. Amongst the diverse bacterial cultures established, *L. reuteri* and *Streptococcus thermophilus* were designated for β-galactosidase making from the permeate-based substrate. The highest enzyme production (6.31 U ml^{-1})

by *L. reuteri* has been testified at a lactose concentration of 6%, initial pH of 5.0–7.5, ammonium phosphate as the nitrogen source, and incubation temperature at 30 °C after 24 h. Whereas in the instance of *S. thermophilus*, the highest β-galactosidase production (7.85 $U\,ml^{-1}$) was attained at 10% lactose concentration of the permeate medium, at initial pH 6.0–6.5, ammonium phosphate as the nitrogen base, and incubation temperature 35 °C after 24 h.[109]

Intracellular β-galactosidase was manufactured from *Streptococcus thermophilus* cultured in whey.[110] The enzyme was refined by various techniques (ammonium sulphate precipitation, dialysis, gel filtration chromatography using Sephadex G-100, and SDSPAGE) and the properties of the purified enzyme such as pH, temperature optima, and kinetic parameters were assessed. *S. thermophilus* A5 strain showed the highest enzyme output of 7.76 $U\,ml^{-1}$ and protein content of 67 $\mu g\,ml^{-1}$ at 40 °C and pH 7.2. The V_{max} and K_m values were established to be 2.8 $IU\,ml^{-1}$ and 3.05 mM, respectively.

The manufacture of β-galactosidase (β-gal) has been accepted in a deproteinized whey ground substrate by *Bifidobacterium animalis* ssp. *lactis* Bb12 and *Lactobacillus delbrueckii* ssp. *bulgaricus* ATCC 11842.[111] Amongst the four procedures (sonication, acetone toluene, SDS chloroform and lysozyme EDTA treatment) established for β-galactosidase extraction, sonication presented the highest enzyme activity (6.80 $U\,ml^{-1}$) for *B. animalis* ssp. *lactis* Bb12 while for *L. delbrueckii* ssp. *bulgaricus* ATCC 11842, lysozyme EDTA handling was found to be the top (7.77 $U\,ml^{-1}$). Between the seven strains of the genus *Kluyveromyces* tried for β-galactosidase activity, *Kluyveromyces marxianus* ATCC 16045 was designated for culture medium optimization on whey and parboiled rice discharge by submerged fermentation.[112] The optimization for manufacture of β-galactosidase was assessed using a factorial proposal.

This enzyme was manufactured using cheese whey and corn steep liquor by *Kluyveromyces marxianus* CBS 6556 utilizing a central composite rotational design (CCRD). The maximum β-galactosidase activity (21.99 $U\,ml^{-1}$) was attained after 24 h with additional cheese whey amount. The outcomes also showed that consuming corn steep liquor as an addition has an encouraging influence on β-galactosidase activity.[113]

Separately from submerged fermentation, solid-state fermentation employing numerous agro-industrial remains for the making of β-galactosidase consuming fungal species such as *Aspergillus, Penicillium, Fusarium, etc.*, has been considered by various investigators. The most widely used media for enzyme making was wheat fiber due to the occurrence of a great hemicellulose amount, total existing sugars, and the accessibility of vital nutrients that act as an inducer for β-galactosidase manufacture in several microscopic classes. A combination of wheat fiber and rice shell in the ratio of 1 : 1 was evaluated for their prospective in enzyme making from *A. flavus*.[114] This combination was effective in generating good amounts of β-galactosidase upon adding glucose (12.5% w/w) and sodium nitrate (1%) as carbon and nitrogen bases, individually. Furthermore, the activity was improved under the optimum settings of moistness (90%), pH (5), and temperature (30 °C). Likewise, using wheat fiber as a medium, highest enzyme making (5661.4 ± 20.9 U ml^{-1}) from *A. tubingensis* halotolerant fungal segregate was detected when de-proteinized whey was recycled as a moisturizing agent.[115] The enzyme action was significantly inclined with the addition of corn steep liquid and magnesium sulphate; thus under the optimized settings, the activity was augmented to 15 936 U ml^{-1}.

Five diverse agro-industrial remains (wheat fiber, soya oil bar, soya meal, whole wheat, and beet paste) have been explored for enzyme making from *Penicillin canescens*.[116] The highest activity (4010 ± 84 U g^{-1}) was attained with wheat fiber, followed by whole wheat (3818 ± 84 U g^{-1}) after 7 and 9 days, respectively. Furthermore, to improve the enzyme making, mixtures of wheat fiber and whole wheat (7 : 3, 5 : 5, and 3 : 7) were taken, which specified raised intensities of enzyme manufacture, possessing an activity of about 5292 ± 111 U g^{-1} under the optimal settings (83% moisture content, 10^6 spores in per gram, and 30 °C).

6.6 Upcoming Scenarios and Industrialized Choice

Nowadays, enzymes have a huge industrial demand including foodstuff, fodder, medical, *etc.* and in order to encounter the customer call, the marketplace for the industrialized enzyme is also growing enormously. The industrialized enzyme market place was prized to be USD 4.2 billion in the year 2014, which is projected to increase to USD 6.2 billion dollars by the year 2020. Amongst the diverse business segments, roughly more than 50% of the request for enzymes rises from foodstuff commerce. The main foodstuff segments comprise sugar and starch handling, bakery, and dairy activities. The most significant food-grade enzymes such as amylase, galactosidases, pectinases, proteases, lipase, *etc.* have a pronounced possible marketplace and call in the food segment. But the foremost obstacle in the employment of these enzymes in the manufacturing segment lies in the non-availability of precise features, rate of production, and shortage of particular enzymes to convey out the manufacturing procedures. Consequently, to decrease the above disputes, new methods such as protein engineering and genetic alteration have been established for the creation of enzymes having the required properties. Also, with the progression in the investigation and improvement, novel biotechnological tools are being and have been established to curtail the rate of enzyme construction. Despite all of this, the constant burden from the ecological organizations to curtail the effluence, the attention has been moved to employing the diverse litters created from the agro-industries for enzyme making, such as Novozymes, the world's foremost enzyme manufacturer that employs renewable bases such as soy gravels and corn starch for enzyme construction. Seeing the benefits of using microbes as enzyme factories, there is a rising necessity for biodiversity samples, in the hunt for novel molecular objects (including greater enzyme-making variability). Not only that, genetic engineering and synthetic biology methods are anticipated to offer strains with improved efficiency and acceptance to cultivation situations. Brazil is a stimulating home to bio-prospect novel microbes, due to its pronounced natural biodiversity and its agriculture-based budget, providing a huge diversity of agro-industrial leftovers to be considered.

6.7 Conclusion

The usage of agro-industrial remains as media for microbial enzyme production is a natural, justifiable way of resolving difficult contamination while lowering expenses in the making of added value resources. The utilization of agro-industrial wastes for microbial fermentation not only improves the process economics, but also can reduce the environmental issues related to their disposal. Although different residues have been exploited and tested for their potential as substrates for microbial fermentation of enzymes, still the dairy and cereal industry litters/by-products have a substantial probability for enzyme construction. Furthermore, amongst the diverse kinds of fermentation (solid state and submerged) tried for enzyme making, a great amount of effort has been made under solid-state conditions by batch fermentation. While numerous microbial bases have been used for enzyme making, novel and effective microscopic bases are desired for the ideal exploitation of agro-industrial litters to attain advanced harvests of the enzymes. The approach is still long and unpaved for microbe-based enzymes, and as their usage increases, so will the growth of connected abilities with balanced rate declining. Finally, the usage of agro-industrial remnants needs to be more active in manufactures not only to decline the prices of the making but also to implement a circular budget that is an eco-friendly method, enormously essential in the existing times.

In summary, the microbial enzyme manufacture using agro-industrial residues is an unusual win-win situation, in which "one guy's waste is another guy's wealth."

Ethical Issues

There are none to be declared.

Conflicts of Interest

The author endorses that this editorial content has no conflict of interest.

Acknowledgements

The writer is obliged to the Department of Science & Technology, DST-SERB New Delhi for the monetary funding. The author is also grateful to Dr Ashok K. Chauhan, Founder President, Dr Atul Chauhan, Chancellor, and Prof. Dr Balvinder Shukla, Vice Chancellor, Amity University – UP, Noida, for their motivation and research facilities.

References

1. M. D. Trevan, Enzyme Production, in *Biotechnology: The biological principles*, ed. M. D. Trevan, S. Boffey, K. H. Goulding and P. Stanbury, Tata McGraw Hill Publishing Company Ltd, New Delhi, 1988, p. 155.
2. E. Rigo, J. L. Ninowa, M. Di Luccio, J. V. Oliveira, A. Polloni, D. Remonatto, F. Arbter, R. Vardanega, D. de Oliveira and H. Treichel, Lipase production by solid fermentation of soybean meal with different supplements, *LWT: Food Sci. Technol.*, 2010, **43**, 1132.
3. A. Salihu, M. Z. Alama, M. I. Ismail AbdulKarim and H. M. Salleh, Optimization of lipase production by Candida cylindracea in palm oil mill effluent based medium using statistical experimental design, *J. Mol. Catal. B: Enzym.*, 2011, **69**, 66.
4. B. Weber and E. A. Stadlbauer, Sustainable paths for managing solid and liquid waste from distilleries and breweries, *J. Cleaner Prod.*, 2017, **149**, 38.
5. V. Mathioudakis, P. W. Gerbens-Leenes, T. H. Van der Meer and A. Y. Hoekstra, The water footprint of second-generation bioenergy: a comparison of biomass feed-stocks and conversion techniques, *J. Cleaner Prod.*, 2017, **148**, 571.
6. R. C. Kuhad, R. Gupta and A. Singh, Microbial cellulases and their industrial applications, *Enzyme Res.*, 2011, 10.
7. A. Ahamed and P. Vermette, Enhanced enzyme production from mixed cultures of *Trichoderma reesei* RUT-C30 and *Aspergillus niger* LMA grown as fed batch in a stirred tank bioreactor, *Biochem. Eng. J.*, 2008, **42**, 41.
8. P. S. Panesar, R. Kaur, G. Singla and R. S. Sangwan, Bio-processing of Agro-industrial Wastes for Production of Food-grade Enzymes: Progress and Prospects, *Appl. Food Biotechnol.*, 2016, **3**, 208.
9. S. I. Mussatto, L. F. Ballesteros, S. F. Martins and J. A. Teixeira, Use of agro-industrial wastes in solid state fermentation processes. in *Industrial waste*, ed. K. Y. Show and X. Guo, Intech, Europe, 2012, p. 121.
10. B. Fatima, Z. Hussain and M. A. Khan, Utilization of agroindustrial waste residues for the production of amylase from *Aspergillus oryzae* IIB-62014, *Br. Biotechnol. J.*, 2014, **4**, 350.
11. C. Cara, E. Ruiz, J. M. Oliva, F. Sáez and E. Castro, Conversion of olive tree biomass into fermentable sugars by dilute acid pre-treatment and enzymatic saccharification, *Bioresour. Technol.*, 2008, **99**, 1869.
12. V. K. Joshi and A. Devrajan, Ethanol recovery from solid state fermented apple pomace and evaluation of physico-chemical characteristics of residue, *Nat. Prod. Radiance*, 2008, 7, 127–132.
13. P. T. William, *Water treatment and disposal*, John Willy, 2005.
14. D. Miljkovic and G. S. Bignami, Nutraceuticals and methods of obtaining nutraceuticals from tropical crops, USA, application number: 10/992.502, 2002.
15. M. R. Wilkins, L. Suryawati, N. O. Maness and D. Chrz, Ethanol production by Saccharomyces cerevisiae and *Kluyveromyces marxianus* in

the presence of orange peel oil, *World J. Microbiol. Biotechnol.*, 2007, **23**(8), 1161.
16. A. Mirzaei-Aghsaghali and N. Maheri-Sis, By-products from fruits and vegetable generation, characteristics and their nutritional value, Proceeding of Third national congress of recycling and reuse of renewable organic resources in agriculture, Isfahan, Iran, 13–15 May, 2008.
17. J. Fernández-López, E. Sendra, E. Sayas-Barberá, C. Navarro and J. A. Pérez-Alvarez, Physico-chemical and microbiological profiles of "salchich" (Spanish dry fermented sausage) enriched with orange fiber, *Meat Sci.*, 2008, **80**, 410.
18. A. Moongngarm, N. Daomukda and S. Khumpika, Chemical compositions, phytochemicals, and antioxidant capacity of rice bran, rice bran layer, and rice germ, *APCBEE Proc.*, 2012, **2**, 73.
19. M. E. Qureshi, M. K. Wegener and T. Mallawaarachchi, The economics of sugar mill waste management in the Australian Sugar Industry: Mill mud case study, 45th Annual Conference of the Australian Agricultural and Resource Economics Society, Adelaide, South Australia, 23–25 January, 2001.
20. S. K. Solomon, Environmental pollution and its management in sugar industry in India: An appraisal, *Sugar Technol.*, 2005, **7**, 77.
21. P. S. Panesar and J. F. Kennedy, Biotechnological approaches for the value addition of whey, *Crit. Rev. Biotechnol.*, 2012, **32**(4), 327.
22. B. S. Shete and N. P. Shinkar, Dairy industry wastewater sources, characteristics & its effects on environment, *Int. J. Current Engg. Technol.*, 2013, **3**(5), 1611.
23. J. S. White, B. K. Yohannan and G. M. Walker, Bioconversion of brewers spent grain to bioethanol, *FEMS Yeast Res.*, 2008, **8**, 1175.
24. S. I. Mussatto, G. J. M. Rocha and I. C. Roberto, Hydrogen peroxide bleaching of cellulose pulp obtained from brewers spent grain, *Cellulose*, 2008, **15**, 641.
25. R. Panesar, S. Kaur and P. S. Panesar, Production of microbial pigments utilizing agro-industrial waste: a review, *Curr. Opin. Food Sci.*, 2015, **1**, 70.
26. J. A. Heredia-Guerrero, A. Heredia, E. Dominguez, R. Cingolani, I. S. Bayer and A. Athanassiou, *et al.*, Cutin from agro-waste as a raw material for the production of bioplastics, *J. Exp. Bot.*, 2017, **68**, 5401.
27. L. Zihare, K. Spalvins and D. Blumberga, Multi criteria analysis for products derived from agro-industrial by-products, *Energy Procedia*, 2018, **147**, 452.
28. A. Nayak and B. Bhushan, An overview of the recent trends on the waste valorization techniques for food wastes, *J. Environ. Manage.*, 2019, **233**, 352.
29. C. K. Venil, L. Dufossé and P. Renuka Devi, Bacterial Pigments: sustainable compounds with market potential for pharma and food industry, *Front. Sustainable Food Syst.*, 2020, **4**, 100.

30. S. Majumdar, T. Mandal and D. Dasgupta Mandal, Production kinetics of β-carotene from *Planococcus* sp. TRC1 with concomitant bioconversion of industrial solid waste into crystalline cellulose rich biomass, *Process Biochem.*, 2020, **92**, 202.
31. R. Sharma and G. Ghoshal, Optimization of carotenoids production by *Rhodotorula mucilaginosa* (MTCC-1403) using agro-industrial waste in bioreactor: a statistical approach, *Biotechnol. Rep.*, 2020, **25**, e00407.
32. N. F. S. Daud, F. M. Said, M. Ramu and N. M. H. Yasin, Evaluation of bio-red pigment extraction from *Monascus purpureus* FTC5357, *IOP Conf. Ser.: Mater. Sci. Eng.*, 2020, **736**, 22084.
33. P. Kaur, G. Ghoshal and A. Jain, Bio-utilization of fruits and vegetables waste to produce β-carotene in solid-state fermentation: characterization and antioxidant activity, *Process Biochem.*, 2019, **76**, 155.
34. S. G. Pandit, K. P. M. Ramesh, M. H. Puttananjaiah and M. A. Dhale, *Cicer arietinum* (Bengal gram) husk as alternative for *Talaromyces purpureogenus* CFRM02 pigment production: bioactivities and identification, *LWT*, 2019, **116**, 108499.
35. R. T. Hilares, R. A. de Souza, P. F. Marcelino, S. S. da Silva, G. Dragone and S. I. Mussatto, *et al.*, Sugarcane bagasse hydrolysate as a potential feedstock for red pigment production by *Monascus ruber*, *Food Chem.*, 2018, **245**, 786.
36. M. A. Haque, V. Kachrimanidou, A. Koutinas and C. S. K. Lin, Valorization of bakery waste for biocolorant and enzyme production by *Monascus purpureus*, *J. Biotechnol.*, 2016, **231**, 55.
37. D. Da Silva Vilar, I. A. Cruz, N. H. Torres, R. T. Figueiredo, L. de Melo and I. T. F. de Resende, *et al.*, Agro-industrial wastes: environmental toxicology, risks, and biological treatment approaches, in *Environmental Contaminants: Ecological Implications and Management*, ed. R. Naresh Bharagava, Springer, Singapore, 2019, pp. 1–23.
38. F. C. Lopes and R. Ligabue-Braun, Agro-Industrial Residues: Eco-Friendly and Inexpensive Substrates for Microbial Pigments Production, *Front. Sustainable Food Syst.*, 2021, **5**, 589414.
39. A. Pandey, C. R. Soccol and D. Mitchell, New developments in solid-state fermentation I. bioprocesses and applications, *Process Biochem.*, 2000, **35**, 1153.
40. G. S. Hoondal, Microbial alkaline pectinases and their industrial applications: A review, *Appl. Microbiol. Biotechnol.*, 2002, **59**, 409.
41. E. S. Martins, D. Silva, R. Da Silva and E. Gomes, Solid state production of thermostable pectinases from thermophilic *Thermoascus aurantiacus*, *Process Biochem.*, 2002, **37**, 949.
42. E. Rosales, S. R. Couto and A. Sanroman, New uses of food waste: application to laccase production by *Trametes hirsute*, *Biotechnol. Lett.*, 2002, **24**, 701.
43. A. De Gregorio, G. Mandalari, N. Arena, F. Nucita, M. M. Tripodo and R. B. Lo Curto, SCP and crude pectinase production by slurry-state fermentation of lemon pulps, *Bioresour. Technol.*, 2002, **83**, 89.

44. D. Silva, K. Tokuioshi, E. da Silva Martins, R. Da Silva and E. Gomes, Production of pectinase by solid-state fermentation with *Penicillium viridicatum* RFC3, *Process Biochem.*, 2005, **40**, 2885.
45. S. R. Patil and A. Dayanand, Production of pectinase from deseeded sunflower head by *Aspergillus niger* in submerged and solid-state conditions, *Bioresour. Technol.*, 2006, **97**, 2054.
46. E. C. Giese, R. F. H. Dekker and A. M. Barbosa, Orange bagasse as a substrate for the production of pectinase and laccase by *Botryosphaeria rhodina* MAMB-05 in submerged and solid state fermentation, *BioResources*, 2008, **3**(2), 335.
47. N. Martin, M. A. U. Guez, L. D. Sette, R. Da Silva and E. Gomes, Pectinase production by a Brazilian thermophilic fungus *Thermomucor indicae seudaticae* n31 in solid state and submerged fermentation, *Mikrobiologiia*, 2010, **79**(3), 321.
48. S. A. Ahmed and F. A. Mostafa, Utilization of orange bagasse and molokhia stalk for production of pectinase enzyme, *Braz. J. Chem. Engg.*, 2013, **30**(3), 449.
49. I. Ahmed, M. A. Zia, M. A. Hussain, Z. Akram, M. T. Naveed and A. Nowrouzi, Bioprocessing of citrus waste peel for induced pectinase production by *Aspergillus niger,* its purification and characterization, *J. Radiat. Res. Appl. Sci.*, 2016, **9**(2), 148.
50. A. M. Embaby, A. A. Masoud, H. S. Marey, N. Z. Shaban and T. M. Ghonaim, Raw agro-industrial orange peel waste as a low cost effective inducer for alkaline polygalacturonase production from *Bacillus licheniformis* SHG10, *SpringerPlus*, 2014, **3**, 1–13.
51. A. Biz, A. T. J. Finkler, L. O. Pitol, B. S. Medina, N. Krieger and D. A. Mitchell, Production of pectinases by solid state fermentation of a mixture of citrus waste and sugarcane bagasse in a pilot-scale packed-bed bioreactor, *Biochem. Engg. J.*, 2016, **111**, 54.
52. V. Poondla, S. K. Yannam, S. N. Gummadi, R. Subramanyam and V. S. R. Obulam, Enhanced production of pectinase by Saccharomyces cerevisiae isolate using fruit and agro-industrial wastes: Its application in fruit and fiber processing, *Biocatal. Agric. Biotechnol.*, 2016, **6**, 40–50.
53. S. Ramachandran, A. K. Patel, K. M. Nampoothiri, F. Francis, V. Nagy, G. Szakacs and A. Pandey, Coconut oil cake-a potential raw material for the production of a-amylase, *Bioresour. Technol.*, 2004, **93**, 169.
54. A. Kunamneni, K. S. Kumar and S. Singh, Response surface methodological approach to optimize the nutritional parameters for enhanced production of -amylase in solid state fermentation by *Thermomyces lanuginosus*, *Afr. J. Biotechnol.*, 2005, **4**(7), 708.
55. H. K. Sodhi, K. Sharma, J. K. Gupta and S. K. Soni, Production of a thermostable α-amylase from *Bacillus* sp. PS-7 by solid state fermentation and its synergistic use in the hydrolysis of malt starch for alcohol production, *Process Biochem.*, 2005, **40**, 525.
56. R. Kammoun, B. Naili and S. Bejar, Application of a statistical design to the optimization of parameters and culture medium for α-amylase

production by *Aspergillus oryzae* CBS 819.72 grown on gruel (wheat grinding by-product), *Bioresour. Technol.*, 2008, **99**, 5602.

57. A. K. Mukherjee, M. Borah and S. K. Rai, To study the influence of different components of fermentable substrates on induction of extracellular α-amylase synthesis by *Bacillus subtilis* DM-03 in solid-state fermentation and exploration of feasibility for inclusion of α-amylase in laundry detergent formulations, *Biochem. Engg. J.*, 2009, **43**, 149.
58. R. Singh, V. Kapoor and V. Kumar, Utilization of agro-industrial wastes for the simultaneous production of amylase and xylanase by thermophilic actinomycetes, *Braz. J. Microbiol.*, 2012, 1545.
59. M. Sahnoun, M. Kriaa, F. Elgharbi, D. Z. Ayadi, S. Bejar and R. Kammoun, *Aspergillus oryzae* S2 alpha-amylase production under solid state fermentation: Optimization of culture conditions, *Int. J. Biol. Macromol.*, 2015, **75**, 73.
60. K. Bhange, V. Chaturvedi and R. Bhatt, Simultaneous production of detergent stable keratinolytic protease, amylase and bio-surfactant by *Bacillus subtilis* PF1 using agro industrial waste, *Biotechnol. Rep.*, 2016, **10**, 94.
61. S. Lanco, O. P. Durive, S. Pérez, Z. D. Montes and N. P. Guerra, Simultaneous production of amylase and proteases by *Bacillus subtilis* in brewery wastes, *Braz. J. Microbiol.*, 2016, **47**, 665.
62. S. R. Shraddha, S. Sehgal, M. Kamthania and A. Kumar, Laccase: Microbial sources, production, purification and potential biotechnological applications, *Enzyme Res.*, 2011, **2011**, 1–11.
63. M. Lorenzo, D. Moldes, S. Rodriguez-Cuoto and A. Sanroman, Improving laccase production by employing different lignocellulosic wastes in submerged cultures of *Trametes versicolor*, *Bioresour. Technol.*, 2002, **82**, 109.
64. G. V. Reddy, P. Ravindra Babu, P. Komaraiah, K. R. R. M. Roy and I. L. Kothari, Utilization of banana waste for the production of lignolytic and cellulolytic enzymes by solid substrate fermentation using two *Pleurotus* species (*P. ostreatus* and *P. sajor-caju*), *Process Biochem.*, 2003, **38**, 1457.
65. S. R. Couto and M. Sanrom, Effect of two wastes from groundnut processing on laccase production and dye decolourisation ability, *J. Food Engg.*, 2006, **73**, 388.
66. E. Winquist, U. Moilanen, A. Mettälä, M. Leisola and A. Annele Hatakka, Production of lignin modifying enzymes on industrial waste material by solid-state cultivation of fungi, *Biochem. Engg. J.*, 2008, **42**, 128.
67. P. Sathish Kumar, K. Murugesan and T. Palvannan, Production of laccase from *Pleurotus florida* using using agro-wastes and efficient decolorization of Reactive blue 198, *J. Basic Microbiol.*, 2010, **50**(4), 360.
68. F. Gassara, S. K. Brar, R. D. Tyagi, R. P. John, M. Verma and J. R. Valero, Parameter optimization for production of ligninolytic enzymes using agro-industrial wastes by response surface method, *Biotechnol. Bioprocess. Engg.*, 2011, **16**(2), 343.

69. A. H. Kittikun, W. Kaewthong and B. Cheirsilp, Continuous production of monoacylglycerols from palm olein in packed bed reactor with immobilized lipase PS, *Biochem. Engg. J.*, 2008, **40**, 116.
70. H. Alkan, Z. Baysal, F. Uyar and M. Dogru, Production of lipase by a newly isolated *Bacillus coagulans* under solid state fermentation using melon waste, *Appl. Biochem. Biotechnol.*, 2007, **136**, 183.
71. M. L. E. Gutarra, E. D. C. Cavalcanti, L. R. Castilho, D. M. G. Freire and G. L. S. Anna Jr., Lipase production by solid-state fermentation cultivation conditions and operation of tray and packed-bed bioreactors, *Appl. Biochem. Biotechnol.*, 2005, **121–124**, 105.
72. J. G. Mala, N. G. Edwinoliver, N. R. Kamini and R. Puvanakrishnan, Mixed substrate solid state fermentation for production and extraction of lipase from *Aspergillus niger* MTCC 2594, *J. Gen. Appl. Microbiol.*, 2007, **53**, 247.
73. N. G. Edwinoliver, K. Thirunavukarasu, R. B. Naidu, M. K. Gowthaman, T. N. Kambe and N. R. Kamini, Scale up of a novel tri-substrate fermentation for enhanced production of *Aspergillus niger* lipase for tallow hydrolysis, *Bioresour. Technol.*, 2010, **101**, 6791.
74. A. Rajan and A. J. Nair, A comparative study on alkaline lipase production by a newly isolated *Aspergillus fumigatus* MTCC 9657 in submerged and solid-state fermentation using economically and industrially feasible substrate, *Turkish J. Biol.*, 2011, **35**, 569.
75. G. V. Coradi, V. L. da Visitação, E. A. de Lima, L. Y. T. Saito, D. A. Palmieri, M. A. Takita, P. de Oliva Neto and V. M. G. de Lima, Comparing submerged and solid-state fermentation of agro-industrial residues for the production and characterization of lipase by *Trichoderma harzianum*, *Ann. Microbiol.*, 2012, **63**(2), 533.
76. B. H. Oliveira, G. V. Coradi, D. Attili-Angelis, C. Scauri, A. H. P. G. Luques, A. M. Barbosa, R. F. H. Dekker, P. O. Neto and V. M. G. Lima, Comparison of lipase production on crambe oil and meal by *Fusarium* sp. (*Gibberella fujikuroi* complex), *Eur. J. Lipid Sci. Technol.*, 2013, **115**, 1413.
77. A. Rajendran and V. Thangavelu, Utilizing agricultural wastes as substrates for lipase production by *Candida rugosa* NCIM 3462 in solid-state fermentation: response surface optimization of fermentation parameters, *Waste Biomass Valor.*, 2013, **4**, 347.
78. P. A. Faisal, E. S. Hareesh, P. Priji, K. N. Unni, S. Sajith, S. Sreedevi, M. S. Josh and S. Benjamin, Optimization of parameters for the production of lipase from *Pseudomonas sp.* BUP6 by solid state fermentation, *Adv. Enzyme Res.*, 2014, **2**, 125.
79. E. A. Silveira, P. W. Tardioli and C. S. Farinas, Valorization of palm oil industrial waste as feedstock for lipase production, *Appl. Biochem. Biotechnol.*, 2016, **179**, 558.
80. J. S. De Lima, R. Cruz, J. C. Fonseca, E. V. Medeiros, M. H. C. Maciel, K. A. Moreira and C. M. de Souza Motta, Production, characterization of tannase from *Penicillium montanense* URM 6286 under SSF using agro

industrial wastes, and application in the clarification of grape juice (*Vitis vinifera* L.), *Sci. World J.*, 2014, **2014**, 1–9.
81. R. Paranthaman, R. Vidyalakshmi and K. Singaravadivel, Comparative study on the suitability of different substrates for tannin acyl hydrolase production using *Aspergillus oryzae*, *J. Pharm. Sci. Res.*, 2009, **1**(4), 36.
82. S. K. Mohan, T. Viruthagiri and C. Arunkumar, Statistical optimization of process parameters for the production of tannase by *Aspergillus flavus* under submerged fermentation, *Biotech*, 2014, **4**(2), 159.
83. R. N. Bhoite and P. S. Murthy, Biodegradation of coffee pulp tannin by *Penicillium verrucosum* for production of tannase, statistical optimization and its application, *Food Bioprod. Process.*, 2015, **94**, 727.
84. N. R. Malgireddy and L. N. R. Nimma, Optimal conditions for production of tannase from newly isolated *Aspergillus terrus* under solid state fermentation, *Eur. J. Biotechnol. Biosci.*, 2015, **3**, 56.
85. F. Amin, H. N. Bhatti and M. Asgher, Partial purification and characterization of an acid invertase from *Saccharum officinarum* L, *Pak. J. Bot.*, 2010, **42**, 2531.
86. S. M. Kotwal and V. Shankar, Immobilized invertase, *Biotechnol. Adv.*, 2009, **27**, 311.
87. T. Mase, E. Hirose, K. Shimizu and E. Uchikawa, Isolation, characterization and application of invertase from *Pseudozyma sp.* 1-8. Sugiyama Jogakuen University, *Res. J.: Hum., Soc. Sci. Nat. Sci. Hen.*, 2009, **40**, 147.
88. I. Ul-Haq and S. Ali, Invertase production from a hyper producing *Saccharomyces cerevisiae* strain isolated from dates, *Pak. J. Bot.*, 2005, **37**(3), 749.
89. C. Uma, D. Gomathi, C. Muthulakshmi and V. K. Gopalakrishnan, Production, purification and characterization of invertase by *Aspergillus flavus* using fruit peel waste as substrate, *Adv Biol. Res.*, 2010, **4**(1), 31.
90. R. Kumar and B. Kesavapillai, Stimulation of extracellular invertase production from spent yeast when sugarcane press-mud used as substrate through solid state fermentation, *SpringerPlus*, 2012, **1**, 1–6.
91. H. Ashraf and Z. H. Bilal, Biosynthesis, partial purification and characterization of invertase through carrot (*Daucus carota* l.) peels, *J. Biochem. Tech.*, 2015, **6**(1), 867.
92. K. Ahmed, S. Munawar and M. A. Khan, Beta-D-fructo-furanosidase production by *Aspergillus niger* IBGE 01 using shaken flask technique of submerged fermentation, *Pure Appl. Biol.*, 2015, **4**(3), 323.
93. R. Gupta, Q. K. Beg and P. Lorenz, Bacterial alkaline protease: molecular approaches and industrial applications, *Appl. Microbiol. Biotechnol.*, 2002, **59**, 15.
94. R. S. Prakasham, C. S. Rao and P. N. Sarma, Green gram husk-an inexpensive substrate for alkaline protease production by *Bacillus* sp. in solid state fermentation, *Bioresour. Technol.*, 2006, **97**, 1449.
95. A. K. Mukherjee, H. Adhikari and S. K. Rai, Production of alkaline protease by a thermophilic Bacillus subtilis under solid state

fermentation (SSF) condition using *Imperata cylindrica* grass and potato peel as low cost medium: Characterization and application of enzyme in detergent formulation, *Biochem. Engg. J.*, 2008, **39**, 353.

96. A. Madhuri, B. Nagaraju, N. Harikrishna and G. Reddy, Production of alkaline protease by *Bacillus altitudinis* GVC11 using castor husk in solid state fermentation, *Appl. Biochem. Biotechnol.*, 2012, **167**, 1199.
97. P. Meena, A. D. Tripathi, S. K. Srivastava and A. Jha, Utilization of agro-industrial waste (wheat bran) for alkaline protease production by *Pseudomonas aeruginosa* in SSF using taguchi (DOE) methodology, *Biocatal. Agric. Biotech.*, 2013, 210.
98. R. R. Desmukh and N. N. Vidhale, Utilization of agro industrial waste for production of protease by *Fusarium oxysporum* in solid state fermentation, *Indian J. Appl. Res.*, 2015, **5**, 611.
99. S. Singh and B. K. Bajaj, Bioprocess optimization for production of thermoalkali-stable protease from *Bacillus subtilis* K-1 under solid state fermentation, *Prep. Biochem. Biotechnol.*, 2016, **46**, 717.
100. A. M. Cal, L. C. Glory, G. Lizama-Uc and E. Ortiz-Vázquez, Naringinase production from filamentous fungi using grapefruit rind in solid state fermentation, *Afr. J. Microbiol. Res.*, 2010, **4**(19), 1964.
101. M. Shanmugaprakash, V. V. Kumar, M. Hemalatha, V. Melbia and P. Karthik, Solid-state fermentation for the production of debittering enzyme naringinase using *Aspergillus niger* MTCC 1344, *Eng. Life Sci.*, 2011, **11**, 322.
102. A. C. Petri, J. B. Buzato, M. A. P. C. Celligoi and D. Borsato, Optimization of the production of α-L-rhamnosidase by *Aspergillus niger* in solid state fermentation using agro-industrial residues, *Br. Microbiol. Res. J.*, 2014, **4**(11), 1198.
103. A. N. Shehata and A. A. A. El Aty, Optimization of process parameters by statistical experimental designs for the production of naringinase enzyme by marine fungi, *Int. J. Chem. Eng.*, 2014, **2014**, 1–10.
104. P. S. Panesar, R. Panesar, R. S. Singh, J. F. Kennedy and H. Kumar, Microbial production, immobilization and applications of β-D-galactosidase, *J. Chem. Technol. Biotechnol.*, 2006, **81**, 530.
105. P. S. Panesar, Production of -D-galactosidase from whey using *Kluyveromyces marxianus*, *Res. J. Microbiol.*, 2008, **3**, 24.
106. A. El-Gindy, Z. Ibrahim and H. Aziz, Improvement of extracellular β-galactosidase production by thermophilic fungi *Chaetomium thermophile* and *Thermomyces lanuginosus*, *Aust. J. Basic Appl. Sci.*, 2009, **3**(3), 1925.
107. S. Bansal, H. S. Oberoi, G. S. Dhillon and R. T. Patil, Production of β-galactosidase by *Kluyveromyces marxianus* MTCC 1388 using whey and effect of four different methods of enzyme extraction on βgalactosidase activity, *Ind. J. Microbiol.*, 2008, **48**, 337.
108. A. M. Gupte and J. S. Nair, β-galactosidase production and ethanol fermentation from whey using *Kluyveromyces marxianus* NCIM 3551, *J. Sci. Ind. Res.*, 2008, **69**, 855.

109. H. A. Murad, R. I. Refaea and E. M. Aly, Utilization of UF permeate for production of β-galactosidase by lactic acid bacteria, *Pol. J. Microbiol.*, 2011, **60**, 139.
110. S. Princely, N. S. Basha, J. J. Kirubakaran and M. D. Dhanaraju, Biochemical characterization, partial purification, and production of an intracellular beta galactosidase from *Streptococcus thermophilus* grown in whey, *Eur. J. Expt. Biol.*, 2013, **3**(2), 242.
111. L. N. Prasad, B. C. Ghosh, F. Sherkat and N. P. Shah, Extraction and characterisation of βgalactosidase produced by *Bifidobacterium animalis* spp. *lactis* Bb12 and *Lactobacillus delbrueckii* spp. *bulgaricus* ATCC 11842 grown in whey, *Int. Food Res. J.*, 2013, **20**, 487.
112. A. R. C. Braga, P. A. Gomes and S. J. Kalil, Formulation of culture medium with agro-industrial waste for βgalactosidase production from *Kluyveromyces marxianus* ATCC 16045, *Food Bioprocess Technol.*, 2012, **5**, 1653.
113. B. L. B. Perini, H. C. M. Souza, M. Kelbert and G. P. Apati, Production of β-galactosidase from cheese whey using *Kluyveromyces marxianus* CBS 6556, *Chem. Eng. Transactions*, 2013, **32**, 991.
114. S. Nizamuddin, A. Sridevi and G. Narasimha, Production of β-galactosidase by *Aspergillus oryzae* in solid-state fermentation, *Afr. J. Biotechnol.*, 2008, 7, 1096.
115. G. G. Raol, B. V. Raol and V. S. Prajapati, Utilization of agro-industrial waste for β-galactosidase production under solid state fermentation using halotolerant *Aspergillus tubingensis* GR1 isolate, *Biotech*, 2015, **5**, 411.
116. A. A. Antoine, M. E. Bedikou, S. Y. A. Anastasie, N. L. Sebastien, D. Jacqueline and T. Philippe, β-galactosidase production by solid state fermentation of wheat bran/whole wheat without any supplement, *World J. Pharm. Pharma Sci.*, 2015, **4**, 196.

CHAPTER 7

Converting Agricultural Waste Biomass Into Value-added Fuels Via Thermochemical Processes

ALA'A H. AL-MUHTASEB,*[a] FARRUKH JAMIL,[b] AHMAD I. OSMAN[c] AND NAWAF ALHAJERI[d]

[a] Department of Petroleum and Chemical Engineering, College of Engineering, Sultan Qaboos University, Muscat, Oman; [b] Department of Chemical Engineering, COMSATS University Islamabad (C.U.I.), Lahore, Pakistan; [c] School of Chemistry and Chemical Engineering, Queens University Belfast, Belfast, Northern Ireland, UK; [d] Environmental Technology Management Department, College of Life Sciences, Kuwait University, P.O. Box 5969, Safat, 13060, Kuwait

*Email: muhtaseb@squ.edu.om

7.1 Introduction

Due to the ever-increasing demand for energy to meet the needs for chemicals and commodities, the size of the global economy has expanded dramatically. With a growing population and a rising standard of living, daily energy consumption is rising, but the supply cannot keep up with the demand.[1,2] The world will face a massive energy crisis if its output is not increased to meet the demand and consumption, particularly in countries where nuclear energy generation is prohibited.[3,4] According to one of the reports published by the United States, the transportation, industrial, and

Green Chemistry Series No. 78
Agri-food Waste Valorisation
Edited by Pankaj Chowdhary and Abhay Raj

Published by the Royal Society of Chemistry, www.rsc.org

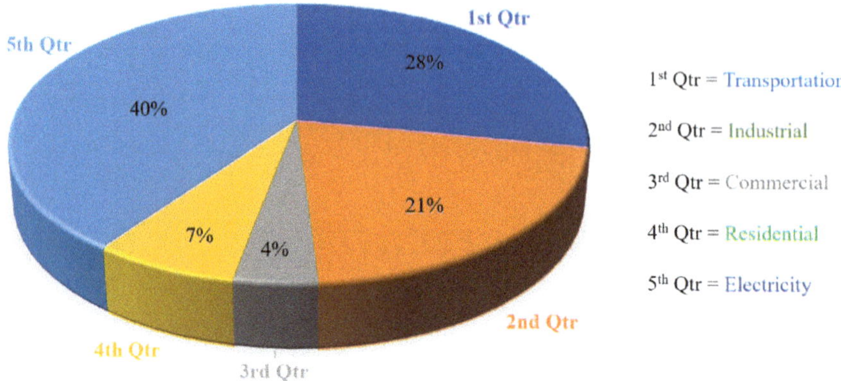

Figure 7.1 Energy consumption by each sector as published by the U.S. energy information.

power sectors consume the most energy, as depicted in Figure 7.1. 100 percent of transport and 80–86 percent of energy on a global scale are provided by nonrenewable resources such as coal, natural gas, and fossil fuels.

At this time, fossil-based resources are essential and promising for meeting the energy needs of various sectors. But the major issues associated with fossil fuel consumption, particularly in vehicles, are environmental damage, climate change, and increased health problems due to carbon dioxide pollution.[5,6] In 2018, 6677 gigatons of greenhouse gases were released into the U.S. environment, according to a report by the Environmental Protection Agency. This distribution of greenhouse gases was attributed to the transportation sector at 28%, electricity production at 27%, industrial activities at 22%, commercial and domestic use at 12%, and agriculture at 10%.[7] Several techniques exist for removing carbon dioxide from the environment, and they are categorized as conventional techniques, negative emission techniques, and radiative forcing geoengineering techniques.[8] First, traditional strategies use decarbonization methods to minimize CO_2 emissions, such as nuclear power, renewable energy, switching to different fuels, efficiency improvements, and carbon capture, storage, and utilization. Most of these technologies have been fully developed and have manageable and acceptable risk levels. Second, the methods that could be used to remove carbon dioxide from the atmosphere are called negative emission technologies. These negative emission techniques include bioenergy carbon capturing and storing, biochar, enhanced weathering, direct air carbon capturing and storing, ocean fertilization, ocean alkalinity enhancement, soil carbon sequestration, afforestation and reforestation, wetland construction and restoration, as well as alternative negative emission utilization and storage methods such as mineral carbonation and using biomass in construction. The third approach is based on the idea that the Earth's radiation balance can be changed by controlling solar and terrestrial radiation.

Table 7.1 A comparative study of fossil fuels and biofuels.

Fossil fuels	Biofuels
• Obtained from geological processes	• Obtained from biological processes
• Are nonrenewable and non-sustainable	• Renewable and sustainable
• High energy per unit of biomass	• Less energy per unit mass
• Environmental unfriendly	• Environmentally friendly
• Production through unsafe techniques	• Production through safer techniques
• Declining day by day	• Growing day by day
• Low investment cost	• Huge investment cost

These methods, known as radiative forcing geoengineering technologies, primarily aim to stabilize or reduce temperature.

In addition, scientists have brought about a significant revolution by developing pre-combustion, oxyfuel combustion, and post-combustion systems that can be utilized for carbon dioxide capture.[9] The acquired CO_2 is compressed and liquefied in pre-combustion systems for storage and transportation applications with minimum power needs. Due to expensive O_2 separation and production, the oxyfuel combustion pathway uses recycled flue gas and an O_2 stream instead of air to burn the carbon-based fuel. However, it is easy to capture and separate CO_2, and oxyfuel combustion is superior to other routes in terms of energy efficiency. The post-combustion pathway captures and separates diluted CO_2 under an oxidizing atmosphere from a combustion system flue gas. CO_2 is compressed into fluid after desulfurization and denitrification of exhaust gas in this system. However, these systems are in an early stage of development and cannot be implemented on a large scale or for commercial purposes.[10,11] Furthermore, fossil fuels are nonrenewable, so fossil-based resources will eventually become scarce and expensive.[12]

As a result, there is a need for such resources for energy generation that are less expensive, renewable, and sustainable and have little to no impact on the environment. All of these characteristics are found in biomass. The primary advantage of using waste-derived biomass as an alternative energy source is that there will be no carbon dioxide emissions when biomass-derived fuels are used in various application areas.[13,14] Table 7.1 compares biomass-derived fuels (biofuels) and fossil-derived fuels (fossil fuels), highlighting their primary benefits and drawbacks.

Several biomass feedstocks have been used as alternative fuel sources for many years. Based on feedstocks and the generation process, fuels are generally categorized as 1st generation, 2nd generation, 3rd generation, and 4th generation.[15] The biomass used to produce this biofuel is discussed in Section 7.2.

7.2 Sources of Biomass Available for Biofuel Production

Figure 7.2 depicts the various types of biomasses utilized in the production of other biofuels. Biofuels comprising the first generation are produced from

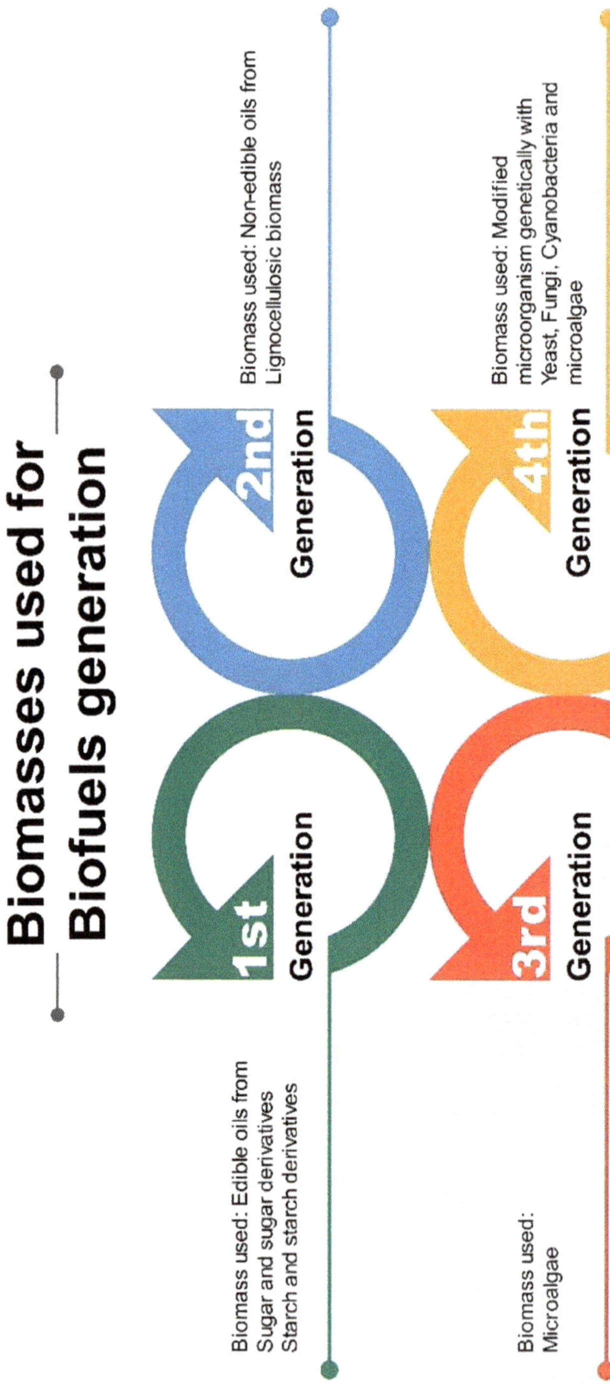

Figure 7.2 Biomass feedstocks and their sources are used to produce biofuels of the first, second, third, and fourth generations.

edible biomass such as starch and sugar, whose production costs are rising.[16,17] The starch used to produce first-generation fuels is derived from wheat, corn, potato, and barley, whereas sugar comes from sugarcane and sugar beets. These biomass resources are also essential for carbon dioxide sequestration, as they consume carbon dioxide for growth. However, there is a concurrent debate between fuel and food.[18] Therefore, large amounts of fertilizers, water, and land are required to meet the fuel and food needs. When producing first-generation biofuels, the availability of fertile land, the competition between food and fuel, the cultivation of crops, cost, transportation, and the chemical composition of biomass must be considered.

The issues associated with the first-generation biofuels have been resolved with the second generation. These biofuels are derived from lignocellulosic biomasses, which are less expensive, renewable, sustainable, and not used as food.[19] Lignocellulosic biomass is one of the most abundant and naturally occurring resources, which has remarkable characteristics for biofuel generation. Nonetheless, second-fuel generation necessitates expensive and, in some cases, impractical unique technology requirements that prevent its commercialization.[20,21] In the past decade, different research groups have introduced third-generation biofuels to eliminate problems associated with the first and second generations.[22] These fuels are derived from marine-based biomasses.[23] Marine biomass provides 20 times more bio-oil than lignocellulosic biomass, has a shorter harvesting period, and has a faster growth rate.[24] In the family of marine-based biomasses, microalgae have the most significant potential for higher yields.[25]

For the fourth-generation biofuels, genetically modified microorganisms such as cyanobacteria, yeast, microalgae, and fungi are utilized. The primary objective of modifying organisms is to create a carbon sink for removing or minimizing carbon and increasing the yield of carbon- and hydrogen-containing compounds (hydrocarbons).[26,27] The advantages of these microorganisms' less complex structure, faster growth rate, and higher fuel or oil content expand their application. In terms of structural components, lipids have a higher energy content than proteins, carbohydrates, and nucleic acids, which is essential for the fourth generation biofuels.[28,29] The advantages and disadvantages of the various biofuel feedstocks and their respective production biomasses are summarized in Table 7.2.

7.3 Agricultural Waste Management for Biofuel Production

As mentioned in Section 7.2, agricultural waste is readily available in large quantities in countries with a high proportion of crop cultivation, *i.e.* agricultural lands. At the end of the growing season, a substantial amount of crop waste is left in the fields.[30–32] This waste should be processed in a planned and sophisticated manner to generate bioenergy or carbon sequestration.[33,34] Typically, in the management of agricultural waste, there

Table 7.2 Summary of different biofuels and biomasses used for production with merits and demerits.

Feedstocks	Oil sources	Merits	Demerits
1st generation biofuels	Edible oils (wheat, corn, potato and barley, sugarcane, and sugar beet, along with others)	• Ecofriendly • Easily available • Renewable, sustainable	• Food *vs.* fuel competition with food crops • Lack of land to cultivate food crops
2nd generation biofuels	Non-edible oils (lignocellulosic biomasses, *e.g.*, sugarcane bagasse, rice husk and straw, corn stalks, wheat straw, and oat straw)	• Easily available • Renewable, sustainable • No competition of food *vs.* fuel	• It needs sophisticated methods to purify • Need for technological developments
3rd generation biofuels	Microalgae	• Renewable • Higher bio-oil yield	• The initial set is not cost-effective
4th generation biofuels	Modified microorganisms (cyanobacteria and yeast)	• Higher cell lipid content • No competing trend between food crops, water, and land usage	• Not viable at a scale of commercial-ization

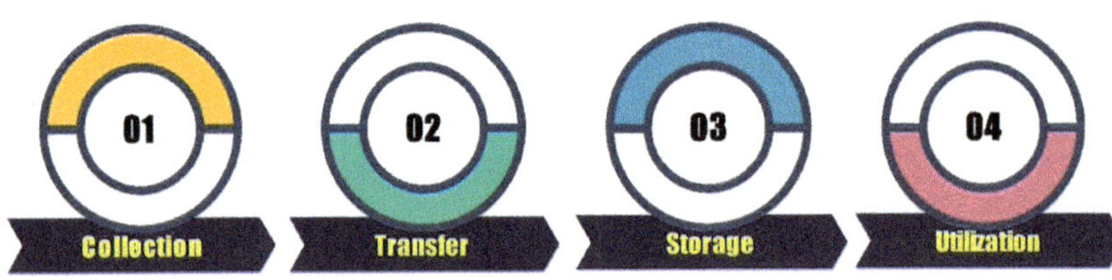

Figure 7.3 The stages involved in managing agricultural waste from the collection site to the processing station.

are four essential steps that research communities must take to manage and process this waste.[35] As shown in Figure 7.3, these steps include collecting agricultural waste, transporting it to the storage and pre-treatment area, and then delivering it to the lab's utilization station.

"Waste collection" refers to the initial capture and gathering of waste from its source to a collection area. For collecting waste, it is necessary to identify the collection methods, location of collection areas, collection schedule, required labour, and essential equipment. Transfer refers to the movement and transport of waste from its origin to storage and pre-treatment areas, followed by the utilization hood. Before reaching final utilization, agricultural waste must pass through several transfer stages.

Transferring waste should involve consistent labour, and the distance between transfer points must be factored into frequency, scheduling, installation, and management expenses.

Storage is the temporary containment of waste. The storage facility of a waste management system allows the manager to oversee the system's schedule and timing. For instance, with adequate storage, waste-spreading activities can be scheduled when they do not interfere with other essential tasks, the weather and field conditions are favourable, and the crop utilizes the waste's nutrients most effectively. The waste management system should determine the storage time, required storage volume, storage facility type, projected size, location, installation cost, and storage process management. Treatment refers to any process designed to reduce the potential for pollution or alter the physical characteristics of waste, such as its moisture and total solid content, to enable more efficient and effective management. Before treatment, the strategy should include an examination of the properties of the waste: determining the desired properties of the treated waste; selecting the type, size, location, and cost of installation of the treatment facility; and calculating the cost of treatment process management. Utilization examples include reusing or recycling waste products. Agricultural wastes can be used as energy, bedding, mulch, organic matter, and plant nutrition. They are marketable if adequately processed. This exhaustive analysis includes the selection of fields, desired equipment, time management from raw materials to finished products, equipment installation, and management costs.

7.4 Bio-refinery Concept and Management

Like a petroleum refinery, a biorefinery system extracts various chemicals, energy, and fuel from crude oil. Feedstocks are the only way to distinguish biorefineries from petroleum refineries. Biorefinery is the utilization of renewable resources to produce valuable products. In contrast, petroleum refineries utilize nonrenewable resources such as fossil fuels.[36,37] Biorefinery is derived from feedstock materials, renewable biomasses, and conversion techniques frequently used to treat and process feedstocks. After feedstock selection, the first stage of a biorefinery's multistage operation typically involves treating biomasses for subsequent operations. The biomass is then treated biologically and chemically.[38–41] The term "biorefinery" has multiple meanings, but all have the same objective. The National Renewable Energy Laboratory (NREL) has developed a typical biorefinery model involving three distinct "platforms" to enhance diverse product pathways: bio-chemical, thermo-chemical, and microbiological. Biorefinery is a system that combines techniques for converting biomass into fuels, electricity, chemicals, and other valuable products. According to the U.S. Department of Energy, biorefinery is the process of converting a variety of biomasses into multiple valuable products.[42] In simple words, biorefining is the long-term conversion of biomass into commercial products and energy.

The products derived from integrated biorefinery facilities fall into two broad categories: energy and material products. Due to their energy content, energy products provide electricity, heat, and transportation services. In contrast, material products are utilized for purposes other than energy generation due to their chemical or physical properties.[43] Some products, such as biohydrogen and bioethanol, can be used as fuels or chemical compounds in chemical synthesis, necessitating an additional distinction when characterizing the product.

Figure 7.4 depicts a biorefinery that accepts one of the many conceivable biomasses as input for processing and conversion to produce various products, including both primary and coproduct valorization.

Biomass and final product selection are crucial to the design of a biorefinery due to their enormous production effects. Initial biomass availability, potential uses, and characteristics must all be considered. There are numerous options available, each with advantages and disadvantages.

A biorefinery can consume biomass from forests, agriculture, aquaculture, and industrial and residential wastes, including wood, short-rotation woody crops, agricultural wastes, and wood wastes. Complete analysis and characterization of raw materials, intermediate, and final products are required to comprehend better how distinct biomass components and reaction products interact at various processing stages. Biorefineries are classified as either lignocellulosic biomass or marine, depending on the biomass feedstock they process.[44,45]

In addition, biorefinery design requires a comprehensive analysis of all components of a sustainable supply chain and the development of a

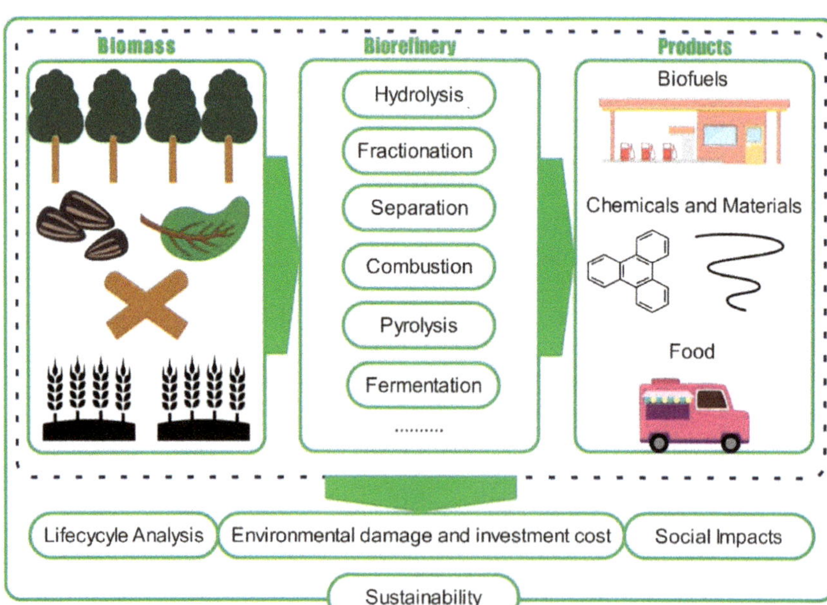

Figure 7.4 Schematic illustration of the biorefinery concept.

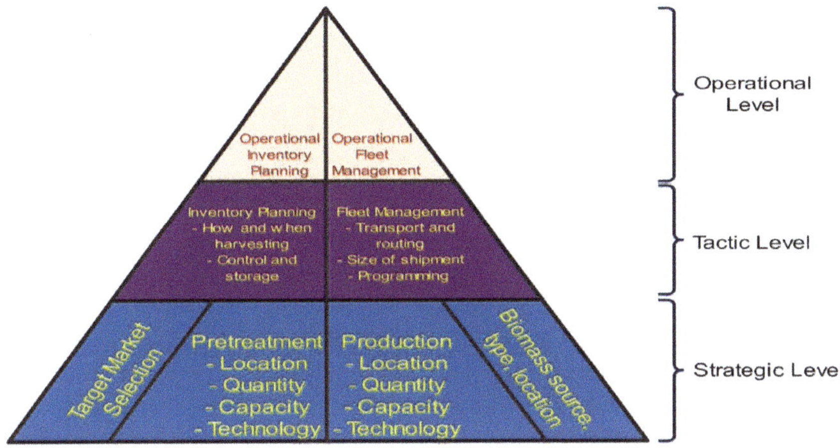

Figure 7.5 Principal decision factors for each biorefinery concept and management level.

realistic model. The design and operations of supply chain networks are also crucial and demanding for the industrialization of biorefineries.[46,47] The supply chain is a comprehensive management system rather than a collection of individual functions. Consequently, the most critical aspect of supply chain management is the development of decision-making models that can accommodate numerous stakeholders and activities linked across the supply chain network.[48,49] The supply chain decision-making process is segmented into strategic, tactical, and operational levels. As shown in Figure 7.5, strategic decisions serve as the basis for tactical and functional choices. Long-term design considerations for the supply chain (SC) are addressed at the strategic level. In contrast, the tactical level oversees the management of monthly (medium-term) decisions that typically span six to twelve months. At the operational level, decisions such as inventory planning (daily inventory control, lack of inventory at distribution sites) and vehicle programming are made.[50]

In conclusion, environmental, economic, and social impacts must be carefully considered when designing a sustainable biorefinery. Waste management from the harvesting area to the storage and processing unit also requires more care.

7.5 Techniques Available to Valorize Agricultural Waste

In the biorefinery process, the techniques available for directly converting biomass feedstock into bioenergy are classified into two major pathways: thermochemical and biochemical. The thermochemical methods often need a significant energy expenditure and the inclusion of a solvent or a catalyst. On the other hand, the biochemical approach has a long cycle time and is less effective in breaking down biomass components. Therefore, incorporating the advantages of both approaches in the manufacturing of biofuels makes merging those two strategies promising. However, hydrothermal routes are employed for pre-treatment to provide the proper biomass feedstock for the subsequent biological pathways to improve the overall process efficiency and ultimate product yields, and *vice versa*.[51]

7.5.1 Thermochemical Techniques

Thermochemical processes decompose plant and waste biomass by thermally decomposing their complex components, with or without catalysts. The procedure entails deconstructing and repurposing an intermediate material into various fuels and chemicals. It enables efficient utilization of a wide range of feedstocks, ensuring year-round supply and the potential for regional energy security. However, changes in the feedstock composition necessitate adjusting the parameters of the technology, resulting in alterations in product distribution and efficiency. Variations in raw materials and operating conditions must be avoided to produce products with desired properties.[52] Researchers focus on mixing wastes from multiple sources generated at variable rates to achieve a constant and predictable feedstock composition. As shown in Figure 7.6, combustion, pyrolysis, and hydrothermal technologies are the three primary categories of biomass and organic waste thermochemical conversion into valuable products.[53]

7.5.2 Biochemical Techniques

In biochemical processes, enzymes and microorganisms extract valuable products from waste biomass.[53] Figure 7.7 depicts a conventional biochemical biomass valorization process.

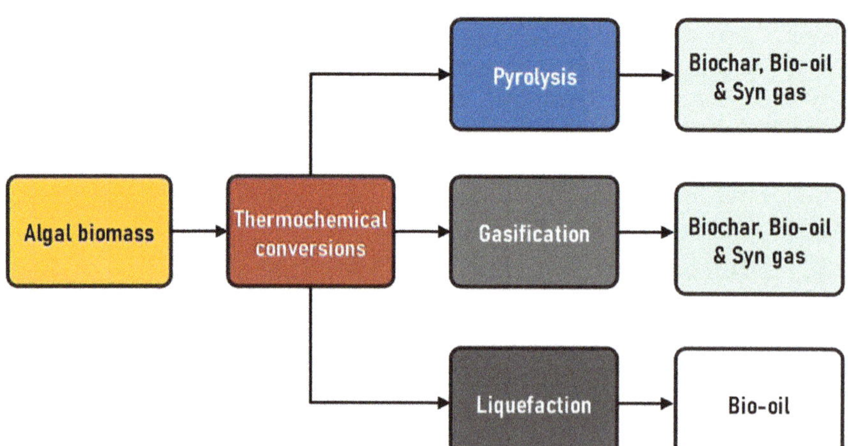

Figure 7.6 Biofuel production *via* the algal biomass conversion processes.

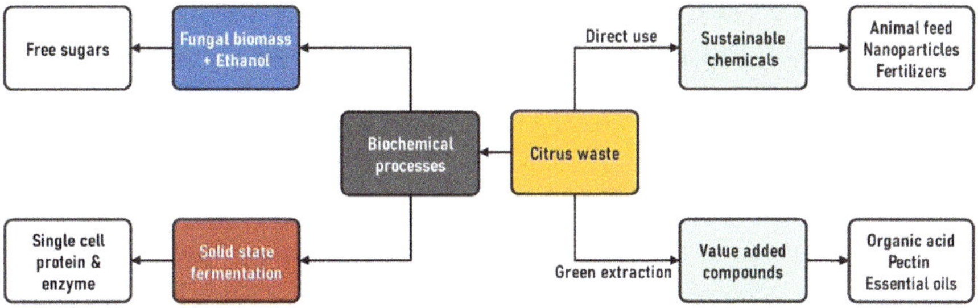

Figure 7.7 Citrus waste valorization for valuable products.

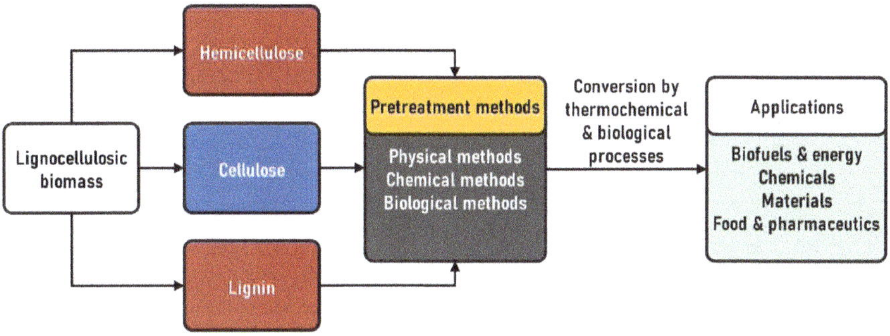

Figure 7.8 Overview of the pre-treatment process of lignocellulosic biomass and its conversion into valuable components.

The key components include biomass pre-treatment, cellulose hydrolysis (saccharification), sugar fermentation, and product recovery (distillation, recovery of solids, and steam and power generation).[54] Economically and technologically, obtaining concentrated sugars from lignocellulosic biomass is a significant issue.[55–57] Waste biomass's various physical and chemical properties pose numerous obstacles to developing cellulose hydrolysis processes by enzymes that degrade cell walls economically and efficiently.[58,59] Developing low-cost monosaccharide manufacturing techniques that overcome the recalcitrance of lignocellulosic plant cell walls is critical to success. Cellulose microfibers are bundled and supported by a lignocellulosic matrix to form plant cell walls. These characteristics make the cell wall an essential structural and defensive component of the plant, making lignocellulose biomass resistant to enzymatic degradation.[60–62] Thermochemical pre-treatment of biomass can help overcome this resistance, and it is a prerequisite for economically viable biochemical conversion systems.[63] Figure 7.8 illustrates an overview of the pre-treatment of lignocellulosic biomass and its conversion into valuable components.

7.5.3 Comparison Between Biochemical and Thermochemical Techniques for Biomass Valorization

According to Verma *et al.*,[64] thermochemical valorization methods provide benefits over biological procedures. Compared to thermochemical methods, biochemical pathways involve microorganisms and enzymes, limiting the feedstocks that may be used. In addition, feedstocks must be treated to be suitable for enzymatic activity (feedstock pre-treatment[55] and washing).[63] Due to these various processes, the technology is not economically viable. Because less retention time is required, thermochemical techniques are more efficient than biochemical techniques. Biological factors impede the productivity of biochemical methods, which explains why this is the case. For instance, the kinetics of cellulose hydrolysis by cellulases demonstrates a decline that diminishes the digestion output. The inability of enzymes and microorganisms to degrade lignin (the structural element of biomass) accounts for 10 to 25 percent of lignocelluloses, making it the second most common natural polymeric material after cellulose,[65] and narrows the range of products obtained. This may change in the future, thanks to advancements in genetically modified microorganisms and enzymes. Recent research has demonstrated that cellulases adsorb non-selectively on lignin structures, decreasing the hydrolysis yield.[66]

Additionally, biochemical processes generate secondary waste that must be treated. Thermochemical methods have a faster cellulose hydrolysis rate regarding the process retention time (a few minutes or seconds *vs.* many hours).[67] The disadvantages of thermochemical processes are greater energy consumption and capital expenditures than those of biochemical processes. In addition, Table 7.3 outlines the advantages and disadvantages of thermochemical techniques (gasification, incineration, pyrolysis, and torrefaction).[68]

7.6 Thermochemical Techniques

Thermochemical conversion is one of the most effective methods for converting biomass into biofuel.[69] This approach is more favourable than the biochemical pathway because it can handle much biomass and is one of the most efficient, quick, and easy conversion methods. In this method, the thermal breakdown of organic matter occurs at very high temperatures, resulting in the degradation of the biomass yield and the formation of bio-oils and gases. As shown in Figure 7.9, this is one of the essential processes, which is classified as follows: torrefaction, pyrolysis, gasification, and the hydrothermal process.[70,71]

7.6.1 Pyrolysis

Pyrolysis is one of the most promising methods for turning biomass into an intermediate liquid product that can be improved into hydrocarbon

Table 7.3 Comparison of the benefits and drawbacks of thermochemical methods.

Process	Operating temp. (°C)	Advantages	Disadvantages
Gasification	550–1600	• Energy recovery with a high efficiency • Decreased emissions • Syngas and biochar development	• Complicated technology • High investment and operating expenses • Extensive gas cleaning of the syngas
Incineration	850–1200	• Up to 90% decrease in volume • Established technologies and infrastructure • Feedstock quality is not important	• Pollution and greenhouse gas emissions • Low energy efficiency – flue gas cleaning
Pyrolysis	500–800	• A process with no waste	• High investment costs
Torrefaction	200–300	• Reduced emissions • Syngas, biofuels, and biochar production • 90% reduction in volume	• Extensive gas cleaning of the syngas • Feedstock sensitivity

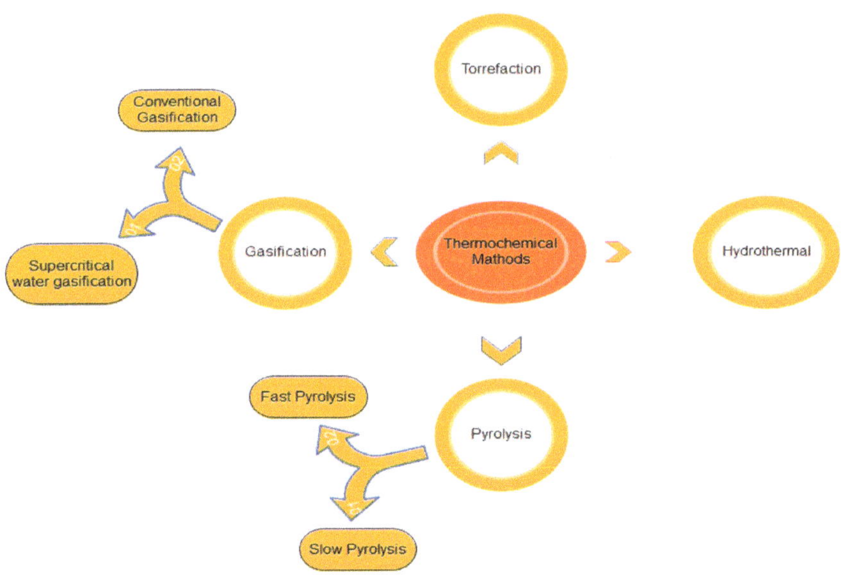

Figure 7.9 Thermochemical conversion processes and the products obtained.

biofuels, oxygen-enriched fuel additives, and petrochemical substitutes. The principal goal of pyrolysis is to heat an organic substance or a material, such as biomass, without oxygen. Moreover, to produce enough heat for biomass degradation during pyrolysis, biomass is frequently pyrolyzed at temperatures of 500 °C or higher. Pyrolysis also does not result in combustion because no oxygen is present in the pyrolysis process. In pyrolysis, the biomass is thermally degraded into charcoal, bio-oil, and other combustible gases.

Additionally, most combustible gases can be condensed to produce pyrolytic or bio-oil. Some permanent gases such as H_2, CO, CO_2, and light hydrocarbons can be employed in the combustion process to provide the necessary heat for pyrolysis. As a result, biomass's three main products of pyrolytic thermal degradation are biochar, bio-oil, and syngas. However, the ratio of these products in the pyrolysis process is influenced by several parameters, including the biomass (feedstock) content and the operational or process conditions. The exciting feature of pyrolysis is that it can connect the agricultural, forest, or wood business to petrochemical processes, preferably by fast pyrolysis, and with different techniques to upgrade the liquid to accomplish the standards of fossil fuel.[72] The pyrolysis technique is subdivided into slow, fast, and slow pyrolysis, depending on the temperature and heating rate provided.[73]

7.6.1.1 Fast Pyrolysis

Fast/flash pyrolysis refers to pyrolysis techniques with a brief residence period of severe heat treatment, generally between 0.5 and 3 seconds at 400–600 °C.[74] In fast/flash pyrolysis, the smaller length of thermal exposure of the organic material (*e.g.*, biomass) increases the significance of heat and mass transfer, phase change, and chemical reaction kinetics.[75] Charcoal production is favoured by long residence durations (a few min to h) and a lower temperature span (200–350 °C). Fluidized bed reactors generally use small particle sizes and higher temperatures to produce rapid heat transmission, reducing char formation. Surprisingly, the poor thermal conductivity of biomass pellets is efficiently used in the ablative reactor, where biomass particles are compressed against a hot surface, generating pyrolysis vapour while exposing the entire inner surface. Fast pyrolysis requires dry biomass (less than 10% moisture), tiny particle sizes (less than 3 mm), short retention durations, controlled temperatures, and quick cooling of pyrolysis vapour. The shortage of predictable kinetics constant for rapid pyrolysis is related to its unsteady state character since biomass complexities need multiphase heat breakdown with significant amounts of volatile chemicals produced (at operating temperatures). All these parameters significantly influence the development of a fast pyrolysis process that must quickly warm biomass to the necessary temperature and cool the products.

7.6.1.2 Slow Pyrolysis

There is a long history of using slow pyrolysis for manufacturing charcoal. It is the most frequent pyrolysis, with a modest heating rate and a high residence duration. Slow pyrolysis involves heating biomass to temperatures between 400 and 500 °C at approximately 0.1 to 1 °C s^{-1} over 5 to 30 minutes. Slow pyrolysis promotes char production, although a tiny amount of gaseous and liquid products are also produced.[76] Slow pyrolysis uses a slower heating rate and a more excellent vapour retention time to provide a desirable environment and give the secondary reactions enough time to proceed. Furthermore, a more extended vapour residence period allows those vapours to be eliminated during the secondary reaction; this eventually leads to the generation of solid carbon-containing biochar.

7.6.2 Gasification

Gasification, as the name implies, is a conversion method that involves partial oxidation of the feedstock to produce a high yield of gaseous products (syngas). Air, oxygen, steam, or a mix may be used as the gasifying medium. The process usually occurs at temperatures ranging from 700 to 1450 °C and is often net exothermic due to incomplete oxidation in the gasifier. There are various steps in the gasification process. The raw material is usually dried before being fed into the gasifier. Still, the remaining moisture fractions can only be entirely dried in the gasification reactor when the biomass particles are burnt to temperatures over 100 °C. Pyrolysis occurs, creating gaseous, liquid (condensable gas, tar), and solid products when the temperature of the particle approaches that of the reactors. Tar cracking, char gasification, incomplete burning, water–CO shift reactions and eliminating the hydrogen and carbon dioxide are among the final reactions.[77] Tar cracking, in particular, is a significant factor since the formation of tars lowers gas production while posing other issues, such as apparatus degradation and pollutants.

Gasification operating variables and the composition of syngas generated from the husk, parchment, pulp, silverskin, and spent coffee ground (SCG) have previously been discussed. The syngas product contains considerable amounts of CO, CO_2, and H_2, with less CH_4. The results show that coffee residue has a high energy potential. The gas-phase yields of the system are increased when the gasification temperature rises. The molar proportion of nitrogen was approximately half; significant nitrogen dilution of the syngas may be avoided by employing oxygen and heat as the gasification medium. The impact may be as strong as increasing the gas's HHV.[78] The syngas may be purified and utilized for vehicle fuel, industrial operations, or the power heat sector. It can be converted to gaseous or liquid fuel using the Fischer–Tropsch process.[79] The eucalyptus sp. is a popular biomass used in Brazil for power production. Densification of eucalyptus stumps after torrefaction gave improved production of gas (1.22 Nl/g daf) with a cold gas efficiency of 65%. The syngas composition was 8% hydrogen, 27% of both CO and carbon dioxide, and 5% CH_4. For more detail see ref. 80. An overall picture of syngas production using the gasification process and its use for different applications are shown in Figure 7.10. Table 7.4 provides a detailed comparison between the general thermochemical conversion techniques.

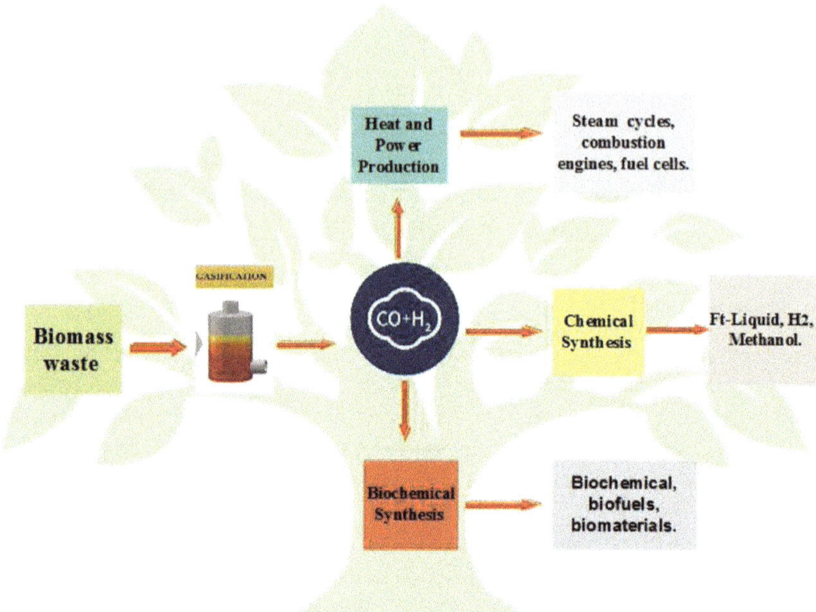

Figure 7.10 Representation of the biomass gasification process and syngas use.

7.6.3 Combustion

Direct combustion is the most basic form of thermochemical conversion under oxidizing circumstances with abundant oxygen. The most common biomass thermochemical conversion pathway is still directly burning raw biomass. Drying is the first step in the combustion process, followed by devolatilization, gas burning, and char combustion. The technique aims for near-total biomass combustion into CO_2 and water with minimum heat loss due to the carbon residue and gaseous hydrocarbons (Figure 7.11). The equipment size may range from small household stoves to huge boilers with hundreds of megawatts of power. The only direct result is heating, which may be utilized to create mechanical work and, from there, energy. Different burning processes are available, for various kinds of biomass. Grate-fired or, provided the fuel is sufficiently rigid and crushable, pulverizing burning suits low moisture biomasses with high carbon contents. At the same time, fluidized bed ignition is suitable for a high moisture content and poor or mixed-quality biomasses. In the literature, there is an incomplete explanation of husk, pulp, and discarded coffee ground direct burning, as well as a thermos gravimetric examination of faulty coffee beans in an oxidizing environment. Because of the lack of established commercial markets, the husk and pulp are mainly burned on the field or utilized as an organic fertilizer. SCG is a solid fuel to produce heat in the soluble coffee

Table 7.4 Comparison of the significant thermochemical conversion processes.

Thermochemical processes	Combustion	Gasification	Pyrolysis
Purpose	Biomass to heat and power conversion	Biomass conversion to high HV gas	Biochar and bio-oil production from biomass
Reaction medium	Air	Air, pure oxygen, and steam	N/A
Temperature	700–1400 °C	500–1300 °C	380–830 °C
Advantages	• The procedure is basic. Current power facilities do not need to be modified to accommodate biomass and coal co-combustion. • Simple and available technology.	• The development of a great variety of chemical compounds, such as methanol and other hydrocarbons, permits quick market adaption. • Maximum conversion can be obtained.	• Liquid fuels are created immediately, and with suitable treatment, they may be processed in traditional refineries. • Produces hydrogen directly.
Disadvantages	• During combustion, NO_x, SO_x, and pollutants are produced. • Feasible only for biomass with a low moisture content. • Requires feedstock pretreatment.	• Tar may obstruct downstream processes and decrease the performance of gasification. • Thermal efficiency decreases drastically for biomass with high moisture.	• Because of its thermally activated nature, pyrolysis requires a lot of energy, and further study is required before it can be used on a large scale. • High char abrasion and erosion to equipment. • Catalyst deactivation due to char.

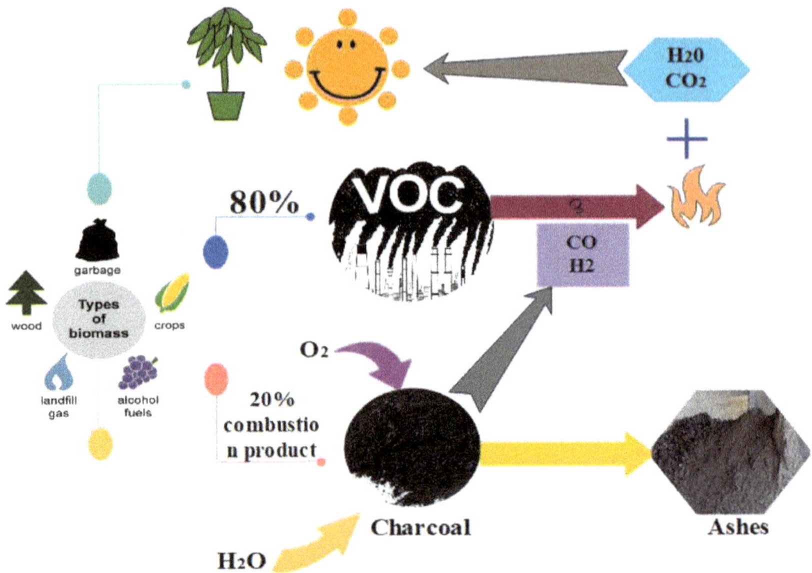

Figure 7.11 Combustion of biomass and the carbon cycle.

manufacturing process. To increase the combustion efficiency, moisture is lowered to around 25–30% before burning.[81] Although wet leftovers such as SCG may be boiled in a proper boiler, maintaining a high enough temperature would need huge combustion chambers, additional surplus air, and extra fuel. It would make wet SCG burning ineffective and costly.

7.7 Conclusions

The worldwide shortage of fossil fuels is a significant issue being pondered by researchers and economists. The most suitable alternatives to replace fossil fuels are renewable energy resources, including solar, wind, geothermal, and biomass energy; it has been discussed and proved that among all renewable resources, biomass is the most suitable for energy production and the transportation sector. Biomass, if combusted directly, for sure most of its energy content will be exhausted towards the environment, so based on this fact, direct combustion is not recommended as an economical technique for biomass utilization. However, pyrolysis of biomass to produce bio-oil followed by further transformation of that bio-oil into potential fuel is considered a suitable thermal technique to valorize biomass.

Similarly, synthesis gas can be obtained from waste biomass by adopting a gasification technique that needs intensive processing conditions in the presence of a limited amount of oxygen. Moreover, one of the most common biofuels, biodiesel, is also obtained from biomass processing. Biodiesel is obtained by mechanical pressing or solvent extraction of oil obtained from biomass followed by transesterification of that bio-oil. Biodiesel is composed of methyl esters. Thus, it can be used as a fuel to replace conventional fuels. Therefore, agricultural waste biomass can be a potential alternative to producing biofuels by adopting thermal conversion techniques.

References

1. A. Naquash, M. A. Qyyum, J. Haider, A. Bokhari, H. Lim and M. Lee, State-of-the-art assessment of cryogenic technologies for biogas upgrading: Energy, economic, and environmental perspectives, *Renewable Sustainable Energy Rev.*, 2022, **154**, 111826.
2. M. K. Yesilyurt, C. Cesur, V. Aslan and Z. Yilbasi, The production of biodiesel from safflower (Carthamus tinctorius L.) oil as a potential feedstock and its usage in compression ignition engine: A comprehensive review, *Renewable Sustainable Energy Rev.*, 2020, **119**, 109574.
3. S. N. Gebremariam and J. M. Marchetti, Economics of biodiesel production: Review, *Energy Convers. Manage.*, 2018, **168**, 74–84.
4. S. Vasistha, A. Khanra, M. Clifford and M. P. Rai, Current advances in microalgae harvesting and lipid extraction processes for improved biodiesel production: A review, *Renewable Sustainable Energy Rev.*, 2021, **137**, 110498.
5. S. S. Rosli, W. N. Amalina Kadir, C. Y. Wong, F. Y. Han, J. W. Lim, M. K. Lam, S. Yusup, W. Kiatkittipong, K. Kiatkittipong and A. Usman, Insight review of attached microalgae growth focusing on support material packed in photobioreactor for sustainable biodiesel production and wastewater bioremediation, *Renewable Sustainable Energy Rev.*, 2020, **134**, 110306.
6. E. Quayson, J. Amoah, S. Hama, A. Kondo and C. Ogino, Immobilized lipases for biodiesel production: Current and future greening opportunities, *Renewable Sustainable Energy Rev.*, 2020, **134**, 110355.
7. U.S. Environmental Protection Agency. Inventory of U.S. Greenhouse gas emissions and sinks, 2020.
8. S. Fawzy, A. I. Osman, J. Doran and D. W. Rooney, Strategies for mitigation of climate change: a review, *Environ. Chem. Lett.*, 2020, **18**, 2069–2094.
9. A. I. Osman, M. Hefny, M. I. A. Abdel Maksoud, A. M. Elgarahy and D. W. Rooney, Recent advances in carbon capture storage and utilization technologies: a review, *Environ. Chem. Lett.*, 2020, **19**, 797–849.
10. S. Fawzy, A. I. Osman, H. Yang, J. Doran and D. W. Rooney, Industrial biochar systems for atmospheric carbon removal: a review, *Environ. Chem. Lett.*, 2021, **19**, 3023–3055.
11. F. Long, W. Liu, X. Jiang, Q. Zhai, X. Cao, J. Jiang and J. Xu, State-of-the-art technologies for biofuel production from triglycerides: A review, *Renewable Sustainable Energy Rev.*, 2021, **148**, 111269.
12. S. Soltanian, C. L. Lee and S. S. Lam, A review on the role of hierarchical zeolites in the production of transportation fuels through catalytic fast pyrolysis of biomass, *Biofuel Res. J.*, 2020, **7**, 1217–1234.

13. V. B. Veljković, M. O. Biberdžić, I. B. Banković-Ilić, I. G. Djalović, M. B. Tasić, Z. B. Nježić and O. S. Stamenković, Biodiesel production from corn oil: A review, *Renewable Sustainable Energy Rev.*, 2018, **91**, 531–548.
14. S. B. Živković, M. V. Veljković, I. B. Banković-Ilić, I. M. Krstić, S. S. Konstantinović, S. B. Ilić, J. M. Avramović, O. S. Stamenković and V. B. Veljković, Technological, technical, economic, environmental, social, human health risk, toxicological and policy considerations of biodiesel production and use, *Renewable Sustainable Energy Rev.*, 2017, **79**, 222–247.
15. A. I. Osman, U. Qasim, F. Jamil, A. A. H. Al-Muhtaseb, A. A. Jrai, M. Al-Riyami, S. Al-Maawali, L. Al-Haj, A. Al-Hinai, M. Al-Abri, A. Inayat, A. Waris, C. Farrell, M. I. A. A. Maksoud and D. W. Rooney, Bioethanol and biodiesel: Bibliometric mapping, policies and future needs, *Renewable Sustainable Energy Rev.*, 2021, **152**, 111677.
16. S. Dawood, A. K. Koyande, M. Ahmad, M. Mubashir, S. Asif, J. J. Klemeš, A. Bokhari, S. Saqib, M. Lee, M. A. Qyyum and P. L. Show, Synthesis of biodiesel from non-edible (Brachychiton populneus) oil in the presence of nickel oxide nanocatalyst: Parametric and optimization studies, *Chemosphere*, 2021, **278**, 130469.
17. M. V. Rodionova, R. S. Poudyal, I. Tiwari, R. A. Voloshin, S. K. Zharmukhamedov, H. G. Nam, B. K. Zayadan, B. D. Bruce, H. J. M. Hou and S. I. Allakhverdiev, Biofuel production: Challenges and opportunities, *Int. J. Hydrogen Energy*, 2017, **42**, 8450–8461.
18. M. U. Qadeer, M. Ayoub, M. Komiyama, M. U. Khan Daulatzai, A. Mukhtar, S. Saqib, S. Ullah, M. A. Qyyum, S. Asif and A. Bokhari, Review of biodiesel synthesis technologies, current trends, yield influencing factors and economical analysis of supercritical process, *J. Cleaner Prod.*, 2021, **309**, 127388.
19. D. LeBauer, R. Kooper, P. Mulrooney, S. Rohde, D. Wang, S. P. Long and M. C. Dietze, BETYdb: a yield, trait, and ecosystem service database applied to second-generation bioenergy feedstock production, *GCB Bioenergy*, 2017, **10**, 61–71.
20. A. Ben Hassen Trabelsi, K. Zaafouri, W. Baghdadi, S. Naoui and A. Ouerghi, Second generation biofuels production from waste cooking oil via pyrolysis process, *Renewable Energy*, 2018, **126**, 888–896.
21. D. K. Westensee, K. Rumbold, K. G. Harding, C. M. Sheridan, L. D. van Dyk, G. S. Simate and F. Postma, The availability of second generation feedstocks for the treatment of acid mine drainage and to improve South Africa's bio-based economy, *Sci. Total Environ.*, 2018, **637–638**, 132–136.
22. S. Nanda, R. Rana, P. K. Sarangi, A. K. Dalai and J. A. Kozinski, A Broad Introduction to First-, Second-, and Third-Generation Biofuels, *Recent Advancements in Biofuels and Bioenergy Utilization*, 2018, pp. 1–25.
23. H. Chowdhury, B. Loganathan, I. Mustary, F. Alam, S. M. A. Mobin, Algae for biofuels: The third generation of feedstock, *Second and Third Generation of Feedstocks*, 2019, pp. 323–344.

24. H. Chowdhury and B. Loganathan, Third-generation biofuels from microalgae: a review, *Curr. Opin. Green Sustainable Chem.*, 2019, **20**, 39–44.
25. J. Chen, J. Li, W. Dong, X. Zhang, R. D. Tyagi, P. Drogui and R. Y. Surampalli, The potential of microalgae in biodiesel production, *Renewable Sustainable Energy Rev.*, 2018, **90**, 336–346.
26. A. Banerjee, C. Banerjee, S. Negi, J.-S. Chang and P. Shukla, Improvements in algal lipid production: a systems biology and gene editing approach, *Crit. Rev. Biotechnol.*, 2017, **38**, 369–385.
27. B. Chen, C. Wan, M. A. Mehmood, J.-S. Chang, F. Bai and X. Zhao, Manipulating environmental stresses and stress tolerance of microalgae for enhanced production of lipids and value-added products–A review, *Bioresour. Technol.*, 2017, **244**, 1198–1206.
28. Y. S. Shin, H. I. Choi, J. W. Choi, J. S. Lee, Y. J. Sung and S. J. Sim, Multilateral approach on enhancing economic viability of lipid production from microalgae: A review, *Bioresour. Technol.*, 2018, **258**, 335–344.
29. K. Azizi, M. Keshavarz Moraveji and H. Abedini Najafabadi, A review on bio-fuel production from microalgal biomass by using pyrolysis method, *Renewable Sustainable Energy Rev.*, 2018, **82**, 3046–3059.
30. M. Arabani and S. A. Tahami, Assessment of mechanical properties of rice husk ash modified asphalt mixture, *Constr. Build. Mater.*, 2017, **149**, 350–358.
31. A. Fareed, S. B. A. Zaidi, N. Ahmad, I. Hafeez, A. Ali and M. F. Ahmad, Use of agricultural waste ashes in asphalt binder and mixture: A sustainable solution to waste management, *Constr. Build. Mater.*, 2020, **259**, 120575.
32. Z. Liu, X. Huang, A. Sha, H. Wang, J. Chen and C. Li, Improvement of Asphalt-Aggregate Adhesion Using Plant Ash Byproduct, *Materials*, 2019, **12**, 605.
33. N. Tripathi, C. D. Hills, R. S. Singh and C. J. Atkinson, Biomass waste utilization in low-carbon products: harnessing a major potential resource, npj Climate and Atmospheric, *Science*, 2019, **2**, 35.
34. M. Amin, T. Murtaza, K. Shahzada, K. Khan and M. Adil, Pozzolanic Potential and Mechanical Performance of Wheat Straw Ash Incorporated Sustainable Concrete, *Sustainability*, 2019, **11**, 519.
35. W. H. B. A. R. S. L. M. Mac Safely, Agriculture waste management systems, United States Department of Agriculture Natural Resources Conservation Service, 2011.
36. F. Cherubini, The biorefinery concept: Using biomass instead of oil for producing energy and chemicals, *Energy Convers. Manage.*, 2010, **51**, 1412–1421.
37. K. Özdenkçi, C. De Blasio, H. R. Muddassar, K. Melin, P. Oinas, J. Koskinen, G. Sarwar and M. Järvinen, A novel biorefinery integration concept for lignocellulosic biomass, *Energy Convers. Manage.*, 2017, **149**, 974–987.
38. V. Ashokkumar, W.-H. Chen, A. A. H. Al-Muhtaseb, G. Kumar, P. Sathishkumar, S. Pandian, F. N. Ani and C. Ngamcharussrivichai,

Bioenergy production and metallic iron (Fe) conversion from Botryococcus sp. cultivated in domestic wastewater: Algal biorefinery concept, *Energy Convers. Manage.*, 2019, **196**, 1326–1334.

39. E. Domínguez, T. Nóvoa, P. G. del Río, G. Garrote and A. Romaní, Sequential two-stage autohydrolysis biorefinery for the production of bioethanol from fast-growing Paulownia biomass, *Energy Convers. Manage.*, 2020, **226**, 113517.
40. A. A. V. Julio, E. A. O. Batlle, A. B. Trindade, S. A. Nebra, A. M. M. Reyes and J. C. E. Palacio, Energy, exergy, exergoeconomic, and environmental assessment of different technologies in the production of bio-jet fuel by palm oil biorefineries, *Energy Convers. Manage.*, 2021, **243**, 114393.
41. E. Najafi, E. Castro and K. Karimi, Biorefining for olive wastes management and efficient bioenergy production, *Energy Convers. Manage.*, 2021, **244**, 114467.
42. A. F. Ferreira, Biorefinery Concept, *Biorefineries*, 2017, pp. 1–20.
43. M. Hingsamer and G. Jungmeier, Biorefineries, *The Role of Bioenergy in the Bioeconomy*, 2019, pp. 179–222.
44. K. Balina, F. Romagnoli and D. Blumberga, Seaweed biorefinery concept for sustainable use of marine resources, *Energy Procedia*, 2017, **128**, 504–511.
45. Z. Usmani, M. Sharma, A. K. Awasthi, T. Lukk, M. G. Tuohy, L. Gong, P. Nguyen-Tri, A. D. Goddard, R. M. Bill, S. C. Nayak and V. K. Gupta, Lignocellulosic biorefineries: The current state of challenges and strategies for efficient commercialization, *Renewable Sustainable Energy Rev.*, 2021, **148**, 111258.
46. N. K. Tsolakis, C. A. Keramydas, A. K. Toka, D. A. Aidonis and E. T. Iakovou, Agrifood supply chain management: A comprehensive hierarchical decision-making framework and a critical taxonomy, *Biosyst. Eng.*, 2014, **120**, 47–64.
47. A. Mortazavi, A. Arshadi Khamseh and P. Azimi, Designing of an intelligent self-adaptive model for supply chain ordering management system, *Eng. Appl. Artif. Intell.*, 2015, **37**, 207–220.
48. M. Eskandarpour, P. Dejax, J. Miemczyk and O. Péton, Sustainable supply chain network design: An optimization-oriented review, *Omega*, 2015, **54**, 11–32.
49. J. Venugopalan, V. S. Sarath, R. J. Pillai, S. A. Krishnan and S. P. Anbuudayasankar, Analysis of Decision Models in Supply Chain Management, *Procedia Eng.*, 2014, **97**, 2259–2268.
50. A. T. Espinoza Pérez, M. Camargo, P. C. Narváez Rincón and M. Alfaro Marchant, Key challenges and requirements for sustainable and industrialized biorefinery supply chain design and management: A bibliographic analysis, *Renewable Sustainable Energy Rev.*, 2017, **69**, 350–359.
51. A. I. Osman, N. Mehta, A. M. Elgarahy, A. Al-Hinai, A. A. H. Al-Muhtaseb and D. W. Rooney, Conversion of biomass to biofuels and life cycle assessment: a review, *Environ. Chem. Lett.*, 2021, **19**, 4075–4118.

52. B. Cantero-Tubilla, Valorization of residues from agriculture and food industries towards biofuels and bioproducts using biochemical and thermochemical technologies, PhD thesis, Cornell University, 2017.
53. K. Tekin, S. Karagöz and S. Bektaş, A review of hydrothermal biomass processing, *Renewable Sustainable Energy Rev.*, 2014, **40**, 673–687.
54. A. U. B. E. D. Larson, Initiative, Biofuel production technologies: status, prospects and implications for trade and development, 2008.
55. J. S. Luterbacher, Q. Chew, Y. Li, J. W. Tester and L. P. Walker, Producing concentrated solutions of monosaccharides using biphasic CO_2–H_2O mixtures, *Energy Environ. Sci.*, 2012, **5**, 6990–7000.
56. L. Liu, J. Sun, M. Li, S. Wang, H. Pei and J. Zhang, Enhanced enzymatic hydrolysis and structural features of corn stover by FeCl3 pre-treatment, *Bioresour. Technol.*, 2009, **100**, 5853–5858.
57. J. S. Luterbacher, J. Y. Parlange and L. P. Walker, A pore-hindered diffusion and reaction model can help explain the importance of pore size distribution in enzymatic hydrolysis of biomass, *Biotechnol. Bioeng.*, 2013, **110**, 127–136.
58. V. Arantes and J. N. Saddler, Access to cellulose limits the efficiency of enzymatic hydrolysis: the role of amorphogenesis, *Biotechnol. Biofuels*, 2010, **3**, 4.
59. P. Bansal, M. Hall, M. J. Realff, J. H. Lee and A. S. Bommarius, Modeling cellulase kinetics on lignocellulosic substrates, *Biotechnol. Adv.*, 2009, **27**, 833–848.
60. D. L. Sills and J. M. Gossett, Using FTIR to predict saccharification from enzymatic hydrolysis of alkali-pretreated biomasses, *Biotechnol. Bioeng.*, 2012, **109**, 353–362.
61. M. Foston and A. J. Ragauskas, Biomass Characterization: Recent Progress in Understanding Biomass Recalcitrance, *Ind. Biotechnol.*, 2012, **8**, 191–208.
62. L. Zhu, J. P. O'Dwyer, V. S. Chang, C. B. Granda and M. T. Holtzapple, Structural features affecting biomass enzymatic digestibility, *Bioresour. Technol.*, 2008, **99**, 3817–3828.
63. S. P. Chundawat, B. Venkatesh and B. E. Dale, Effect of particle size based separation of milled corn stover on A.F.E.X. pre-treatment and enzymatic digestibility, *Biotechnol. Bioeng.*, 2007, **96**, 219–231.
64. M. Verma, S. Godbout, S. K. Brar, O. Solomatnikova, S. P. Lemay and J. P. Larouche, Biofuels Production from Biomass by Thermochemical Conversion Technologies, *Int. J. Chem. Eng.*, 22012, **012**, 1–18.
65. D. Watkins, M. Nuruddin, M. Hosur, A. Tcherbi-Narteh and S. Jeelani, Extraction and characterization of lignin from different biomass resources, *J. Mater. Res. Technol.*, 2015, **4**, 26–32.
66. A. O. Converse, H. Ooshima and D. S. Burns, Kinetics of enzymatic hydrolysis of lignocellulosic materials based on surface area of cellulose accessible to enzyme and enzyme adsorption on lignin and cellulose, *Appl. Biochem. Biotechnol.*, 1990, **24**, 67.
67. J. M. Prado, R. Vardanega, G. C. Nogueira, T. Forster-Carneiro, M. A. Rostagno, F. Maugeri Filho and M. A. A. Meireles, Valorization of Residual

Biomasses from the Agri-Food Industry by Subcritical Water Hydrolysis Assisted by CO2, *Energy Fuels*, 2017, **31**, 2838–2846.

68. M. C. Samolada and A. A. Zabaniotou, Comparative assessment of municipal sewage sludge incineration, gasification and pyrolysis for a sustainable sludge-to-energy management in Greece, *Waste Manag.*, 2014, **34**, 411–420.
69. Y. H. Chan, K. W. Cheah, B. S. How, A. C. M. Loy, M. Shahbaz, H. K. G. Singh, N. R. Yusuf, A. F. A. Shuhaili, S. Yusup, W. Ghani, J. Rambli, Y. Kansha, H. L. Lam, B. H. Hong and S. L. Ngan, An overview of biomass thermochemical conversion technologies in Malaysia, *Sci. Total Environ.*, 2019, **680**, 105–123.
70. W. H. Chen, B. J. Lin, M. Y. Huang and J. S. Chang, Thermochemical conversion of microalgal biomass into biofuels: a review, *Bioresour. Technol.*, 2015, **184**, 314–327.
71. D. Chiaramonti, M. Prussi, M. Buffi, D. Casini and A. M. Rizzo, Thermochemical Conversion of Microalgae: Challenges and Opportunities, *Energy Procedia*, 2015, **75**, 819–826.
72. A. V. Bridgwater, Review of fast pyrolysis of biomass and product upgrading, *Biomass Bioenergy*, 2012, **38**, 68–94.
73. S. Gulshan, Fast pyrolysis of biomass in a fluidized bed for production of bio-oil and upgradation by ex-situ catalytic bed, 2019.
74. D. Meier and O. Faix, State of the art of applied fast pyrolysis of lignocellulosic materials—a review, *Bioresour. Technol.*, 1999, **68**, 71–77.
75. A. V. Bridgwater, A. J. Toft and J. G. Brammer, A techno-economic comparison of power production by biomass fast pyrolysis with gasification and combustion, *Renewable Sustainable Energy Rev.*, 2002, **6**, 181–246.
76. A. Demirbas and G. Arin, An Overview of Biomass Pyrolysis, *Energy Sources*, 2002, **24**, 471–482.
77. P. Basu, Torrefaction, *Biomass Gasification, Pyrolysis and Torrefaction*, 2013, pp. 87–145.
78. N. L. Dina Bacovsky, M. Ognissanto and M. Worgetter, Status of advanced biofuels demonstration facilities in 2012, A report to IEA Bioenergy task 39, 2013.
79. L. Wang, C. L. Weller, D. D. Jones and M. A. Hanna, Contemporary issues in thermal gasification of biomass and its application to electricity and fuel production, *Biomass Bioenergy*, 2008, **32**, 573–581.
80. F. Pinto, J. Gominho, R. N. André, D. Gonçalves, M. Miranda, F. Varela, D. Neves, J. Santos, A. Lourenço and H. Pereira, Improvement of gasification performance of eucalyptus globulus stumps with torrefaction and densification pre-treatments, *Fuel*, 2017, **206**, 289–299.
81. M. A. Silva, S. A. Nebra, M. J. Machado Silva and C. G. Sanchez, The use of biomass residues in the Brazilian soluble coffee industry, *Biomass Bioenergy*, 1998, **14**, 457–467.

CHAPTER 8

Agri-food Waste to Biofuels: Current Trends and Challenges

HARITH SYAHMI ZAIMI,[a] MUHD ARSHAD AMIN,[a] HAFIZA SHUKOR,*[b] NOOR FAZLIANI SHOPARWE[c] AND MUAZ MOHD ZAINI MAKHTAR[d]

[a] Faculty of Chemical Engineering Technology, Universiti Malaysia Perlis, 02600 Arau, Perlis, Malaysia; [b] Centre of Excellence for Biomass Utilization, Faculty of Chemical Engineering Technology, Universiti Malaysia Perlis, 02600 Arau, Perlis, Malaysia; [c] Faculty of Bioengineering and Technology, Universiti Malaysia Kelantan, Jeli Campus, 17600 Jeli, Kelantan, Malaysia; [d] Bioprocess Technology Division, School of Industrial Technology, Universiti Sains Malaysia, Gelugor 11800, Pulau Pinang, Malaysia
*Email: hafizashukor@unimap.edu.my

8.1 Introduction

In 2021, it was anticipated that the world would generate 2.24 billion tonnes of solid waste or 0.79 kg per person every day. Rapid population growth and urbanisation are predicted to raise annual waste creation by 73% from 2021 levels to 3.88 billion tonnes in 2050.[1,2] This rapid growth leads to the accumulation of a range of wastes, which can be attributed to the shift in lifestyles of the current generation, and also necessitates the pursuit of continuous energy generation.[2] In order to provide an acceptable framework for waste management and to safeguard our energy security, crucial steps must be taken. A huge number of developing countries are currently implementing waste valorization as an alternative renewable strategy. This strategy is based on the notion of reusing, recycling, and composting waste

Green Chemistry Series No. 78
Agri-food Waste Valorisation
Edited by Pankaj Chowdhary and Abhay Raj

Published by the Royal Society of Chemistry, www.rsc.org

materials from a variety of sources, which are then turned into useful industrial products such as high-value chemicals and fuels, thus reducing waste and ultimately addressing environmental problems.[3]

In addition, according to Maina *et al.*, the majority of developing countries are moving toward the concept of a circular bio-economy, which supports a cleaner approach to energy generation and mitigates environmental issues for a more balanced approach to sustainability.[4] The linchpin of this concept is ensuring that waste and pollution are eliminated while preserving the products and materials in use and allowing them to renew in a natural system.[2,5] These preserved items will then be recycled as raw materials for the later production of bioenergy. While this concept maintains worldwide demand, it results in the generation of a large amount of waste as the rate of production increases.[4] As a result, this waste will not be accumulated for disposal but will be reused within the system and will continue to circulate as long as it prevents the need for new resources in the manufacture of new products while also promoting waste reduction.[1,3]

Therefore, a circular bio-economy is a possible way to address the waste issue as it is circular in nature. To further grasp the concept of waste valorization, many recent studies emphasized specifically on food waste valorization due to the vast amount of waste accumulated annually. Chin *et al.*, for instance, highlighted an innovative strategy to waste valorization that centred on the composting of lignocellulosic food waste.[6] This was achieved by merging two distinct technologies: co-composting various types of lignocellulosic waste and inoculating the co-compost with a combination of microorganisms with the purpose of achieving quick composting.[2,6] By combining these technologies, it was feasible to reduce the amount of time and area necessary to compost waste, according to the findings. Consequently, this provides a viable long-term solution for solid waste management that simultaneously solves the issue of solid biowaste disposal.[4–6]

In addition, currently waste valorization is mostly emphasized on food waste which provides a premise for the generation of biofuels such as biodiesel and bioethanol.[7] This is due to the properties of the food wastes that are produced in large quantities, inedible, and currently underutilized but have the potential to be hydrolyzed to functional beneficial products such as lipids, carbohydrates, amino acids, and phosphate.[3,5] Through their laboratory experiment, Karmee *et al.* found that bakery and mixed food wastes contain substantial amounts of lipids (about 30%) and carbohydrates (approximately 49%).[7] As a result, these hydrolysates can be further transformed into biofuels, with the obtained lipids being transformed into biodiesel and carbohydrates into bioethanol.[5] In the same context, a good illustration of the huge potential of waste valorization for creating a sustainable society is the use of lipids and carbohydrates derived from food waste for the creation of biofuels.[4] Given that food waste is known to be a zero-cost resource, this chapter will further shed light on food waste valorization treatment and hydrolysis in sustainable biofuel generation and will also discuss the current trends and problems while taking future

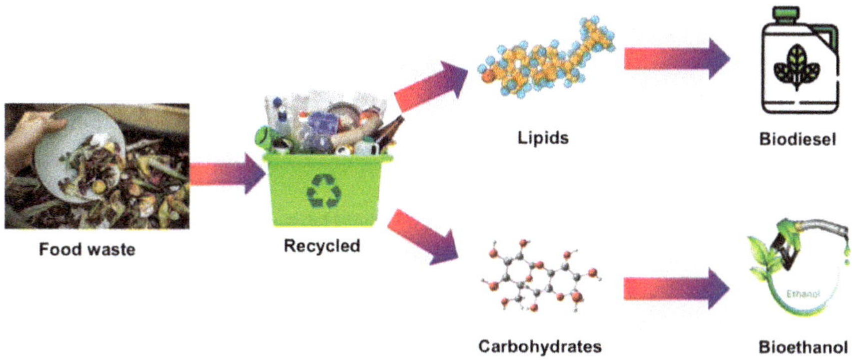

Figure 8.1 Overall scheme of food waste recycling into biofuels.

perspectives into consideration. Figure 8.1 shows the overall scheme of food waste recycling into biofuels.

8.2 Variation of Agri-food Waste

Generally, agri-food waste is originally derived at any point of the food supply chain which includes various types of stages such as production, post-harvest, industrial processing, distribution, and domestic processing.[8] The volume created at each stage varies, with up to 51% of agri-food waste arising from industrial processing such as food processing companies.[6,7] Furthermore, other types of agri-food wastes that are generated can be separated into several categories such as households, food service, and wholesale and retail.[8] As stated in Section 8.1, these agri-food wastes have the potential to be recovered and transformed into useful products due to their high-value properties, such as polysaccharides, proteins, carbohydrates, polyphenolic components, and many more.[3,5,8] Moreover, these non-edible wastes can be used as an alternative feedstock to biofuel production, since the conventional method employs edible feedstock, such as soybean, rapeseed, and canola oils for biodiesel, and sweet sorghum juice, sugarcane juice, barley, and corn for bioethanol and biobutanol.[7,9] Consequently, this can minimize reliance on edible feedstocks and enhance food security for a sustainable future. In the same context, rather than over accumulating waste and causing environmental deterioration, these wastes can be exploited for processes such as valorization, which provides the premise for obtaining unused resources at a low cost and simultaneously promotes environmental health.[3–5]

Numerous recent studies demonstrate the efficacy of the agri-food waste valorization approach.[9] For instance, Tropea *et al.* demonstrated the viability of employing domestic waste such as used cooking oil for biodiesel production as a result of the organic rich nature of food waste, which consists primarily of fat.[10,11] Furthermore, agri-food waste utilization with the intention of creating a circular bioeconomy is also gaining appealing interest in European countries whereby potato peel waste which was regarded as having "no value" was turned into bioethanol by incorporating environmentally benign biocatalytic methods.[8,10] Other accomplishments can be seen in the study of Rahmati *et al.*, a study that recovered oil from instant noodle waste whereby from 99 g of noodle waste, 82 g of purified starch and 5 g of oil were obtained.[12] The acquired oil-free starch residue was used for the simultaneous saccharification and fermentation process for the production of bioethanol.[10,12]

Moreover, according to Parthasarathy *et al.*, oil palm waste is currently utilized for domestic purposes in Malaysian processing mills.[13] This not only benefits the mills in terms of steam and electricity generation, but also encourages a circular economy because the process is carbon neutral due to the use of a substantial proportion of waste instead of coal, which is essential for reducing greenhouse gas (GHG) emissions.[14,15] In addition, numerous studies have demonstrated the effectiveness of using oil palm waste for biobutanol production.[7,8,13,16] Similarly, waste from fruits and vegetables, such as Indian water chestnut, sweet potato, jackfruit, and pineapple, has been widely incorporated into the bioethanol production process, which employs an acid hydrolysis approach and yields fermentable sugar.[17,18] Table 8.1 shows the summary of several achievements through valorization of the agri-food waste approach.

8.3 Impact of Agri-food Waste Generation from the Environmental, Economic, and Society Points of View

Globally, food waste accumulation has increased significantly over the years and according to the Food and Agriculture Organization (FAO), the current amount of agri-food waste generated is about 1.3 billion tons which is equivalent to about 990 billion USD.[24] As a result, if there is no mitigation strategy in place to avert this scenario, the wastes generated will have a wide range of impacts on numerous sectors, particularly from an environmental, economic, and societal perspective, which will be described in greater detail in this chapter. In an environmental context, food waste led to a significant loss of fresh water and groundwater, as the agricultural sector relies heavily on water, which accounts for up to 69% of global water withdrawals.[25] This issue can be demonstrated further whereby, for instance, dumping two pounds of meat is equivalent to a loss of 49 000 L of water that was also used to produce that meat.[5,24,25] In addition, as the rise in global population can be linked to an increase in food production, land utilization grows in order to meet the need for food manufacturing, with half of all liveable land used for agriculture. Since land is commandeered for food production, this necessitates deforestation, which contributes to an increase in GHG emissions and waste generation.[24,26]

Apart from that, as mentioned previously, the majority of wasted food is deposited in landfills, resulting in excessive waste accumulation.[4,5] This not only contributes to global warming and climate change, but also mandates the management of these wastes, since landfills will run out of space and be

Table 8.1 Summary of several achievements through valorization of the agri-food waste approach.

Type of agri-food waste	Achievements	Ref.
Instant noodle waste	• Extracted oil from instant noodle waste was incorporated into simultaneous saccharification and fermentation process in order to **produce bioethanol**	12
Oil palm waste	• The waste was utilized *via* direct combustion for **steam and electricity generation** for processing mills	13
	• **Biobutanol was produced** by utilizing oil palm kernel shells (OPKS) *via* slow pyrolysis	19
Fruit and vegetable waste	• Potato peel waste was **turned into bioethanol** by incorporating environmentally benign biocatalytic methods	10
	• Utilization of tomato wastes as a functional structure for **colourants, antioxidants**, or other components with positive health effects was possible.	17 and 18
	• Indian water chestnut, sweet potato, jackfruit, and pineapple were also widely utilized in the acid hydrolysis-based **bioethanol production** process, yielding fermentable sugar.	17 and 18
	• Orange peel waste proves its viability as a medium to support growth and lipid production by oleaginous yeasts which can be incorporated into **biodiesel production**	20 and 21
	• Due to the content of cellulose (50%) and hemicellulose (40%) in vegetable waste, it became a viable option for **the production of bioethanol** which was executed *via* the hydrolysis method in the presence of microorganisms	22
	• Apple pomace which is a type of waste obtained after apple processing was found to be feasible for **biobutanol production** which was readily fermented by *Clostridium beijerinckii* CECT 508 in 96 h	23
Domestic waste	• Household waste such as used cooking oil is known as organic-rich waste that contains primarily fats and is suitable for **biodiesel production**	10 and 11

unable to store the waste for a long period of time.[25] Andrea Sebrini highlighted in his study that the decomposition of this waste results in the production of methane, a greenhouse gas that is 21 times more potent than carbon dioxide.[24,27] In addition, he stated that food waste contributes up to 7% of the world's total greenhouse gas emissions.[27] Not only that, the distribution of food waste for disposal purposes from one location to another in the entire supply chain produces an increase in the amount of fuel consumed, which in turn increases greenhouse gas emissions.[24,25,27]

Moreover, food waste not only impacts the environment but also results in economic loss. This is further confirmed by the fact that, according to economic estimates, worldwide food waste amounts to nearly one trillion dollars per year, the majority of which comes from developed nations.[24,27]

In addition, the FAO reported that almost one-third of all food produced for human use is lost or wasted.[27] Not only that, Woon *et al.* highlighted in their investigations the intertwined relationship between producers and consumers, in which the large quantity of food waste generated is directly related to the price level of food, which is a result of pricing policy impacting demand.[28] From a societal point of view, this means that those with low incomes who cannot afford to spend more on food are at the greatest risk. Besides, due to the current trend of unexpected policy behaviour on food waste management, it is expected that global calories wasted at the consumer level will nearly double by 2049.[27,28]

To further underline the societal impact of food waste generation, Andrea Sebrini added in his study that the global increase in food production is still not sufficient to feed the entire population, since billions of people still suffer from hunger and malnutrition.[27,29] In addition, this means that just a portion of the population has access to adequate food, while the disadvantaged have no opportunity of obtaining quality food and may even go hungry. Besides, there is a connection between the issue of global hunger and an ethical and moral dimension within the broader idea of global food security, which mandates the management of this nutritional issue in today's society.[29,30] On a positive notion, the current approach to mitigating these issues is focused on converting the food waste generated, which contains numerous useful components, into food additives to reduce malnutrition and hunger. For instance, mango by-products such as peels and seeds are successfully used by Nigeria for the diet of infants and adults since they increase the content of protein and antioxidants.[30] Figure 8.2 depicts the interwoven interaction among environmental, economic, and social aspects.

8.4 Current Agri-food Waste Management

As the population and production of municipal solid waste, including food waste, continue to grow year after year, the majority of countries have started to realise that a good waste management system is required to handle the daily quantity of municipal solid wastes created. A paradigm shift from a conventional food waste disposal model to a sustainable food waste management system is reflected in the government's commitment to food waste management. Due to Malaysia's constrained food waste management funds, food waste management and treatment solutions are regarded as being less successful.[31] In 2009, the Ministry of Housing and Local Government of Malaysia and the Ministry of the Environment of Japan collaborated to plan and propose the National Strategic Plan for Food Waste Management in Malaysia (NSPFWMM), which aims to instil in the public a good habit of disposing of food waste, including food waste segregation for reducing greenhouse gas emissions and reducing land utilization.[32] Other than that, the National Solid Waste Management (2001–2019), National Recycling Program (1999–2004), and Waste Minimization Master Plan (2004) are just a few of the national initiatives that have been made public. However, the food waste treatment strategy has not

Figure 8.2 The interwoven interaction among environmental, economic, and social aspects.

been as successful as other initiatives because the current waste management practises have not improved.[33]

The Malaysian waste management system currently uses a limited number of technologies, including five common treatment techniques that have been widely used in developing nations which are animal feeding, composting or organic fertilisers, anaerobic digestion, incineration, and landfills. Indirectly, there are two categories of techniques for handling food waste: those that are more environmentally friendly and those that are less feasible. According to the data on current food waste treatments in developing nations, the most prevalent food waste treatment method at the moment is dumps/landfills, which are used for food waste treatment at a rate of over 89%, while composting is the second most popular food waste treatment method, with a rate ranging from 1% to 6%.[34]

8.4.1 Landfill and Incineration

Open dumping or landfills are the main food waste treatment methods employed in all developing countries. Landfilling is still the preferred technique of waste management in many countries, including Malaysia, despite being a horrible alternative because it is affordable and requires little technical expertise to dispose of food waste. Leachate, groundwater contamination, the potential discharge of hazardous gases, and odour are further environmental problems brought on by landfills. Both organic and solid wastes contribute significantly to these problems. Due to the decreasing amount of landfill space and rising expenses associated with finding and building new landfills, the government is forced to discover alternative ways to meet the demand for rubbish disposal.[35] Hence, reducing the amount of waste and the need for landfill space through incineration is effective. As a result, numerous nations, notably Singapore and the United States, employ this technique.[34] However, incineration of waste has a high potential for releasing dioxin gases, which are highly toxic and can endanger human health.[36] Additionally, it calls for expensive equipment and highly sophisticated procedures to limit gas emission leftovers.

8.4.2 Eco-friendly Techniques for Handling Food Waste

Anaerobic digestion, animal feeding, and composting are the more popular and environmentally friendly ways to handle food waste. These are the ideas for a sustainable food waste management system that Malaysia could improve in the future as opposed to landfilling or incineration, both of which have negative environmental effects.

8.4.2.1 Anaerobic Digestion

Anaerobic digestion of food waste is a biological process that converts the putrescible organic substrate into biogas through four major steps: hydrolysis, acidogenesis, acetogenesis, and methanogenesis. This technique of treatment places a strong emphasis on waste minimization, recycling, and recovery. This technology is supported by its affordability, environmental friendliness, and use of renewable fuel. Due to the creation of valuable biogas as well as the by-product of this process known as digestate, which are organic particles with a high nutrient content, it can be used as a compost or fertiliser.[37] Since 2005, it has been widely used for food waste treatment in the European Union and many Asian developed countries.[38] In contrast, it is noted that anaerobic digestion is still not widely used as a significant treatment strategy for food waste control in developing nations. Many institutions and NGOs have developed various anaerobic digesters on a domestic and commercial scale in China and India to advance the anaerobic digestion technology for food waste treatment.

8.4.2.2 Animal Feeding

Food waste is a nutrient-dense feedstock for animals in general. As a result, food waste can be used to supplement animal feed. It is estimated that 1 tonne of dry food waste could replace the same amount of maize grain in order to meet the protein needs of a given animal.[39] Since bone meal, eggshells, and sea shells are significant sources of calcium, phosphorus, and other minerals as dietary supplements, they contribute the majority of food waste to animal feed. As the loading and separation of food waste from municipal waste are rarely practised in underdeveloped nations, using food waste as a feedstock for animal feeding is not appropriate there. Except for the more widely used practices of harvesting cadaver bone and eggshell, not all food waste in Malaysia is eligible for use as animal feed for livestock, according to the FEED ACT (Act 697).[32] Meanwhile, there are local laws that support utilizing food waste to feed animals in several nations where there is a strong demand for animal feeding, such as Japan, South Korea, and Taiwan, where it accounts for 33%, 80%, and 71.1% of total food waste generation, respectively.[34]

8.4.2.3 Composting

Utilizing composting to handle food waste is by far the simplest method of getting rid of home and commercial kitchen waste. In essence, it entails food waste breaking down in garden soil. The latter then takes all the nutrients in and grows more fertile. Composting is the process of biologically breaking down organic waste under regulated aerobic conditions to produce a solid humus-like substance. Composting has been receiving more interest than the other waste treatment methods due to its advantages at both an economic and environmental level[36] but the Malaysian government has used vermicomposting as a key national strategy to use food waste to create bio-fertilizers. However, there are still considerable inefficiencies in the manufacture of compost due to unpurified waste feedstock, which comes from the insufficient source segregated food waste system in the majority of poor nations. As a result, the market for composting is small, and food waste composts must compete with a variety of chemical fertilisers, which presents problems for the management and capital expenditures of composting plants.

8.5 Valorization of Agri-food Waste as Biofuel Resources

At present, food waste is being brought up as a worldwide issue connected to greenhouse gas emissions. The food waste problem is anticipated to worsen in Asian nations during the next 25 years, coinciding with economic advancement, population expansion, and urbanisation.[40] Additionally, if the energy component of food waste is not salvaged for use, it is lost when it is buried in landfills.[41]

Biofuels are a great example of how microorganisms can be used to provide sustainable energy while reducing the need for non-renewable energy sources. In order to produce biofuels, a suitable feedstock must be chosen, which is a crucial step. Utilizing waste materials such as food waste and inedible feedstock has many advantages for biofuels. These advantages include lower raw material costs, a lack of dependence on arable land, and a defence against competition from the food market.[40] Therefore, using food waste to make biofuels could help stop the growing food *versus* fuel issue from happening. Figure 8.3 illustrates the advantages of producing biofuels from food waste.

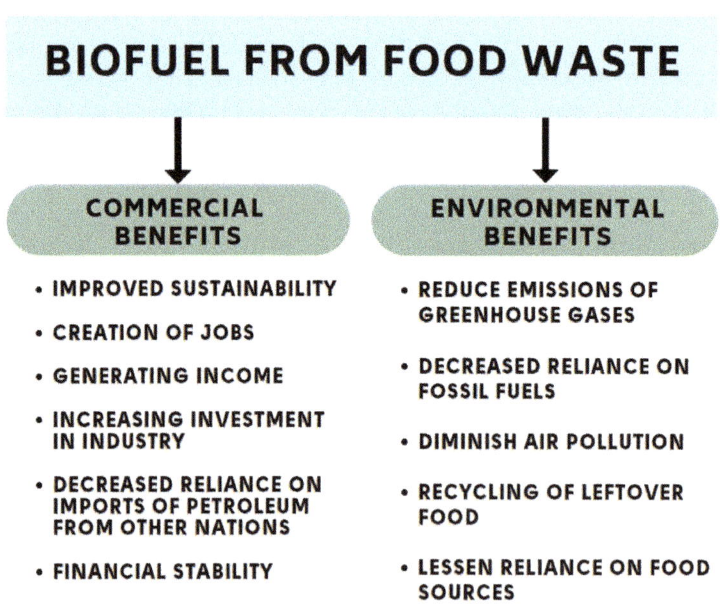

Figure 8.3 Benefits of biofuels from food waste.

8.5.1 Thermochemical Processes

Using biomass as biofuels is a viable way to get over the aforementioned challenges through thermochemical conversion. Thermochemical conversion produces biofuels with high quality and densified energy content by using heat and chemical processes.[42] The most popular thermochemical conversion techniques for turning lignocellulosic and non-lignocellulosic biomass into biofuels include pyrolysis and gasification.

8.5.1.1 Pyrolysis and Gasification

Pyrolysis is the high-temperature, anaerobic thermochemical breakdown of organic materials. Pyrolysis is a preferable alternative to combustion because of its lower process temperatures and lower pollutant emissions.[43] Waste is thermally decomposed during pyrolysis through a series of reactions that include dehydration, depolymerization, decarboxylation, charring, isomerization, and aromatization. These reactions take place concurrently and sequentially. Three fractions-bio-oils, charcoal, and biogas are created during this process, and their separation allows for their use as a liquid fuel.[44] The bio-oil portion can be utilized in furnaces and as a fuel for vehicles. The bio-oil must be improved if it is to be used as a transportation fuel in order to get rid of some organic components that can make the fuel unstable and corrosive due to its high oxygen concentration.[45] According to Wang *et al.*, a continuous fast microwave-assisted pyrolysis system successfully extracted bio-oil from *Camellia oleifera* shell and rice straw (31.86%) employing these materials.[46] In addition to bio-oil, non-condensable byproducts from pyrolysis, such as syngas, can be collected and refined to produce biofuels or biochemicals. Synthetic biofuel can then be produced by further enhancing the hydrogen component of syngas.[43]

While pyrolysis and gasification both rely on the thermal breakdown of organic matter to produce hydrocarbons, gasification involves partial oxidation of biomass at a higher temperature, often between 800 and 1200 °C.[47] Gasification is thought to have a less environmental impact than landfilling or incineration of food waste and requires less space to operate. In addition, it generates fuel gas that is thought to be a clean fuel and may be used in a burner, boiler, or steam turbine. On the downside, the fuel gas may have minute amounts of undesirable elements such as fine particles, organic pollutants such as dioxins, and acidic gases such as nitrogen oxides, hydrogen chloride, and sulphide.[48]

8.5.2 Biochemical Processes

The two main biochemical processes for recovering intrinsic energy from biowastes such as food waste in the form of liquid and gaseous energy carriers, respectively, are fermentation and anaerobic digestion. Depending on the total solid concentration or operating temperatures, different anaerobic digestion applications are conceivable.[44]

8.5.2.1 Bio-hydrogen and Bio-methane

Due to their high calorific values – 143 $kJ\,g^{-1}$ for hydrogen and 55 $kJ\,g^{-1}$ for methane, respectively – hydrogen and methane are frequently employed in the chemical and process industries.[49] Since hydrogen does not emit carbon dioxide into the environment when it burns, it is regarded as a clean energy source. Unlike other fuels, the combustion of methane and hydrogen does not produce nitrous oxide or sulphur dioxide, which are the main causes of air pollution. On the other hand, carbon dioxide is still produced when methane is burned.[50] Its combustion produces no greenhouse gases. In general, a variety of methods, including photobiological processes, microbial electrolysis, acidogenic fermentation, dark fermentation, and photo fermentation, can be used to produce biohydrogen. These methods can also be combined. Due to its better potential for producing hydrogen and ability to decompose organic wastes without relying on the presence of light, dark fermentation is seen to be one of the most promising ways.[51] Other than that, biomethane has also been used to make biofuels, but the success of the combined production of biodiesel and biomethane, which produced the maximum fuel (about 2400 MJ using 30 kg of biodiesel and 35 kg of biomethane/1000 kg of wet food waste), stands out.[52] Biomethane fermentation could take place in both mesophilic and thermophilic environments. The yields and contents of methane produced under thermophilic conditions were higher than those under mesophilic conditions.

Additionally, there is a biogas mixture known as bio-hythane that contains both biomethane and biohydrogen as well as carbon dioxide. It is a mixed biogas combining hydrogen, methane, and carbon dioxide with an average composition of 10% H_2, 30% CO_2, and 60% CH_4.[53] The direct use of this mixture with the current vehicle engines and the fact that it has been made commercially available as vehicle fuels in the US and India are the key reasons why bio-hythane is anticipated to be advantageous economically as a future energy resource.[54] Due to their clean nature, high fuel value, and improved heat efficiency, both H_2 and CH_4 may be upgraded properly to replace fossil fuels and be used as gaseous transport biofuels, which offers a means to create a greener economy and makes engines more likely to function with minimum energy.

8.5.2.2 Bio-ethanol

Currently, the primary source of fuel for transportation is fossil-derived fuel, which contributes significantly to greenhouse gas emissions. Among the many possibilities, bioethanol is one of the most promising liquid biofuels for replacing gasoline and reducing its harmful effects on the environment and human health.[55] This biofuel can be created by fermenting carbohydrates, a biological process that is quick and has several benefits, including the high energy output and useful byproducts. Because of its vast range of uses, ethanol has been gaining more attention and demand globally. About 62% of the ethanol produced worldwide is produced only by Brazil and the United States with the production capacity reaching 100 billion litres in order to meet the enormous demand in 2015.[40] Using the right microbial strains, such as *Saccharomyces cerevisiae*, sugary food wastes can be directly transformed into bioethanol. Before the ethanolic fermentation process can begin with starchy food waste, a hydrolysis step is necessary to obtain sugars (glucose). In a study, fermentable sugars were created from food waste collected from food courts at a Malaysian university using a modified sequential acid-enzymatic hydrolysis technique. *S. cerevisiae* then converts the liberated sugars into bioethanol with a conversion efficiency of 85.4% and a yield of 0.42 $g\,g^{-1}$.[56]

8.5.2.3 Bio-diesel

In many other nations, including the U.S. and Europe, biodiesel is used as a fuel. Chemically, biodiesel is made up of methyl esters of fatty acids (FAME). Due to the severe environmental effects of petroleum-fueled diesel engines and the mounting pressure on fast-depleting petroleum resources, biodiesel is one of the viable alternatives to conventional fuels and is steadily gaining prominence.[57] In comparison with petroleum diesel, biodiesel has a number of benefits, including a decrease in soot and sulphur emissions as well as a non-contributory effect on the atmospheric CO_2 level. Because of its high viscosity and appropriateness for use in modern diesel engines, it may also extend engine life.[58] Currently, the majority of feedstocks used in the manufacturing of biodiesel include soybean oil, cottonseed oil, vegetable oils, and animal fats. The use of non-food resources such as food waste is still in its infancy.[40]

Before isolating the lipid-free water, additional water from the mixture of food waste must be removed by heating after the lipid-containing particles have been removed using solvent extraction or centrifugation. The transesterification reaction of the extracted lipid (acid or basic catalytic reaction) and integrated microbial lipids (enzymatic lipase, bacterial, fungi, and microalgae conversion) is the crucial step in the efficient conversion of food waste into biodiesel. In this reaction, the nutrient in the food waste serves as a support system for microorganisms under well-controlled cultivation system growth before being subsequently converted into biodiesel.[59]

8.5.2.4 Acetone–Butanol–Ethanol (ABE)

Due to its distinct properties, including lower instability, reduced toxicity, good combustibility, low vapour pressure, and so on, biobutanol, also known as butyl alcohol, may be used as a transportation fuel by being blended with gasoline.[60] Due to its better qualities of having a larger energy content, blending with gasoline at high percentages, having a low affinity for water, and having lower volatility than ethanol, acetone–butanol–ethanol

(ABE) has attracted a lot of attention as an advanced liquid biofuel. It is anticipated that petrochemical products will be replaced by biobutanol from ABE fermentation when it comes to blending with diesel and gasoline fuels.[61] Corn starch and molasses were formerly used as substrates in the manufacture of ABE by common solventogenic *Clostridium* sp. However, the comparatively expensive cost of the substrate is one of the most significant barriers to commercial-scale ABE production.[62]

Acetone, butanol, and ethanol are produced in around a 3 : 6 : 1 ratio during ABE fermentation, with total product yields of about 20 $g L^{-1}$.[63] To increase the yield of sugar before the conversion process, acidic or enzymatic pretreatment is essential in the upstream phase, while through batch, fed-batch, continuous, and cell immobilised reactors, the strategy for a downstream ABE fermentation aims to increase the production and productivity of the solvent while concurrently lowering the cytotoxic effect on the cells. A successful study on the production of ABE incorporated steps aimed at a thorough use of both apple pomace soluble sugars and acid or alkali hydrolyzed sugars. The results demonstrated the superiority of using apple pomace's water-soluble sugars and hydrolyzed sugars in combination to produce ABE.[64]

8.6 Current Challenges Associated with Biofuel Production

Food waste is a low-value resource that can be used to produce liquid biofuels. The promise of advanced biofuels is dependent on technologies that are both economically and environmentally sound. Biofuel demand is rapidly increasing globally due to its environmentally sustainable and appealing benefits. Nonetheless, numerous challenges must be addressed. Because this is a new research area, a thorough understanding and discussion of various aspects of food waste will aid in overcoming the limitations.

Food loss cannot be overemphasized in food waste. In fact, it represents a loss of land, water, energy, and other resources. Poor food waste management has exaggerated serious issues affecting the economy, environment, and society over the last decade. As a result, it is clear that transitioning from a well-established fossil-based to a bioeconomy will be difficult.

8.6.1 Unorganized Industry and Waste Separation

Food waste collection remains difficult due to the unorganized nature of the industry. The general perception of food waste is that it should be discarded. Food waste is frequently mixed with other solid municipal wastes in many places.[31] A public awareness campaign is required to emphasize the importance of food waste. Food waste collection should be coordinated by urban planning and housing departments, as well as food industries. Volunteers should be encouraged to come every day to transport food waste to community collection facilities. In keeping with this, specialised personnel and transport equipment should be used to collect and deliver food waste to biorefineries. Such efforts will make the collection of food waste much faster and easier. For further processing and utilization, proper separation and sorting methods of food wastes from non-biological wastes are required. Because food waste is diverse and complex, the separation strategy may vary depending on the type of food waste.

8.6.2 Nonrenewable and Unconventional Resource

Food waste is non-renewable, and its makeup is mostly influenced by geography and dietary preferences.[40] Launching significant industries for the recycling of food waste will require a constant supply of enormous quantities of food waste. Maintaining a significant industry based on food waste in this situation is not practical. Therefore, large restaurants and food parks can be connected to small and medium biofuel producing facilities. Additionally, this will lower the cost of transporting the food wastes. Furthermore, its chemical makeup and water content must be identified before it can be used as a resource for the production of biofuels. In contrast to typical feedstocks, food waste is a complicated material. Therefore, it is important to have a proper characterization procedure in place to identify the various food waste types.

8.6.3 Techno-economic Challenges

The technological difficulties in each treatment train are still present. The cost of enzymes, process heat load, and energy requirements for distillation have been identified as the three main bottlenecks in the manufacturing of bioethanol. On the other hand, there are still a number of issues with the manufacturing of biodiesel that are related to the catalyst's nature (acidic or alkaline), which places restrictions on both the economy and the environment.[65] If bioethanol and biodiesel production are combined within a biorefinery that simultaneously values the produced residues, the latter issue might be resolved. This could increase the process efficiency and improve environmental performance. Furthermore, the primary downside of pretreatment techniques was that they led to the production of specific microbe inhibitors. Several strategies could be used to improve the tolerance of microbes to such inhibitors, including genetic and metabolic engineering, which is further divided into various aspects such as *in situ* detoxification, efflux pumps, stress responses, and membrane engineering. Genes for stress tolerance can also be screened for. Other than that, there are several undesired and dangerous elements found in biogas. The toxic and extremely corrosive gases affect heat and power plants and metal components.[66] The generation of biogas is impacted by H_2S because it releases toxic pollutants, corrodes the biogas purification system, and reduces its quantity and quality. Due to the contaminants in the biogas, it often requires pre-treatment to increase methanol yields and post-treatment to remove H_2S. These procedures cost money and require a lot of energy.

8.7 Future Prospectives of Biofuel Production from Agri-food Waste

Several recent studies have raised the possibility of using green energy produced from food waste to operate green vehicles in the near future.[67] These difficulties have spurred important developments in the study and production of fuels and energy produced from biomass. Biofuels are therefore anticipated to be most valuable in this regard to solve such issues in a very sustainable manner. Biofuels have been recognised as the most practical options for lowering carbon dioxide emissions in the transportation industry.[68] Additionally, biofuels can be simply made from locally accessible indigenous resources. The study also identified crucial elements that could be crucial for the sustainability evaluation of future green energy vehicle technologies with the use of food waste as an alternative resource to currently available fossil fuels.

Utilizing this low-value resource to create high-value goods should be the primary goal of future legislation. For instance, it is important to create sustainable valorization methods to produce biofuels such as biodiesel, bioethanol, and biooil. Such initiatives are helpful for a variety of reasons, including making the country energy independent and strengthening the local economy.

8.8 Conclusions

Food waste is a non-valuable and inedible resource. As a result, it can be used to build a biofuel manufacturing process that is both efficient and affordable. The development of novel food valorization techniques for the conversion of food waste into biofuels and value-added goods is the current focus of academic and industrial research. Around the world, there is a lot of research being done to increase the production of biofuels. Without a doubt, the research and business around biofuels are expanding quickly, but they cannot yet totally replace fossil fuels. Several combined engineering and biological techniques are still needed to optimise the production of biofuels on a commercial scale. Future research should concentrate on the viability of producing biofuels from food waste on a wide scale using economical, chemical, and biochemical processes. Furthermore, in order to create a sustainable biofuel economy, it is essential to comprehend how the production of biofuels will be impacted by future climate changes. Therefore, biofuels as a replacement for fossil fuels in the future will undoubtedly be a major energy supplier in a sustainable fashion with the ability to boost supply security. To develop more effective strategies for food waste management in achieving the biofuel targets, contribution to greenhouse gas emission reduction, and solid food waste minimization goals, it is essential to incorporate the related technology and proposed integrated process handling food waste.

Acknowledgements

The authors would like to acknowledge the support from the Fundamental Research Grant Scheme (FRGS) under grant number FRGS/1/2021/TK0/UN-IMAP/02/31 from the Ministry of Higher Education Malaysia.

References

1. S. S. Lam, Y. F. Tsang, A. T. Quitain, S. E. Hosseini and D. V. N. Vo, Editorial: Energy and Resource Valorization of Biomass and Waste Toward Sustainable Environment via Thermochemical and Biological Application, *Front. Energy Res.*, 2021, **4**(1), 23–28.
2. S. Isah and G. Ozbay, Valorization of Food Loss and Wastes: Feedstocks for Biofuels and Valuable Chemicals, *Front. Sustainable Food Syst.*, 2020, **4**(6), 4–10.
3. E. Imbert, Food waste valorization options: Opportunities from the bioeconomy, *Open Agric.*, 2017, **2**(1), 195–204.
4. S. Maina, V. Kachrimanidou and A. Koutinas, A roadmap towards a circular and sustainable bioeconomy through waste valorization, *Curr. Opin. Green Sustainable Chem.*, 2017, **8**(1), 18–23.
5. S. Otles, S. Despoudi, S. Bucatariu and C. Kartal, Food waste management, valorization, and sustainability in the food industry, in *Food Waste Recovery: Processing Technologies and Industrial Techniques*, 2015, ch. 1, pp. 12–20.
6. K. L. Chin, P. S. H'ng, E. W. Chai, P. S. Khoo, C. L. Lee and W. Z. Go, Valorization of Lignocellulosic Food Industry Waste in Malaysia by Accelerated Co-composting Method: Changes in Physicochemical and Microbial Community, *Waste Biomass Valorization*, 2020, **11**(9), 4871–4884.
7. S. K. Karmee and C. S. K. Lin, Valorisation of food waste to biofuel: current trends and technological challenges, *Theor. Chem. Acc.*, 2014, **2**(1), 4–7.
8. X. Turon, J. Venus, M. Arshadi, M. Koutinas, C. S. K. Lin and A. Koutinas, Food waste and byproduct valorization through bio-processing: Opportunities and challenges, *BioResources*, 2014, **9**(4), 5774–5777.
9. P. Gosavi, Y. Chaudhary and A. Durve-Gupta, Production of biofuel from fruits and vegetable wastes, *Eur. J. Biotechnol. Biosci.*, 2017, **5**(3), 2321–9122.
10. A. Tropea, Food Waste Valorization, *Fermentation*, 2022, **8**(4), 12–24.
11. Sahar, S. Sadaf, J. Iqbal, I. Ullah, H. N. Bhatti, S. Nouren and M. Iqbal, Biodiesel production from waste cooking oil: An efficient technique to convert waste into biodiesel, *Sustainable Cities Soc.*, 2018, **41**(5), 220–226.
12. S. Rahmati, W. Doherty, D. Dubal, L. Atanda, L. Moghaddam, P. Sonar and K. Ostrikov, Pretreatment and fermentation of lignocellulosic

biomass: Reaction mechanisms and process engineering, *Reaction Chem. Eng.*, 2020, **5**(11), 2017–2047.

13. P. Parthasarathy, M. Alherbawi, M. Shahbaz, H. R. Mackey and G. Mckay, Conversion of oil palm waste into value-added products through pyrolysis: a sensitivity and techno-economic investigation United States dollars, *Biomass Conversion Biorefinery*, 2022, **3**(2), 23–35.
14. J. S. Teh, Y. H. Teoh, H. G. How, T. D. Le, Y. J. J. Jason, H. T. Nguyen and D. L. Loo, The potential of sustainable biomass producer gas as a waste-to-energy alternative in Malaysia, *Sustainability*, 2021, **13**(2), 20–34.
15. K. L. Ong, G. Kaur, N. Pensupa, K. Uisan and C. S. K. Lin, Trends in food waste valorization for the production of chemicals, materials and fuels: Case study South and Southeast Asia, *Bioresour. Technol.*, 2018, **2**(8), 100–112.
16. M. A. Hassan, M. A. Ahmad Farid, Y. Shirai, H. Ariffin, M. R. Othman, M. H. Samsudin and M. Y. Hasan, Oil Palm Biomass Biorefinery for Sustainable Production of Renewable Materials, *Biotechnol. J.*, 2019, **14**(6), 1–8.
17. V. Melini, F. Melini, F. Luziatelli and M. Ruzzi, Functional ingredients from agri-food waste: Effect of inclusion thereof on phenolic compound content and bioaccessibility in bakery products, *Antioxidants*, 2020, **9**(12), 1–29.
18. J. Hadj Saadoun, G. Bertani, A. Levante, F. Vezzosi, A. Ricci, V. Bernini and C. Lazzi, Fermentation of agri-food waste: A promising route for the production of aroma compounds, *Foods*, 2021, **10**(4), 8–16.
19. L. M. Terry, C. Li, J. J. Chew, A. Aqsha, B. S. How, A. C. M. Loy and J. Sunarso, Bio-oil production from pyrolysis of oil palm biomass and the upgrading technologies: A review, *Carbon Resour. Convers.*, 2021, **4**(10), 239–250.
20. E. Carota, M. Petruccioli, A. D'Annibale, A. M. Gallo and S. Crognale, Orange peel waste-based liquid medium for biodiesel production by oleaginous yeasts, *Appl. Microbiol. Biotechnol.*, 2020, **104**(10), 4617–4628.
21. M. D. Kumar, S. Kavitha, and J. R. Banu, Valorization of food waste for biodiesel production, in *Food Waste to Valuable Resources: Applications and Management*, 2020, ch. 4, pp. 8–20.
22. M. Moneruzzaman Khandaker, U. Aliyu Abdullahi, M. Dogara Abdulrahman, N. Afiza Badaluddin and K. Suryati Mohd, Bio-Ethanol Production from Fruit and Vegetable Waste by Using *Saccharomyces cerevisiae*, *Bioethanol Technol.*, 2021, **4**(11), 44–56.
23. M. Hijosa-Valsero, A. I. Paniagua-García and R. Díez-Antolínez, Biobutanol production from apple pomace: the importance of pretreatment methods on the fermentability of lignocellulosic agro-food wastes, *Appl. Microbiol. Biotechnol.*, 2017, **101**(21), 8041–8052.
24. A. Friman and N. Hyytiä, The Economic and Welfare Effects of Food Waste Reduction on a Food-Production-Driven Rural Region, *Sustainability*, 2022, **14**(6), 4–12.

25. H. Kahiluoto, M. Kuisma, J. Havukainen, M. Luoranen, P. Karttunen, E. Lehtonen and M. Horttanainen, Potential of agrifood wastes in mitigation of climate change and eutrophication – Two case regions, *Biomass Bioenergy*, 2011, **35**(5), 1983–1994.
26. *Responsible consumption and production*, ed. W. Leal Filho, A. M. Azul, L. Brandli, P. G. Özuyar and T. Wall, Springer International Publishing, Cham, 2020.
27. A. Seberini, Economic, social and environmental world impacts of food waste on society and Zero waste as a global approach to their elimination, *SHS Web Conf.*, 2020, 7(4), 3–10.
28. K. S. Woon, I. M. C. Lo, S. L. H. Chiu and D. Y. S. Yan, Environmental assessment of food waste valorization in producing biogas for various types of energy use based on LCA approach, *Waste Manage.*, 2016, **5**(2), 290–299.
29. E. Lopez Barrera and T. Hertel, Global food waste across the income spectrum: Implications for food prices, production and resource use, *Food Policy*, 2021, **9**(8), 8–23.
30. C. Torres-León, N. Ramírez-Guzman, L. Londoño-Hernandez, G. A. Martinez-Medina, R. Díaz-Herrera, V. Navarro-Macias and C. N. Aguilar, Food Waste and Byproducts: An Opportunity to Minimize Malnutrition and Hunger in Developing Countries, *Front. Sustainable Food Syst.*, 2018, **2**(9), 1–17.
31. A. A. Dhim, A. A. Kadir, M. H. Ibrahim, S. Halim, N. A. Sarani, M. I. H. Hassan, N. J. A. Hamid, N. N. H. Hashar and N. F. N. Hissham, Overview on food waste management and composting practice in Malaysia, *AIP Conf. Proc.*, 2021, 2339.
32. W. J. Lim, N. L. Chin, A. Y. Yusof, A. Yahya and T. P. Tee, Food waste handling in Malaysia and comparison with other Asian countries, *Int. Food Res. J.*, 2016, 1–6.
33. S. W. A. Ghafar, Food Waste in Malaysia: Trends, Current Practices and Key Challenges, *FFTC Agricultural Policy Articles*, 2017, https://ap.fftc.org.tw/article/1196.
34. N. B. D. Thi, G. Kumar and C. Y. Lin, An overview of food waste management in developing countries: Current status and future perspective, *J. Environ. Manage.*, 2015, **157**, 220–229.
35. D. Cudjoe and M. S. Han, Economic feasibility and environmental impact analysis of landfill gas to energy technology in African urban areas, *J. Cleaner Prod.*, 2021, **284**, 125437.
36. V. Arumugam, M. H. Ismail, T. A. Puspadaran, W. Routray, N. Ngadisih, J. N. Wahyu Karyadi, B. Suwignyo and H. Suryatmojo, Food Waste Treatment Methods and its Effects on the Growth Quality of Plants: A Review, *Pertanika J. Trop. Agric. Sci.*, 2022, **45**(1), 75–101.
37. P. Kumaran, D. Hephzibah, R. Sivasankari, N. Saifuddin and A. H. Shamsuddin, A review on industrial scale anaerobic digestion systems deployment in Malaysia: Opportunities and challenges, *Renewable Sustainable Energy Rev.*, 2016, **56**, 929–940.

38. C. Negri, M. Ricci, M. Zilio, G. D'Imporzano, W. Qiao, R. Dong and F. Adani, Anaerobic digestion of food waste for bio-energy production in China and Southeast Asia: A review, *Renewable Sustainable Energy Rev.*, 2020, **133**, 110138.
39. K. Murugesan, K. R. Srinivasan, K. Paramasivam, A. Selvam and J. Wong, Conversion of Food Waste to Animal Feeds, *Current Developments in Biotechnology and Bioengineering: Sustainable Food Waste Management: Resource Recovery and Treatment*, 2021, pp. 305–324.
40. S. Dhiman and G. Mukherjee, Present scenario and future scope of food waste to biofuel production, *J. Food Process Eng.*, 2021, **44**(2), 94–135.
41. K. L. Ong, G. Kaur, N. Pensupa, K. Uisan and C. S. K. Lin, Trends in food waste valorization for the production of chemicals, materials and fuels: Case study South and Southeast Asia, *Bioresour. Technol.*, 2018, **248,** 100–112.
42. H. C. Ong, W. H. Chen, Y. Singh, Y. Y. Gan, C. Y. Chen and P. L. Show, A state-of-the-art review on thermochemical conversion of biomass for biofuel production: A TG-FTIR approach, *Energy Convers. Manage.*, 2020, **209**, 112634.
43. K. Jones and E. U. Agrocycle, Agri-food waste valorisation, *Food Sci. Technol.*, 2019, **33**(4), 60–63.
44. F. O. Kassim, C. L. P. Thomas and O. O. D. Afolabi, Integrated conversion technologies for sustainable agri-food waste valorization: A critical review, *Biomass Bioenergy*, 2022, **156**, 106314.
45. B. R. Patra, S. Nanda, A. K. Dalai and V. Meda, Slow pyrolysis of agro-food wastes and physicochemical characterization of biofuel products, *Chemosphere*, 2021, **285**, 131431.
46. S. Elkhalifa, T. Al-Ansari, H. R. Mackey and G. Mckay, Food waste to biochars through pyrolysis: A review, *Resour., Conserv. Recycl.*, 2019, **144**, 310–320.
47. F. H. Pour and Y. T. Makkawi, A review of post-consumption food waste management and its potentials for biofuel production, *Energy Rep.*, 2021, **7**, 7759–7784.
48. V. S. Sikarwar, M. Zhao, P. S. Fennell, N. Shah and E. J. Anthony, Progress in biofuel production from gasification, *Prog. Energy Combust. Sci.*, 2017, **61**, 189–248.
49. D. Bolzonella, F. Battista, C. Cavinato, M. Gottardo, F. Micolucci, G. Lyberatos and P. Pavan, Recent developments in biohythane production from household food wastes: A review, *Bioresour. Technol.*, 2018, **257**, 311–319.
50. G. K. Dinesh, R. Chauhan and S. Chakma, Influence and strategies for enhanced biohydrogen production from food waste, *Renewable Sustainable Energy Rev.*, 2018, **92**, 807–822.
51. A. Akhbari and S. Ibrahim, Bioenergy recovery from food waste through dark fermentation direction, 2022, DOI: 10.21203/rs.3.rs-1965745/v1.
52. H. Guo, Y. Zhao, A. Damgaard, Q. Wang, H. Wang, T. H. Christensen and W. Lu, Quantifying global warming potential of alternative biorefinery

systems for producing fuels from Chinese food waste, *Waste Manage.*, 2021, **130**, 38–47.

53. B. David, B. Federico, C. Cristina, G. Marco, M. Federico and P. Paolo, Biohythane Production From Food Wastes, *Biohydrogen*, 2019, 347–368.
54. R. A. A. Meena, J. Rajesh Banu, R. Yukesh Kannah, K. N. Yogalakshmi and G. Kumar, Biohythane production from food processing wastes – Challenges and perspectives, *Bioresour. Technol.*, 2020, **298**, 122449.
55. S. Sinha and P. Tripathi, Trends and challenges in valorisation of food waste in developing economies: A case study of India, *Case Stud. Chem. Environ. Eng.*, 2021, **4**, 100162.
56. H. Kazemi Shariat Panahi, M. Dehhaghi, G. J. Guillemin, V. K. Gupta, S. S. Lam, M. Aghbashlo and M. Tabatabaei, Bioethanol production from food wastes rich in carbohydrates, *Curr. Opin. Food Sci.*, 2022, **43**, 71–81.
57. Y. H. Tan, M. O. Abdullah, J. Kansedo, N. M. Mubarak, Y. S. Chan and C. Nolasco-Hipolito, Biodiesel production from used cooking oil using green solid catalyst derived from calcined fusion waste chicken and fish bones, *Renewable Energy*, 2019, **139**, 696–706.
58. A. Syafiuddin, J. H. Chong, A. Yuniarto and T. Hadibarata, The current scenario and challenges of biodiesel production in Asian countries: A review, *Bioresource Technol. Rep.*, 2020, **12**, 100608.
59. M. Carmona-Cabello, I. L. García, A. Papadaki, E. Tsouko, A. Koutinas and M. P. Dorado, Biodiesel production using microbial lipids derived from food waste discarded by catering services, *Bioresour. Technol.*, 2021, **323**, 124597.
60. I. Veza, M. F. Muhamad Said and Z. A. Latiff, Recent advances in butanol production by acetone-butanol-ethanol (ABE) fermentation, *Biomass Bioenergy*, 2021, **144**, 105919.
61. N. E. Poe, D. Yu, Q. Jin, M. A. Ponder, A. C. Stewart, J. A. Ogejo, H. Wang and H. Huang, Compositional variability of food wastes and its effects on acetone-butanol-ethanol fermentation, *Waste Manage.*, 2020, **107**, 150–158.
62. P. K. Sarangi and S. Nanda, Recent developments and challenges of acetone-butanol-ethanol fermentation, *Recent Advancements in Biofuels and Bioenergy Utilization*, 2018, pp. 111–123.
63. C. Zhang, Z. Ling and S. Huo, Anaerobic fermentation of pretreated food waste for butanol production by co-cultures assisted with in-situ extraction, *Bioresource Technol. Rep.*, 2021, **16**, 100852.
64. Q. Jin, N. Qureshi, H. Wang and H. Huang, Acetone-butanol-ethanol (ABE) fermentation of soluble and hydrolyzed sugars in apple pomace by Clostridium beijerinckii P260, *Fuel*, 2019, **244**, 536–544.
65. E. M. Barampouti, S. Mai, D. Malamis, K. Moustakas and M. Loizidou, Liquid biofuels from the organic fraction of municipal solid waste: A review, *Renewable Sustainable Energy Rev.*, 2019, **110**, 298–314.

66. S. Dutta, M. He, X. Xiong and D. C. W. Tsang, Sustainable management and recycling of food waste anaerobic digestate: A review, *Bioresour. Technol.*, 2021, **341**, 125915.
67. D. Kour, K. L. Rana, N. Yadav, A. N. Yadav, A. A. Rastegari, C. Singh, P. Negi, K. Singh and A. K. Saxena, Technologies for Biofuel Production: Current Development, Challenges, and Future Prospects, in *Prospects of Renewable Bioprocessing in Future Energy Systems*, Springer International Publishing, 2019, pp. 32–49.
68. N. B. D. Thi, C. Y. Lin and G. Kumar, Waste-to-wealth for valorization of food waste to hydrogen and methane towards creating a sustainable ideal source of bioenergy, *J. Clean. Prod.*, 2016, 122.

CHAPTER 9

Valorization of Agricultural Residues Generated from Corn/Maize: Acquiring Valuables from Waste

TALAT ILYAS,[a] PRATYUSH TRIPATHI,[a] VAGMI SINGH[b] AND HARSH VARDHAN SINGH*[a]

[a] Plant-Microbe Interaction and Rhizosphere Biology Lab, ICAR-National Bureau of Agriculturally Important Microorganisms, Kushmaur, Maunath Bhanjan-275 103, India; [b] Seed Quality Laboratory, Genetics & Plant Breeding Division, CSIR-Central Institute of Medicinal and Aromatic Plants, Lucknow, 226015, India

*Email: drharsh2006@rediffmail.com

9.1 Introduction

Agricultural productivity has escalated considerably over the previous five decades, reaching approximately about 23.7 million tonnes of sustenance daily, and will continue to increase in the times ahead to meet the requirements of the expanding global population.[1,2] According to Leisner, it has been anticipated that by 2050, the population of the world will be about 9.1 billion people. Therefore, an increase in the production of agricultural wastes is predictable as agricultural production rises.[3] It has been estimated that the worldwide waste generation from several agricultural sources is expected to be around 140 billion metric tonnes annually since the production of agricultural solid waste has been increasing at a rate of 7.5% per year.[4,5] Agricultural waste

Green Chemistry Series No. 78
Agri-food Waste Valorisation
Edited by Pankaj Chowdhary and Abhay Raj

Published by the Royal Society of Chemistry, www.rsc.org

is frequently burnt and dumped in exposed places in developing countries, resulting in pollution of the environment and considerable acceleration of global warming.[6] Therefore, it is necessary to develop novel approaches to manage agro-wastes in such a way that they should not cause ecological inconvenience but can also be generated into valuable products through valorization in the utmost environmentally friendly way. The process of turning waste into affluence focuses on the value-added conversion of agricultural wastes into bioproducts, which are favorable owing to their eco-friendliness and benefaction to the economies. El-Ramady states that crop leftovers from harvests such as maize, rice, wheat, and sugarcane consist of cobs, stems, husks, bagasse, shells, *etc.* as possible sources of agricultural wastes. These huge wastes are typically burned off or placed on farms to decompose naturally. The pollutants from agro-industrial processes are also frequently discharged into the environment without being further processed.[7] All these methods have an adverse effect on the environment in different ways, including tainting of water and air by burning, spreading of infections, and the formation of rodents' and insects' habitat, along with the deterioration of the region's natural surroundings. To solve these difficulties connected to the improper disposal of agricultural wastes, an innovative scheme for utilizing them is crucial for environmental sustainability. These massive wastes could be processed into bioproducts that can be utilized to produce paper, bioethanol, biogas, for the generation of electricity, compost, and organic fertilizers, for the removal of heavy metals, added as an addition to cement or can be valorized by microorganisms for the production of enzymes and feed additives (Figure 9.1).

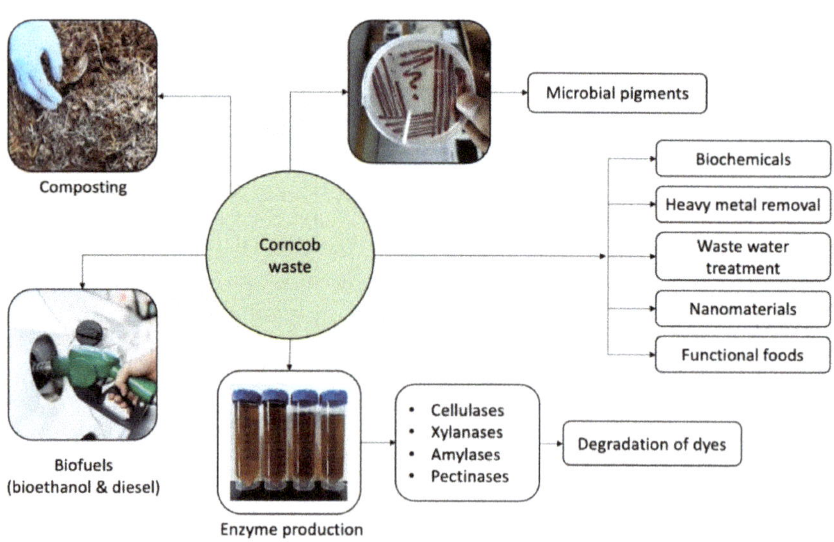

Figure 9.1 Various strategies for corncob waste management.

The potential benefits of these wastes have not been realized in the majority of developing countries due to the lack of technology to turn them into valuable products. One of the crops that produce a substantial amount of waste is corn (*Zea mays*), generating corncobs, maize chaff, and other plant biomass, throughout the chain of its processing. About 200 million tonnes of cobs are produced as a result of corn cultivation each year; however, they are not significantly used on a wide scale.

9.2 Corncob as a Source of Agricultural Waste

Maize (*Zea mays*), commonly known as corn, is the staple crop with the highest global production. Currently, nearly 1147.7 million MT of maize is being produced together by over 170 countries from an area of 193.7 million hectares with an average productivity of 5.75 t per ha. India ranks fourth in area and seventh in yield among nations that cultivate maize, accounting for around 4% of the global maize area and 2% of the total production.[8] In India, the maize area reached 9.2 million hectares in 2018–19.

Due to the large production of maize, the producer nations now confront a significant environmental concern from the leftover residues. Corncob, a corn by-product, is the central portion of maize, where the grains develop. In the production of maize, corn grains are detached from corncobs either mechanically or manually. The corncob, a portion that is left over after the grains have been extracted, has been labeled as trash since its economic worth has not been fully appreciated. According to Li and Chen, maize cultivation may result in the production of up to 203 million tonnes of garbage annually.[9] Maize has been listed as one of the key crops creating waste by Prasad, with its cobs and leaves being the main wastes generated.[10] The most prevalent practices for getting rid of agricultural waste in many developing nations are dumping in landfills or burning on fields after harvesting. Burning may produce significant amounts of greenhouse gases, air pollution, and other particulates, whereas landfilling encourages the growth of diseases, pollutes groundwater through leachate seepage, and pollutes the air through the discharge of unpleasant odors. The growing production of agricultural wastes is responsible for several serious environmental issues, such as pathogen proliferation, groundwater contamination, and greenhouse gas emissions.[11,12]

Corncobs can be recycled by composting to produce organic compost or by microbial fermentation to make high-end eco-friendly products. Corncob makes up about 27–30% of maize trash. It is a lignocellulosic substance that contains a significant amount of cellulose (69.2%), hemicellulose (22.8%), and lignin (8%) respectively. In many industrialized nations, corn residues are used as raw materials for the production of energy or in the creation of ceramic materials.[13] It might be used as a feedstuff or as a feed substrate supplement for animals,[14] as well as a profound source of carbon for the progression of several bacteria and fungi, for the development of essential enzymes having pharmaceutical and nutraceutical values. According to a study conducted by Seesuriyachan, corncob could be used in a variety of industrial processes since it is a relatively cheap and easily accessible biomass. Therefore, there is promising potential for biotechnologically valuing corncob, which is growing by the day. Within the scope of the circular and bioeconomy, corncob, a prominent agricultural waste, is exploited as a substrate to manufacture high-end goods by microbial valorization.[15]

9.3 Valorization of Corncob

Microbes utilize an extensive range of substrates for growth by the formation of enzymes that help in breaking down complex substrates into simpler forms that can be ingested by them. This is made possible by the metabolic diversity of microbes. Nowadays, new metabolic engineering methods are being used to improve microbial metabolism so they can use a greater variety of complicated and unique substrates to create diverse bioproducts.[16] Numerous studies have been documented stating bacteria have been used in corncob for creating a variety of bioproducts (Table 9.1).

9.3.1 Enzyme Production

Fermentation is a desirable method that can be used to value corncob and produce valuable compounds such as enzymes. Through the use of biological agents, the process of fermentation inclines to purify and increase the nutritional value of corncobs.[17] Submerged fermentation (SmF) and solid-state fermentation are the two possible forms (SSF). However, SSF has a higher potential than SmF for producing enzymes from corncob residues. The detoxification and nutritional improvement of corncobs, particularly when filamentous fungi are present, are greatly aided by SSF, which also increases the output of enzymes. Furthermore, the physiochemical characteristics of corncob residues inherit the tendency to promote solid-phase cultivation for the complete implementation of the solid-state fermentation technique.[18] Enzymes are macromolecules that support significant biochemical processes in living things along a clearly defined route.[19] According to Singh, enzymes are produced by all living things, but due to their high manufacturing yield, simplicity in genetic modification, affordability, durability, improved activity over a broad range of temperatures and pH, and environmental friendliness, microbial enzymes are favored for research in industrial applications.[20] Microbial enzymes are being used by several industries, such as those in pharmaceutical, waste management, food and animal feed, detergent, leather, and textile sectors.[21,22]

Industrial microbial enzymes are sold in a highly competitive worldwide market. Industrial enzymes had a market value of USD 5.6 billion in 2019 and were expected to increase 6.4% the annual growth rate from 2020 to 2027. Microbial enzymes have a wide range of uses, but their usage is restricted because of their high manufacturing costs, which result in an increased importation rate, particularly in poor nations. However, research has

Table 9.1 Production of value-added byproducts from corncob using different microorganisms.

S. no.	Value-added products	Microorganisms used	Substrate	Mechanism/summary	Ref.
1.	Corncob, corncob nanocellulose, and functionalized corncob	—	Corncob	Methyl orange removal	5
2.	Xylanase	*Penicillium purpurogenum*	Corncob	Solid state fermentation	38
3.	Xylanase	*Penicillium purpurogenum*	Corncob	—	32
4.	Pectinase	*Penicillium purpurogenum*	Corncob	Submerged fermentation	32
5.	Xylanase	*Aspergillus niger* and *Trichoderma longibrachiatum*	Corncob	Both submerged and solid-state fermentation	41
6.	Pectinase	*Geotrichum candidum*	Corncob	Submerged fermentation	46
7.	Amylase	*Aspergillus niger*	Corncob	Solid state fermentation	44
8.	Cellulase	*Aspergillus niger*	Corncob	Usage of basal media	33
9.	Pectinase	*Aspergillus niger*	Corncob	Submerged fermentation	33
10.	Cellulase	*Trichoderma reesei*	Corncob	Submerged fermentation	31
11.	Bioethanol	*Coprinopsis cinerea* and *Trametes hirsuta*	Corncob	An ethanol concentration of 8.02 $g\,L^{-1}$ is obtained	53
12.	Bioethanol	*Candida glabrata*	Corncob	Ethanol concentrations of 31.32 and 21.92 $g\,L^{-1}$ are obtained	50
13.	Bioethanol	*Candida tropicalis*	Corncob	An ethanol concentration of 25.3 $g\,L^{-1}$ is obtained	52
14.	Bioethanol	*Saccharomyces cerevisiae*	Corncob	Fed batch fermentation; ethanol yields of 69.2 and 84.7 $g\,L^{-1}$ are obtained	49
15.	Biodiesel	*Aspergillus awamori*	Corncob	33.3% lipid productivity is obtained	54
16.	Monascus pigment	*Talaromyces atroroseus*	Corncob	Pigment production was higher than the control	59
17.	Mycelium pigment	*Neurospora* sp.	Corncob	Pigment absorbance was higher (0.64)	57
18.	Natural pigment	*Monascus purpureus*	Corncob	Yield of 25.42 OD U per g pigment is obtained	58
19.	D-Lactic acid	*Sporolactobacillus inulinus*	Corncob	Fed batch fermentation; yield of 0.85 $g\,g^{-1}$ glucose is obtained	62
20.	Lactic acid	*Bacillus* sp.	Corncob	Fed batch fermentation; a concentration of 74.7 $g\,L^{-1}$ is obtained	61

Table 9.1 (*Continued*)

S. no.	Value-added products	Microorganisms used	Substrate	Mechanism/summary	Ref.
21.	Lactic acid	*Rhizopus oryzae*	Corncob	Yield of 772 $g\,kg^{-1}$ lactic acid is obtained	62
22.	Lactic acid	*Acremonium thermophilus*	Corncob	Yield of more than 25 $g\,L^{-1}$ lactic acid	60
23.	Lactic acid	*A. thermophilus* and *Rhizopus* sp.	Corncob	Yield of approximately 24 $g\,L^{-1}$ is obtained	60
24.	Neutral red and methylene blue	*Pleurotus ostreatus*	Corncob	—	64
25.	Brilliant green dye	*Brevibacillus parabrevis*	Corncob biochar		65
26.	Pulp and paper mill wastewater	White rot fungi	Corncob	COD and removal of color were best at pH 4.5–5	69
27.	Biosorption of lead and copper ions	Natural fungi growing on corncob	Corncob	Biosorption of Pb and Cu was 4.29 and 1.76 $mg\,g^{-1}$, respectively	70
28.	Cellulose-based nanoparticles (CPNs)	—	Corncob	Drug delivery system	71
29.	Silver nanoparticles	—	Corncob (fungal xylanase from corncob)	Dye degradation and anticoagulant and antioxidant activities	72
30.	Gold nanoparticles	—	Corncob (fungal xylanase)	Antimicrobial, antioxidant, and thrombolytic activities	72
31.	Gold–silver alloy nanoparticles	—	Corncob	Antimicrobial, antioxidant, and thrombolytic activities	72
32.	Silica/corncob nanocomposites	—	Corncob	Methylene blue removal	76
33.	Corncob silica supported nanoscale zero-valent iron	—	Corncob	Hexavalent chromium removal	76
34.	Silica nanoparticles	—	Corncob	Arsenic contamination removal	75
35.	Corncob biochar-based nanocomposites	—	Corncob	Improved growth and yield of plants	74

focused on using inexpensive substrates that would enable cost-effective manufacturing. According to Kanengoni, corncobs are a very nutrient-dense lignocellulosic material that promotes the development of microorganisms and provides an affordable substrate for the synthesis of enzymes.[23] A pre-treatment is required to improve the use of sugar and enzyme accessibility before it is used as the substrate for fermentation or other biotechnological processes. Grinding, steam explosion, hydrothermal treatment, sodium hydroxide usage, and ultra-sonication are a few pre-treatment methods.[24–26] These techniques can be physical/mechanical, chemical, and biological.[27]

9.3.1.1 Cellulases

Cellulases are economically significant because they are the second-largest industrial enzymes and have several applications in the food, beverages, wine, textile, laundry and detergent, agricultural, animal feed, paper, cosmetics, and environmental management sectors.[28,29] Cellulases are in charge of catalyzing processes in which cellulose is hydrolyzed into glucose and cello-oligosaccharides. Despite the enormous potential uses of cellulase, one of its main drawbacks is the high cost of commercial cellulose, which drives up the production costs.[30] As a result, there is increasing interest in using lignocellulosic materials, which are abundant and inexpensive in nature, to replace cellulose. As a lignocellulosic material, corncob can be a great source of cellulose. *Trichoderma reesei* ZU-02 was found to produce cellulase from corncob in a study by Liming and Xueliang.[31] Without any prior treatment, milled and sieved corncobs yielded a significant output of cellulase. In another study, cellulase production from *A. niger* isolated from corncob waste was assessed.[32,33] *Penicillium purpurogenum* has also been shown to produce cellulase from corncobs. In 2020, a study stated the isolation and identification of the thermophilic bacteria *Bacillus aerius* from the Indian thermal springs with the aim of manufacturing a thermostable cellulase, which was subsequently used for corncob saccharification after it had been de-lignified using a combined approach of hydrodynamic cavitation reaction paired with laccase generated by an agitating laccase-producing fungus.[34] According to Soeka,[35] *Bacillus subtilis* A8 produced cellulase by utilizing corncob at pH 7.0 and a temperature of 50 °C resulting in an activity of 7.0 U ml^{-1}. According to a study by Ire,[36] among nine fungal isolates, one strain of *Penicillium* sp. produced a high concentration of cellulase when cultivated on a corncob. Santos reported *Penicillium* sp. FSDE15 produced cellulases utilizing corncob and wheat bran. After 216 hours of fermentation using a mixture of 50% wheat bran and 50% corncob as substrates, cellulase and FPase activities of 21.11 and 1.29 U g^{-1} were achieved respectively.[37]

9.3.1.2 Xylanases

Due to the numerous commercial uses of xylanases, research in recent years has focused on how to produce them efficiently on a large scale while reducing the cost of the enzymes by using high-yielding microbes and inexpensive substrates. The hemicellulolytic portion of lignocellulose is broken down by a group of enzymes known as xylanases (EC 3.2.1.8), which results in the production of xylose.[38] They have been discovered to be pertinent in several industrial applications, including the production of bread, wine, textiles, livestock, juice extraction, biorefineries, and paper.[39] Because of the high cost of the purified xylan utilized as the substrate for the commercial manufacturing of xylanases, the production is unprofitable. However, xylan, which is made up of 1,4-linked D-xylopyranosyl residues, is a crucial component of hemicellulose. Corncobs have the largest concentration of xylan (around 40%) of all the lignocellulosic agro-wastes, making them a good substitute substrate for xylanase synthesis.[40] Some fungal isolates used corncob for the production of xylanases by submerged and solid-state fermentation. Corncob was pulverized and sieved before being degraded by the fungal isolates for xylanase production, improving the juice clarification and rising of dough in bakery applications, demonstrating their significance in the food industry.[41] According to Sunkar, *Penicillium purpurogenum* produced xylanase from corncob in SSF at a rate of 1097 6.76 U g^{-1} under ideal circumstances created using the Taguchi approach.[38] Additionally, *Aspergillus* sp. such as *A. niger, A. flavus*, and *A. fumigatus* isolated from soil were reported by Ire *et al.* to produce xylanase on corncobs.[42] The fungal isolates were modified using UV and based on the quantity and quality of the enzymes produced in SmF; *Aspergillus niger* was selected as the top xylanase-producing strain. Using ammonium sulfate precipitation, xylanases from two strains were partially purified.

9.3.1.3 Amylases

Amylase is a key industrial enzyme used in the pharmaceutical, textile, food, detergent, and paper industries.[43] Alpha-amylases break down the α-1,4-glycosidic linkages in polysaccharides to produce oligosaccharides with a variety of lengths. As a result, the creation of amylases is required for the conversion of starch to oligosaccharides.[18] One of the low-cost lignocellulolytic materials that have been described as α-amylase production substrates is corncobs. The competence of *A. niger* in generating α-amylase using various types of agro-waste as substrates was evaluated by Aliyah, in SSF.[44] The substrates used were sugarcane bagasse, rice husk, and maize cob. The substrate with the highest amylase activity was corncob.

9.3.1.4 Pectinases

Pectinases are crucial for a large number of industrial processes, including oil extraction from seeds, the production of pigments, fruit and vegetable processing, fermentation of coffee, tea, and cocoa, bleaching of paper, and the treatment of wastewater.[45] Utilizing corncob as a substrate, a variety of bacteria, filamentous fungi, and yeasts have been used to produce pectinases. In a study by Ejaz, pectinase synthesis by immobilized and free *Geotrichum candidum* resulted in 0.554 and 0.215 IU ml^{-1}, respectively.[46] Pectinase production on corncobs by *Penicillium purpurogenum* was also reported by Mardones.[32]

9.3.2 Generation of Bioenergies

The issues posed by the rising energy demand in emerging nations have made it necessary to look for ecological alternatives of raw materials that may be used to satisfy such demands. Renewable energy sources are receiving more attention as a result of the rising global energy demand, diminishing fossil fuel supplies, and worries about climate change. These have sparked interest in the field of renewable energy research, a field in which bioresources have been thoroughly investigated.

9.3.2.1 Bioethanol

To battle the energy crisis and address environmental issues, lignocellulose, a by-product derived from several agricultural operations, remains a feasible substrate for bioconversion to bioethanol.[47] A large amount of agricultural industrial waste produced results in the abundance of fertile land and has served as inspiration for finding solutions to energy problems. Corncob is recognized as a favorable substrate for the generation of bioenergy because of its high cellulosic and hemicellulosic amounts, wide availability, and low cost. The implication of ethanol as a harmless and healthy conveyance fuel is considered a response to the depletion of air pollution as well as fuel reserves. In recent years, the conversion of lignocellulosic materials into glucose and other fermentable sugars has been viewed as an effective technique for the production of ethanol.[48] Cellulases and other hydrolytic enzymes enzymatically hydrolyze cellulose to release sugars (saccharification), which may then be processed to produce ethanol from lignocellulosic materials such as corncobs. Commercial production of ethanol from lignocellulosic materials must produce more than 4% (v/v) of ethanol in the fermentation broth to be financially sustainable.[49] According to Boonchuay,[50] the desiccated compact portion of cellulose-rich corncob was produced after the synthesis of xylooligosaccharides employed as a substrate for simultaneous saccharification and fermentation (SSF) and separate hydrolysis and fermentation (SHF) of cellulose-rich corncob to create bioethanol using thermotolerant *Candida glabrata* KY618709. In a study conducted by Chen, corncob was effectively hydrolyzed with *T. reesei* and *A. niger* cellulases.[51]

The cellulosic hydrolysate in corncob fermented with *Saccharomyces cerevisiae* results in the production of 45.7 g L^{-1} of ethanol in 18 hours. Another study by Cheng, reported the bioethanol production from non-detoxified and aerobically fermented corncobs through fermentation and saccharification using *Candida tropicalis*.[52] Using the fed-batch SSF process, ethanol may be made from high-dry matter corncobs, according to Zhang.[49] The effects of neutral fungal laccase (pH 6.5) from *Coprinopsis cinerea* and acidic fungal laccase rLacA (pH 4.0) from *Trametes hirsuta* as controls on the generation of bioethanol from alkaline fungal laccase were evaluated by Liu.[53]

9.3.2.2 Biodiesel

With the use of corncob waste liquor and direct and indirect transesterification techniques, Subhash and Mohan explored the biocatalytic potential of *Aspergillus* sp. for the generation of biodiesel.[54] The fuel qualities of the created biodiesel were comparatively good, with a density of 0.8342 $g\,cm^{-3}$ among others. Most recently, *Rhodotorula taiwanensis* AM2352 was utilized by Miao to ferment corncob hydrolysate and create lipid, which was then used to create biodiesel.[55] A total of 55.8 g of oil, or up to 81.5% of biodiesel, was produced by the yeast from 1 kg of corncob. Numerous studies have been carried out on using corncobs to make lipids with the goal of increasing biodiesel output.[56]

9.3.3 Production of Microbial Pigments

Natural pigments could be a significant alternative to the possibly dangerous synthetic colors. Chutrtong investigated the pigment production of *Neurospora* sp., extensively utilized as a model in genetics research, on a variety of substrates including dry corncob, corn, and potato dextrose agar.[57] The mycelium pigment was extracted from the organism's substrates using ethyl alcohol, and the absorbance was observed at 450.5 nm using a spectrophotometer. Additionally, the possibility of corncob as a viable substrate for *Monascus purpureus* in solid-state fermentation to create natural pigments was also investigated by Velmurugan.[58] With corncob as the substrate, pigment production from the fermented substrate was achieved, outperforming other agricultural leftovers studied under optimal circumstances. According to Morales-Oyervides, *Talaromyces atroroseus* GH2 can produce pigments using corncob as a low-cost substrate. To release the sugars from corncob, an acid pre-treatment was used.[59] Liquid with a high xylose content was evaluated as a pigment formation fermentation medium with and without nutrient input. It was determined that *Monascus homologous* pigments such as rubropunctatin and monascorubrin made up the majority of the pigments.

9.3.4 Production of Lactic Acid

Lactic acid was reported to be produced from corncobs by Miura[60] in 2004. Cellulase was created using the bacterium *Acremonium thermophilus*, and corncob was then hydrolyzed with it. About 55 $g\,L^{-1}$ of sugars were hydrolyzed from 100 $g\,L^{-1}$ of dried corncobs under the optimal circumstances of temperature, pH, enzyme concentration, and substrate (corncob) concentration. Additionally, this enzymatic hydrolysate produced more than 25 $g\,L^{-1}$ of L-lactic acid. *Rhizopus* sp. MK-96-1196 and *A. thermophilus* were also used to produce L-lactic acid in SSF. The outcomes demonstrated that from 100 $g L^{-1}$ of untreated raw corncob, more than 24 $g L^{-1}$ of L-lactic acid was generated. In the manufacture of L-lactic acid by *Bacillus* sp., Wang showed how to employ corncob molasses, waste possessing a high xylose concentration.[61] Corncob molasses yielded a concentration of L-lactic acid of 74.7 $g L^{-1}$ in the fed-batch fermentation. The synthesis of lactic acid by *Sporolactobacillus inulinus* utilizing corncob remnants was described by Bai.[62] Cellulase from commercial sources was first used to hydrolyze corncobs enzymatically. The hydrolysate supernatant was subsequently concentrated between 100 and 150 $g L^{-1}$ of total reducing sugar. When the hydrolysate supernatant was used in fed-batch fermentation, it resulted in a high concentration of D-lactic acid of 107.2 $g L^{-1}$, under optimal conditions. Furthermore, the purity of D-lactic acid was observed to be 99.2%.

9.3.5 Degradation of Dyes

Nowadays, significant amounts of dyes are being used in various sectors such as the pharmaceutical, food, cosmetic, textile, printing, and leather industries. Due to this reason, brightly colored effluents have been released, which have an impact on water's transparency and gas solubility in the water bodies. They are also dangerous due to their toxicity and carcinogenicity. Therefore, it is crucial to get rid of dyes before dumping them into natural waterways and corncob plays a significant part in this process.[63] According to Yu, using corncob substrate derived from the growth of *Pleurotus ostreatus*, it was feasible to remove the hazardous neutral red dye from aqueous solutions.[64] Additionally, corncob was also used for the adsorption of methylene blue dye from aqueous solutions with a pH range from 4 to 12. The thermodynamic characteristics collected further revealed that the dye adsorption was endothermic and spontaneous in nature. Another study conducted by Giri showed the use of biochar made from corncob with *Brevibacillus parabrevis* for the removal of brilliant green dye from industrial effluents.[65]

9.3.6 Feed Technology

Utilizing corncobs in feed technology is one technique to lessen the impact corncobs have on the environment. There is a need for adequate alternative sources because the production of animal feed in the past has relied on relatively expensive sources including cereals, fish wastes, and plant proteins.[17] The increasing human demand for grains and the need to broaden the variety of feed components accessible to animal producers have encouraged research into the use of fibrous, affordable, unconventional feedstuffs prevalent in nature. Paynor discovered an endophytic fungus connected with bamboo using corncobs in feed technology.[66]

Studies on the usage of fibrous, inexpensive feedstuffs that are plentiful in nature have been prompted by the rising human demand for grains and the need to expand the range of feed ingredients available to animal producers. Olagunju *et al.* used *Lachnocladium* spp. to pre-treat corncobs for fermentation.[67] They then evaluated the content of the fermented corncobs' nutrients and anti-nutrients. Significant increases in protein, ash, and mineral components were seen after incubating for 6 days, while notable decreases in fiber and anti-nutrient levels were observed. Additionally, *Trichoderma viride*, *Trichoderma reesei*, *Aspergillus oryzae*, and *Rhizopus oligosporus* were utilized by Rostika and Safitri to ferment corncob, which was then added as a supplement to fish feed. The study's findings indicated that the highest growth rate for java barbs, 2.01%, was produced by a fish diet comprising corncob that had undergone a 5% fermentation process.[68]

9.3.7 Removal of Heavy Metals and Wastewater Treatment

Corncob has been utilized for heavy metal biosorption and wastewater treatment. Zhang showed how to immobilize white rot fungus using corncob as support, which was subsequently used to treat wastewater from pulp and paper mills after primary treatment.[69] About 1 cm pieces of corncob were cut, cleaned with ethanol and distilled water, and then inoculated with fungi. To evaluate the color and chemical oxygen demand (COD), the immobilized cells were introduced to sterilized wastewater and incubated for 5 days at 30 °C in an air bath shaker. The COD and color removal were reportedly impacted by the pH of the wastewater treatment system. Additionally, the speed of the air bath shaker was also shown to be a conditional variable in most COD and color removal. Jonglertjunya reported the ability of indigenous fungus found on corncobs to remove heavy metals from solutions by absorbing copper and lead ions.[70]

9.3.8 Nanotechnology Applications

A nanotechnology system is one of the recognized options for reducing the substantial amount of corncob trash produced each year. With an emphasis on the green synthesis method, which uses safe and fewer harmful proteins to create nanoparticles for diverse purposes, there is rising recognition of the significance of nanotechnology in a variety of aspects of human existence. In the green synthesis of metal nanoparticles, a variety of biomolecules from plants, arthropods, microorganisms, enzymes, and agro-wastes have been investigated. However, the management of corncob using a nanotechnological method might take one of two forms: directly manufacturing nanoparticles from corncob or employing its by-products in the production of nanoparticles with numerous prospective uses. There have been reports on the production of nanoparticles obtained from corncob with typical diameters ranging from 24 to 200 nm and possible uses in the delivery of medication.[71] It has been reported that fungal xylanases generated by

A. niger and *Trichoderma longibrachiatum via* corncob valorization are used to synthesize silver and gold nanoparticles.[72] The obtained nanoparticles showed exceptional antibacterial, antioxidant, and anticoagulant properties. Additionally, the synthesized silver nanoparticles possessed the properties of degradation of dyes. In a study reported, cellulose micro/nano-fibrils were synthesized containing lignin from corncob.[73] To gradually supply plants with micro- and macronutrients, Lateef *et al.* described the usage of a nano-composite made from biochar from corncobs.[74] Investigations on the corncob biochar-based nano-composite showed that it was more successful in terms of water retention and absorbance. Furthermore, studies revealed a slow nutrient release pattern beneficial for improving plant yield and growth. The removal of arsenic contamination was carried out using a spectrophotometric method, and a report by Balasubramaniam described the use of corncobs in the manufacture of silica nanoparticles. The results showed that an elevated concentration of silica nanoparticles led to the elimination of about 96% of arsenic.[75] Additionally, Kamal observed that silica/corncob nano-composites showed 98% removal effectiveness when it came to removing methylene blue.[76]

9.4 Future Trends for the Bioeconomy Development

The potential applications of biomass from corncobs in the production of bioproducts and enzymes, treatment of wastewater, degradation of dyes, and generation of bioenergy have been effectively emphasized in this study. In addition to these, further non-biological potential uses have also been discussed for construction materials, reinforcing materials, and the cement and concrete industries. Year-round availability of corncobs is necessary for the efficient maximization of this agricultural waste, which calls for the creation and installation of modern storage facilities where the hemicellulosic content will remain constant even while being stored. Additionally, to match up with the aforementioned numerous aspects through which the utilization of corncob can be maximized, plant health and biomass yield per hectare area need to be enhanced. Through genetic engineering, to get rid of tough polymers such as lignin and characteristics such as a higher cellulose content and lower lignin levels, corncob might be taken into concern. However, such a study will need to take into account how the enhancement may affect the maize grains.

A potential study topic that needs to be prioritized is optimization to enhance current outcomes and maybe produce more stable bioproducts. According to Morrison and Golden, multifaceted studies on synergies, potential dangers, and prospects must be conducted to fully utilize agricultural feedstock for the industrial bioeconomy.[77] Additionally, there are still issues that need to be resolved to fully utilize the potential of microorganisms in the valorization of corncob; therefore, finding suitable solutions in this area should be a future research priority. It is important to properly address issues including choosing an effective pre-treatment method for processing corncobs, choosing an appropriate catalyst and perhaps recovering the catalyst, and preventing the production of phenolic chemicals in lignin. The chances of bio-conversion of lignin included in the corncob biomass into usable bioproducts should also be taken into account, as well as other potential bioconversion procedures that would maximize the biotechnological usage of corncobs. The efficient valorization of corncob to produce diverse bioproducts may also benefit from the use of genetically modified strains that contain gene cartridges.

9.5 Conclusion

It has been demonstrated that corncob is a source of crude hemicellulosic material that offers sustainable substitutes for the development of economically significant value-added bioproducts. The manufacture of industrial enzymes, biochemicals, pigments, functional foods, and nanotechnology applications are just a few examples of how bioproducts are applied in these fields. Additionally, corncob can be used as a component in animal feeds whose nutritional content has been enhanced through microbial fermentation, in the treatment of wastewater, in the breakdown of dyes, and in the production of dyes. To enable application at an industrial scale, a reliable bioprocess system must be created. The huge potential for the microbial method of corncob biomass valorization presents a variety of opportunities for biorefineries and businesses to generate distinguishing advantages from the products made from corncob biomass, thereby increasing the potential for corncob in biotechnology management and the supply of a sustainable and environmentally friendly environment in impoverished countries. To meet the goals of sustainable development, it is crucial to use the valorization of corncob to create high-end goods utilizing green and eco-friendly procedures for environmental sustainability and to promote the bioeconomy.

Abbreviations

MT	Metric tons
SmF	Submerged fermentation
SSF	Solid state fermentation
SHF	Separate hydrolysis and fermentation
COD	Chemical oxygen demand

Acknowledgements

The authors sincerely thank Director, ICAR-NBAIM, Mau, for providing scientific and technical support during the preparation of the manuscript. The authors gratefully acknowledge the ICAR-NBAIM and the Indian Council of Agricultural Research, Ministry of Agriculture and Farmers Welfare, Government of India, for providing financial support to the study.

References

1. C. Blattner, *J. Agric. Food Syst. Community Dev.*, 2020, **9**(3), 1–6.
2. D. Sarkar, S. K. Kar, A. Chattopadhyay, Shikha, A. Rakshit and V. K. Tripathi, *Ecol. Indic.*, 2020, **115**, 106412.
3. C. P. Leisner, *Plant Sci.*, 2020, **293**, 110412.
4. L. M. Mohlala, M. O. Bodunrin, A. A. Awosusi, M. O. Daramola, N. P. Cele and P. A. Olubambi, *Alex Eng. J.*, 2016, **55**(3), 3025–3036.
5. I. O. Adejumo and O. A. Adebiyi, *Strategies of sustainable solid waste management*, Intech Open, 2020.
6. B. Wang, F. Dong, M. Chen, J. Zhu, J. Tan and X. Fu, *Proc. Environ. Sci.*, 2016, **31**, 12–17.
7. H. El-Ramady, A. El-Henawy, M. Amer, A. E. D. Omara, T. Elsakhawy and H. Elbasiouny, *Egypt. J. Soil Sci.*, 2020, **60**(4), 349–364.
8. FAOSTAT, Food and Agriculture Organization of the United Nations, Statistical database, 2020.
9. S. Li and G. Chen, *J. Cleaner Prod.*, 2020, **251**, 119669.
10. M. Prasad, R. Ranjan, A. Ali, D. Goyal, A. Yadav and T. B. Singh, *Contam. Agric.*, 2020, 271–287.
11. H. Qi, Y. Zhao, X. Zhao, T. Yang, Q. Dang and J. Wu, *Bioresour. Technol.*, 2020, **299**, 122596.
12. R. Wang, Y. Zhao, X. Xie, T. A. Mohamed, L. Zhu and Y. Tang, *Bioresour. Technol.*, 2020, **295**, 122175.
13. S. Y. Kpalo, M. F. Zainuddin, L. A. Manaf and A. M. Roslan, *J. Cleaner Prod.*, 2021, **284**, 124745.
14. B. O. Adigun, F. I. Jegede and S. O. Tunmilayo, *Int. J. Ceram. Eng. Sci.*, 2020, **2**(1), 17–21.
15. P. Seesuriyachan, A. Kawee-Ai and T. T. Chaiyaso, *Bioresour. Technol.*, 2017, **241**, 537–544.
16. A. J. Borchert, W. R. Henson and G. T. Beckham, *Curr. Opin. Biotechnol.*, 2022, **73**, 1–13.
17. O. O. Oduguwa, M. O. Edema and A. O. Ayeni, *Bioresour. Technol.*, 2008, **99**(6), 1816–1820.
18. R. Ravindran, S. S. Hassan, G. A. Williams and A. K. Jaiswal, *Bioengineering*, 2018, **5**(4), 93.
19. J. A. Elegbede and A. Lateef, *Microbial Nanobiotechnology. Materials Horizons: From Nature to Nanomaterials*, Springer, Singapore, 2021, pp. 185–232.
20. R. S. Singh, T. Singh and A. Pandey, *Adv. Enzyme Technol.*, 2019, 7.
21. I. A. Adelere and A. Lateef, *Kuwait J. Sci.*, 2016, **43**(3), 118–127.
22. I. A. Adelere and A. Lateef, *Springer Series on Polymer and Composite Materials*, Springer International Publishing, Switzerland, 2019, pp. 123–162.
23. A. T. Kanengoni, M. Chimonyo, B. K. Ndimba and K. Dzama, *Asian-Australas. J. Anim. Sci.*, 2015, **28**(12), 1669–1679.
24. S. Kumar, J. S. Upadhyaya and Y. S. Negi, *BioResources*, 2010, **5**(2), 1292–1300.

25. J. A. Elegbede and A. Lateef, *Waste Biomass Valor.*, 2018, **9**(8), 1273–1287.
26. X. Zhang, M. Yang, Q. Yuan and G. Cheng, *ACS Sustainable Chem. Eng.*, 2019, **7**(20), 17166–17174.
27. P. Leite, J. M. Salgado, A. Venâncio, J. M. Domínguez and I. Belo, *Bioresour. Technol.*, 2016, **214**, 737–746.
28. V. A. Kumar, R. S. C. Kurup, C. Snishamol and G. N. Prabhu, *Green Bio-Process.*, 2019, 323–343.
29. V. Kumar, P. P. Sandhu, V. Ahluwalia, B. B. Mishra and S. K. Yadav, *Bioresour. Technol.*, 2019, **291**, 121931.
30. A. Singh, S. Bajar, A. Devi and D. Pant, *Bioresour. Technol. Rep.*, 2021, **14**, 100652.
31. X. Liming and S. Xueliang, *Bioresour. Technol.*, 2004, **91**(3), 259–262.
32. W. Mardones, E. Callegari and J. Eyzaguirre, *Mycology*, 2019, **10**(2), 118–125.
33. S. B. Oyeleke, O. A. Oyewole, E. C. Egwim, B. E. N. Dauda and E. N. Ibeh, *Bayero J. Pure Appl. Sci.*, 2012, **5**(1), 78–83.
34. M. Ganesan, R. M. Mathivani Vinayakamoorthy, S. Thankappan, I. Muniraj and S. Uthandi, *Biotechnol. Biofuels*, 2020, **13**, 124.
35. Y. S. Soeka and Sulistiani, *Conf. Ser.: Earth Environ Sci.*, 2019, **308**(1), 012033.
36. F. S. Ire, A. O. Okoli and V. Ezebuiro, *Microbiolgy*, 2018, **8**(2), 1–10.
37. F. A. Santos, L. C. T. de Carvalho-Gonçalves, A. L. de Carvalho Cardoso-Simões and S. F. de Melo Santos, *Bioresour. Technol. Rep.*, 2021, **14**, 100648.
38. B. Sunkar, B. Kannoju and B. Bhukya, *Front. Microbiol.*, 2020, **11**, 772.
39. B. S. Chadha, B. Kaur, N. Basotra, A. Tsang and A. Pandey, *Bioresour. Technol.*, 2019, **277**, 195–203.
40. A. Knob, D. Fortkamp, T. Prolo, S. C. Izidoro and J. M. Almeida, *BioResources*, 2014, **9**(3), 5738–5773.
41. J. A. Elegbede, A. Lateef, M. A. Azeez, T. B. Asafa, T. A. Yekeen and I. C. Oladipo, *IET Nanobiotechnol.*, 2018, **12**(6), 857–863.
42. F. S. Ire, I. J. Chima and V. Ezebuiro, *Biocatal. Agric. Biotechnol.*, 2021, **31**, 101869.
43. O. Magalhães and M. Souza, *Braz. J. Microbiol.*, 2010, **41**, 850–861.
44. A. Aliyah, G. Alamsyah, R. Ramadhani and H. Hermansyah, *Energy Procedia*, 2017, **136**, 418–423.
45. A. Ahmed and M. Sohail, *J. King. Saud. Univ. Sci.*, 2020, **32**(1), 955–961.
46. U. Ejaz, A. Ahmed and M. Sohail, *Biocatal. Agric. Biotechnol.*, 2018, **14**, 450–456.
47. A. Devi, A. Singh, S. Bajar, D. Pant and Z. U. Din, *J. Environ. Chem. Eng.*, 2021, **9**(5), 105798.
48. M. Gáspár, T. Juhász, Z. Szengyel and K. Réczey, *Process Biochem.*, 2005, **40**(3–4), 1183–1188.
49. M. Zhang, F. Wang, R. Su, W. Qi and Z. He, *Bioresour. Technol.*, 2010, **101**(13), 4959–4964.
50. P. Boonchuay, C. Techapun, N. Leksawasdi, P. Seesuriyachan, P. Hanmoungjai and M. Watanabe, *Bioresour. Technol.*, 2018, **256**, 399–407.

51. M. Chen, L. Xia and P. Xue, *Int. Biodeterior. Biodegrad.*, 2007, **59**(2), 85–89.
52. K. K. Cheng, J. Wu, Z. N. Lin and J. A. Zhang, *Biotechnol. Biofuels*, 2014, **7**(1), 166.
53. S. Liu, H. Liu, C. Shen, W. Fang, Y. Xiao and Z. Fang, *J. Ind. Microbiol. Biotechnol.*, 2021, **48**(1–2), 013.
54. G. V. Subhash and S. V. Mohan, *Biomass Convers. Biorefinery*, 2015, **5**(2), 215–226.
55. Z. Miao, X. Tian, W. Liang, Y. He and G. Wang, *Renewable Energy*, 2020, **161**, 91–97.
56. Y. H. Chang, K. S. Chang, C. F. Lee, C. L. Hsu, C. W. Huang and H. D. Jang, *Biomass Bioenergy*, 2015, **72**, 95–103.
57. J. Chutrtong, *Proc. Soc. Behav. Sci.*, 2015, **197**, 797–800.
58. P. Velmurugan, H. Hur, V. Balachandar, S. Kamala-Kannan, K. J. Lee and S. M. Lee, *J. Biosci. Bioeng.*, 2011, **112**(6), 590–594.
59. O. L. Morales, S. J. P. Ruiz, J. C. Oliveira, G. M. J. Sousa, M. T. K. Morales and A. Albergamo, *Biochem. Eng. J.*, 2020, **161**, 107698.
60. S. Miura, T. Arimura, N. Itoda, L. Dwiarti, J. B. Feng and C. H. Bin, *J. Biosci. Bioeng.*, 2004, **97**(3), 153–157.
61. L. Wang, B. Zhao, B. Liu, B. Yu, C. Ma and F. Su, *Bioresour. Technol.*, 2010, **101**(20), 7908–7915.
62. Z. Bai, Z. Gao, J. Sun, B. Wu and B. He, *Bioresour. Technol.*, 2016, **207**, 346–352.
63. S. R. Couto, *Biotechnol. Adv.*, 2009, **27**(3), 227–235.
64. X. Yu, G. Zhang, C. Xie, Y. Yu, T. Cheng and Q. Zhou, *BioResources*, 2011, **6**(2), 936–949.
65. B. S. Giri, S. Gun, S. Pandey, A. Trivedi, R. T. Kapoor and R. P. Singh, *Bioengineered*, 2020, **11**(1), 743–758.
66. K. Paynor, E. S. David and M. J. G. Valentino, *Mycosphere*, 2016, **7**(2), 139–147.
67. A. Olagunju, E. Onyike, A. Muhammad, S. Aliyu and A. S. Abdullahi, *Afr. J. Biochem. Res.*, 2013, **7**(11), 210–214.
68. R. Rostika and R. Safitri, *APCBEE Proc.*, 2012, **2**, 148–152.
69. A. Zhang, G. Wang, G. Gong and J. Shen, *Ind. Crops Prod.*, 2017, **109**, 538–541.
70. W. Jonglertjunya, *Chiang Mai J. Sci.*, 2008, **35**(1), 69–81.
71. S. Kumar, Y. S. Negi and J. S. Upadhyaya, *Adv. Mater. Lett.*, 2010, **1**(3), 246–253.
72. J. A. Elegbede, A. Lateef, M. A. Azeez, T. B. Asafa, T. A. Yekeen and I. C. Oladipo, *Waste Biomass Valor.*, 2020, **11**(3), 781–791.
73. J. Luo, Y. Su, J. Chen, X. Wang and J. Liu, *Cellulose*, 2021, **28**(8), 4671–4684.
74. A. Lateef, R. Nazir, N. Jamil, S. Alam, R. Shah and M. N. Khan, *Environ. Nanotechnol. Monit Manag.*, 2019, **11**, 100212.
75. T. Balasubramaniam, A. H. A. Bakar, M. N. A. Uda, U. Hashim, N. A. Parmin and A. Anuar, *IOP Conf. Ser.: Mater. Sci. Eng.*, 2020, **864**(1), 012187.
76. K. Kamal, M. S. Attia, N. Ammar and E. M. Abou-Taleb, *Int. J. Dev.*, 2019, **8**(1), 81–93.
77. B. Morrison and J. S. Golden, *BioResources*, 2015, **10**(3), 4411–4440.

CHAPTER 10

Advanced Food Waste Valorization Techniques for Bioenergy Production: A Path in the Direction of Environmental Sustainability

A. N. ANOOPKUMAR,[a] EMBALIL MATHACHAN ANEESH,[a] ARAVIND MADHAVAN,[b] PARAMESWARAN BINOD,[c] MUKESH KUMAR AWASTHI,[d] MOHAMMED KUDDUS,[e] ASHOK PANDEY,[f,g,h] LAYA LIZ KURIAKOSE[i] AND RAVEENDRAN SINDHU*[i]

[a] Centre for Research in Emerging Tropical Diseases (CRET-D), Department of Zoology, University of Calicut, Malappuram, Kerala, India; [b] Rajiv Gandhi Center for Biotechnology, Jagathy, Thiruvananthapuram – 695 014, Kerala, India; [c] Microbial Processes and Technology Division, CSIR-National Institute for Interdisciplinary Science and Technology (CSIR-NIIST), Trivandrum – 695 019, Kerala, India; [d] College of Natural Resources and Environment, Northwest A & F University, Yangling, Shaanxi – 712 100, China; [e] Department of Biochemistry, University of Hail, Kingdom of Saudi Arabia; [f] Centre for Innovation and Translational Research, CSIR-Indian Institute for Toxicology Research (CSIR-IITR), 31 MG Marg, Lucknow – 226 001, India; [g] Centre for Energy and Environmental Sustainability, Lucknow-226 029, Uttar Pradesh, India; [h] Sustainability Cluster, School of Engineering, University of Petroleum and Energy Studies, Dehradun-248 007, India; [i] Department of Food Technology, TKM Institute of Technology, Kollam – 691 505, Kerala, India
*Emails: sindhurgcb@gmail.com; sindhufax@yahoo.co.in; sindhur@tkmit.ac.in

Green Chemistry Series No. 78
Agri-food Waste Valorisation
Edited by Pankaj Chowdhary and Abhay Raj

Published by the Royal Society of Chemistry, www.rsc.org

10.1 Introduction

The problems risen towards the environment by the growing population in recent years have been witnessed as one of the major challenging issues over the globe. The dramatically increasing population rate accelerates the need for basic elements including food, which is escorted by the bulk generation of diverse waste. The generation of food waste biomass also underwrites the enhanced cost of waste disposal along with severe environmental consequences. Many of the investigations in recent years have also explained the link between food waste and water pollution, air pollution, climate change, biodiversity loss and soil degradation. In addition to this, the generation of food waste has also likely to be allied with nutritional loss.[1]

The following elements prominently belong to the food waste category: fruit and vegetable waste, spoiled foods, leftovers from hotels, crops left in the field and leftovers on the plate from restaurants and homes. It is extremely difficult to completely evade the food waste biomass; however, it is applicable to diminish the quantity of wasted food. Hence, the approaches or technologies for food waste valorization are known to illustrate significant contributions towards the recent research targeting sustainable development, bioenergy production and environmental sustainability.[1,2] The extreme homogeneity of food waste along with the concept of biorefinery illustrates their potential for the production of biofuels, bio-based materials and platform chemicals.

Mainly the following four kinds of approaches were primarily implemented for bioenergy production from food waste: biochemical conversion, thermochemical conversion, transesterification and microbial fuel cells. Among them, food waste biomass conversion through microbial fuel cells has been proven as one of the best approaches for food waste valorization. The treatment in this regard significantly contributes to the production of biomethane, biohythane and biogas.[3] Biomethane and biohydrogen can be produced primarily through the fermentation process (single or two-stage). Anaerobic digestion and dark fermentation can be used to produce biohythane, which is a mixture of hydrogen and methane. The applicability of light-dependent techniques including biophotolysis and photo-fermentation for hydrogen production has previously been discussed.[4]

The role of microbial electrolysis cells and dark fermentation as light-independent approaches for hydrogen production has also been discussed in the aforesaid study. Based on the findings from such studies, it was evident that bioelectricity, biobutanol and biodiesel can be generated in a variety of approaches and contribute to a circular economy. The formation of by-products including biofertilizers[5] from the treatment can again be used for the agricultural industry thereby offering a path in the direction of environmental sustainability through recycling.

10.2 Food Waste Valorization

The valorization of food waste usually involves the conversion of food waste into essential products including fuels, chemicals and materials (Table 10.1).

Table 10.1 Benefits of products obtained from food waste.

Food waste	Product and use
▪ Cooking oils ▪ Bakery waste ▪ Beet sugar ▪ Cheese whey	• PHA • Applications in the medical field, food packaging, as a fuel, and fine chemicals
▪ Mixed food waste	• Biohydrogen • As a fuel
▪ Corn ▪ Wheat	• PHB • Applications in the medical field and food packaging
▪ Carbohydrates ▪ Citrus wastes	• Ethanol • Drink products, chemical synthesis of solvent biofuels and disinfectants
▪ Bread wastes	• Succinic acid • Production of the following chemicals: polybutylene succinate, succinic esters, pyrrolidinones and polyamides
▪ Bread wastes	• Lactic acid • PLA production and production of personal care products
▪ Food waste	• Butanol • Alternative to petrol
▪ Starch waste	• Fumaric acid • Production of the following chemicals: tetrahydrofuran and unsaturated polyester resins
▪ Cooking oils	• Glycerol • Health and pharmaceutical products
▪ Tomatoes	• Lycopene • Anticarcinogen and antioxidant
▪ Mixed food waste	• Bioplastics • For multipurpose uses
▪ Corn	• PLA • Applications in the medical field, compost bags, disposable tableware and fabrics
▪ Mixed food waste	• Methane • Alternative to LNG and LPG
▪ Corn ▪ Straw ▪ Wheat	• PHB • Applications in the medical field, biocomposites
▪ Vegetable oils	• Plant sterols • Anti-inflammatory and immune enhancer, anticarcinogen, combat atherosclerosis
▪ Cooking oils	• Glycerol • Sweeteners, health and pharmaceutical products and food products

The conventional approaches for food waste management primarily focused on incineration or landfills. The selection of landfill approaches for food waste management may also be allied with severe drastic impacts on the environment and surroundings. For instance, anaerobic degradation of food waste through aforesaid approaches, especially the landfills, results from the emission of volatile fatty acids, methane and ammonia[6] followed by a lower amount of dissolved oxygen (DO).

The high potential of the conventional approaches to contaminate the environment and the surroundings forced the scientific community to explore the advantageous effects of recent technologies. Incineration also releases volatile organic compounds along with particulate matter which impacts the environment including the human population and the health of various animals inhabited in the particular area. The high energy consumption along with the emission of greenhouse gas (GHG)[7] was also addressed as a prominent negative impact instigated by incineration.

Food waste valorization encompasses primarily the first and second-generation processes. The former approach focused on the complete utilization of food waste biomass for the production of various essentials including compost, energy and animal feed and the latter approach encompasses diverse forms of fractionated utilization.[8] In order to get various forms of industrial products including commodity products, fine chemicals and functional food, many studies have suggested the possibilities of integrating the advanced conversion procedures followed by adapted recovery for specific elements.

10.2.1 Bio-fuel Production

There exists an augmented public awareness concerning the degradation of environmental quality because of the mismanagement of food waste. In order to rectify such concerns, recent studies have implemented various technologies to convert food waste biomass into effective and useful products.[9] For instance, bio-based technologies such as microbial fermentation, aerobic digestion and anaerobic digestion have been prominently used for the production of gaseous (bio-hydrogen and bio-methane) and liquid (bio-diesel and bio-ethanol) bio-fuels from food waste materials. There exists a lot of research concerning food waste treatment using physical methods such as incineration and landfills. But recent studies have addressed the practical implementation of anaerobic digestion and aerobic composting from the same perspective. The most prominent benefit of implementing the above-said techniques is the potential of the same to offer maximum yields and minimum emissions along with the ability to manage the high moisture content.

10.2.1.1 Bio-alcohol (Ethanol) and Bio-methane Production

Ethanol is currently noted as an effective alternative fuel in the current scenario since it shows the same kinds of characteristic features exhibited by petroleum. The high-octane number of ethanol helps the same to accomplish the same kind of anti-knock effect as exhibited by petroleum. For these reasons, the scientific community has focused on the production of bio-alcohol through saccharification. The aforesaid process principally converts the carbohydrates or starch to various reduced forms of sugars with the aid of the following enzymes: glucoamylase, α-amylase and β-amylase. It has been considered the first step of ethanol production from various lignocellulosic substrates including sugarcane, rice and potato. After this stage, the fermentation process begins with microorganisms such as *Saccharomyces cerevisiae*. The zymase and invertase enzymes formed by the yeast cells aid in the conversion of reduced forms of sugars into CO_2 and crude ethanol. In addition to this, the other groups of microbial populations including bacteria and fungi have also been implemented for the fermentation process to get the same kind of output. The use of *Pichia rhodanensis* and *Zymomonas mobilis* for the production of bioethanol through the conversion of pentose sugars has been previously reported. Furthermore, the production of bio-butanol using the *Clostridium acetobutylicum* bacteria from food waste has received much more significance than the production of bioethanol through the same process since the bio-butanol possessed improved combustion efficiency, lower vapor pressure followed by a higher energy density. The complex structure of the lignocellulose in food waste made the process much harder than others, simultaneously affecting the bio-ethanol yield. Hence, to rectify such concerns, various pre-treatment methods using heat, enzymes and alkali have been implemented. However, the following parameters are also known to exhibit significant influence on the specific growth rate of the microbial population thereby affecting the production of bio-ethanol.

Carbon dioxide, methane and trace amounts of impurities including methyl mercaptans, hydrogen sulphide, hydrogen, nitrogen and oxygen are the major constituents of biogas. To enhance the energetic value of the biofuel, various processes such as pressure swing absorption, membrane separation and chemical absorption can be implemented. The technique for the production of biogas primarily involves the initial hydrolysis followed by fermentation and finally the conversion of the various substrates to methane. Each of the steps (hydrolysis, aceto-acidogenesis and methanogenesis) was influenced by the action of the microbial population.[10]

The influence of various surrounding parameters, especially the temperature, has been witnessed by Nayak and Bhushan, 2019. For instance, 25–35 °C of temperature along with a pH of 5.2–6.3 has been required for the optimal performance of the microbial populations used in the first stages such as the hydrolysis and the acidogenesis stage. Likewise, 30–60 °C of temperature followed by a pH of 6.7–7.5 is required to maintain the optimal performance of the microbial populations in the methanogenesis stage. In addition to the environmental factors, the physical as well as chemical characteristic features of the food waste, including the nutrient contents, volatile solids, high lipid content and moisture, are also known to exhibit significant impact on the anaerobic digestion performance.

In order to rectify the above-said concerns and to enhance the biogas yield, co-substrates can be added during the anaerobic digestion processes. For instance, the co-digestion of food waste material in anaerobic mesophilic reactor settings with the aid of cow manure was previously shown to increase the yield by up to 26% compared to the individual processes. Clearly, the aforesaid process thus promises a significant choice that not only enhances the biogas yield but also stabilizes the entire major process.

It was also evident that the following two-stage digestion processes in separate reactors were found to be much more efficient than the single-stage digestion procedure: hydrolysis, acido-acetogenesis and methanogenesis (Figure 10.1). A study by Grimberg *et al.*[11] validated the aforementioned perspective and revealed that the two-stage digestion process for food waste treatment is more efficient than single-stage digester systems. In addition to the various parameters, the design of the reactor would be an important factor, since a better reactor should have the ability to attain high loading, stabilize the methanogenesis stage and immobilize the microbial consortia. The anaerobic plug flow reactor (APFR), continuous stirred tank reactor (CSTR) and anaerobic sequencing batch reactor (ASBR) are the major conventional reactors used so far.[12] Due to the drawbacks allied with them, recent studies have exhibited a shift to the following approaches to replace the conventional strategies: anaerobic baffled reactor (ABR), anaerobic sludge blanket reactor (USAB) and anaerobic contact reactor (ACR). Likewise, the use of anaerobic membrane bioreactors (AnMBRs) has greatly increased in recent years due to the following benefits: cost-effective approach, high effluent quality, free from pathogens and nutrient-rich yields (phosphorus and nitrogen). The ability of anaerobic membrane bioreactors (AnMBRs) to exhibit augmented performance while dealing with toxic constituents is an added advantage of the same. However, membrane fouling followed by poor efficiency under lower temperatures is the major drawback allied with the anaerobic membrane bioreactors (AnMBRs).[13]

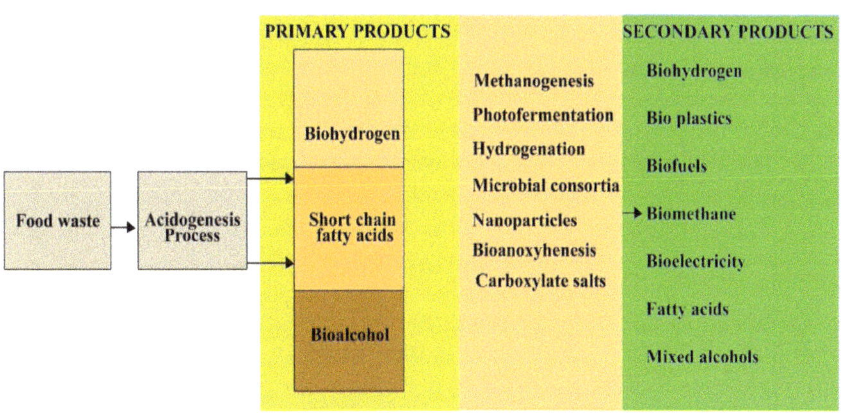

Figure 10.1 Various major approaches and the products obtained from food waste.

A more recent study by Jung *et al.*[14] analysed the two-stage dynamic membrane bioreactor (DMBR) system for biomethane production under mesophilic conditions from food waste materials and found that about 79% energy was recovered along with a high rate of biomethane production. Deheri and Acharya[15] reported the possibilities of augmented biomethane generation with the aid of NaOH and $CaO_2 + CaCO_3$. According to them, about 67.24% of methane content was produced by the methanogenic reactor when NaOH was used as the alkali reagent. The applicability of the two-stage anaerobic fermentation process for the co-production of biomethane from food waste and chicken manure was reported by Liu *et al.*[16]

10.2.2 Bio-oil/Bio-char Generation

The utilization of food waste to produce various bioproducts, especially bio-oil, through the pyrolysis and microwave-assisted approaches has gained much more attention in the current scenario.[17] When providing the following optimum conditions of pyrolysis, the bio-oil yield was rapidly enhanced up to 30.24 wt%: 30 min, 400 °C and 50 mL min^{-1} of nitrogen. A study[18] optimized (7 s and 550 °C) the microwave-assisted response surface methodology and obtained 42 wt% of bio-oil. An enzyme-assisted hydrothermal treatment for the co-production of bio-oil and hydrochar from food waste was reported by Kaushik *et al.*[19] and according to their findings, it was perceived that the pre-treatment with a specific enzyme ratio (carbohydrase : protease : lipase in 1 : 2 : 1) verified to produce the aforesaid products with a maximum yield. To enhance the yield in an energetically favourable way, recent investigations have analysed the practical implementation of certain models in context with the techno-economic analyses of HOT and found that the use of enzymes for the pre-treatment enhances the product quality with expensive nature. The central research institute of the electric power industry in Japan has developed an advanced oil-extraction technique with high energy-saving properties. The advanced oil-extraction technique principally uses an environmentally friendly solvent (liquefied dimethyl ether (DME)) to extract the bio-oil from food waste.[20] As mentioned in the former sections, the microwave-assisted approaches yielded the highest rate of bio-oil (50 wt%).[21] They also compared the bio-oil properties to the many biodiesel standards such as IS 15607:2016, ASTM D 6751 and EN 14214:2012 and recommended that it is essential to upgrade the same to achieve the standard values.

10.2.3 Bio-hydrogen Production

Many strategies are being implemented for the production of bio-hydrogen from food waste. Direct and indirect bio-photolysis, photo-fermentation and hybrid systems followed by dark fermentation are the major approaches primarily used for the production of bio-hydrogen from food waste. The implications of bio-hydrogen production from food waste through dark fermentation are found to be still limited in many situations. In this regard, a study by Jarunglumlert *et al.*[22] discussed the up-to-date status of advanced technologies with special inference on the factors influencing the proper production condition, the practicability of scaled-up production and production potential followed by economic value analysis. According to them, the predominant and suitable food waste for bio-hydrogen production is carbohydrate-rich food waste which is always effortlessly biodegradable. If the food waste contains very low carbohydrates, then additives constituting a high carbon content can be added to augment the bio-hydrogen production yield. Their findings suggested that the following conditions can be implemented for fermentation to get a high bio-hydrogen production yield: a temperature of 50–55 °C and an initial pH of 6–8. The benefits of two-stage production in terms of more energy production have also been suggested by them. The influence of initial pH on the bio-hydrogen production yield was previously studied in certain investigations and it was found that a high hydrogen production rate of 2.90 mmolH_2 per day was achieved at a pH of 5.5.[10]

Many of the studies discussed the possibilities of using dark fermentation of food waste which is becoming a more eminent strategy for biohydrogen production. Acidogenesis and hydrolysis are the early stages of anaerobic digestion. The following microbial species are known to play a significant role in biohydrogen production: *Clostridium saccharobutylicum, Caldicellulosiruptor thermocellum, Clostridium pasteurianum, Thermoanaerobacterium thermosaccharolyticum, Enterobacter aerogenes, Caldicellulosiruptor saccharolyticus, Clostridium butyricum* and *Bacillus thermozeamaize*.[23] The autonomous nature followed by the functional ability in context with the capability to convert any type of carbohydrate-consisted material made the strategy a prominent one for biohydrogen production in the current scenario. The different co-products including volatile fatty acids obtained during the aforesaid stage can also be used for various industrial purposes.[23]

10.2.4 Nanoparticles and Biofuel Production

The use of nanoparticles for protein immobilization, biosensors and biofuel production has enhanced considerably in recent years. Nanoparticles can have a prominent impact on microbial energy metabolism for the production of biohydrogen by shifting electrons proficiently under aerobic conditions. Because of the quantum size effects as well as the surface, nanoparticles primarily stimulate biohydrogen fabrication.[24] Table 10.2 enlists the different types of nanoparticles along with their ability to assist biohydrogen production. The influence of the metal nanoparticles towards the hydrogenase activity on biohydrogen production has been widely reported and for these reasons, researchers are recently focused on nanoparticle-based approaches. For instance, the nanoparticle mixer consisting of Ni and Fe nanoparticles in diverse concentrations exhibited remarkable improvement in the biohydrogen production yield.[25] Likewise, the combined influence of Fe_2O_3 and NiO nanoparticles on the biohydrogen production with molasses through the anaerobic sludge has been reviewed by Maroušek.[26]

Table 10.2 Nanoparticles and biohydrogen production.

Nanoparticles	Biohydrogen production yield (%)	pH and temperature (°C)
Ni and Fe	200	7.0 and 37
Activated carbon	100	6.5 and 25
Fe_2O_3, NPs, NiO and NPs	62	5.5 and 37
CNTs	100	6.5 and 25
TiONPs	50–100	7.0–8.0 and 30–35
CuNPs	3.5	7 and 37
SiO_2NPs	5.1–120	4.2–7.6 and 28–30
AgNPs	67.6	8.5 and 35
PbNPs	1.48–2.48	7 and 37
Fe_2O_3NPs	4.8–63.1	5.5–7.0 and 35–60
NiNPs	0.90–23.0	5.5–7.0 and 37–55

10.3 Environmental Perspectives

The above-said strategies including the dark fermentation system usually result in the formation of certain byproducts including volatile fatty acids and it forced the industry to perform additional treatments to combat the environmental damage. Mainly, because of the high water solubility of the volatile fatty acids, the release of such elements or complexes in water results in the distribution of wastewater to the surroundings and finally it will contaminate the groundwater. However, volatile fatty acids are always found to be beneficial chemical elements and hence can be explored for manufacturing microbial lipids and biopolymers, reducing sulfate and sulfur. The volatile fatty acids formed can be used for the manufacture and preparation of various bioproducts, and it would be noted as one of the major alternative paths to diminish the upsurged volume of waste formed when biofuel is produced. Various approaches are available to treat the volatile fatty acids in wastewater including the adsorption process, membrane separation and chemical coagulation.[27] But, due to the cost-effectiveness and easier use, anaerobic digestion was found to provide significant benefits over them. The production of biomethane from wastewater through anaerobic fermentation always offers a way towards reduced environmental pollution. But, the long operating conditions of the same for volatile fatty acids are not stable in many situations. A previous study has investigated the effects of anaerobic fermentation under the aforesaid condition using a conductive carbon cloth and found that the addition of carbon cloths could assist the anaerobic digestion and prevent it from failing.[28] It also enhanced the biomethane production by up to 200–260% and simultaneously increased the COD removal efficiency, when compared with a control group lacking carbon cloth.

10.4 Conclusion

Worries concerning the energy demand and energy security along with the impact of CO_2 releases from current fuel usage extremely contribute towards climate change and are the predominant driving factors forcing to find alternative energy resources. Biofuels are recognized as the most efficient ways for reducing environmental damage and hence recent years have witnessed the use of food waste as a prominent resource for bioenergy production. Food waste is always noted as a global problem that instigates social, detrimental, environmental and economic impacts. The conversion of food waste material into advantageous bioproducts, especially biofuels, can prove to be an environmentally sustainable path for food waste disposal. Many of the factors including the temperature, pH, use of the strains of microbial population and various conditions have influenced the yield of bioproducts from food waste. Recent studies explored the applicability of nanoparticles for biofuel production. Collectively, recent investigations have indicated that many of the advanced strategies are found to be promising for the conversion of food waste into sustainable bioproducts.

References

1. B. Tsegaye, S. Jaiswal and A. K. Jaiswal, *Foods*, 2021, **10**, 1174.
2. A. R. United Nations, *United Nations Gen. Assem. Resolut.*, 2015.
3. M.-L. T. Nguyen, P.-C. Hung, T.-P. Vo, C.-H. Lay and C.-Y. Lin, *Int. J. Hydrogen Energy*, 2021, **46**, 11313–11324.
4. C. Hitam and A. Jalil, *Biomass Convers. Biorefin.*, 2020, 1–19.
5. P. R. Rout, R. R. Dash, P. Bhunia, E. Lee and J. Bae, *Sustainable Energy Technol. Assess.*, 2021, **48**, 101620.
6. H. Fisgativa, A. Tremier and P. Dabert, *Waste Manage.*, 2016, **50**, 264–274.
7. G. Capson-Tojo, M. Rouez, M. Crest, J.-P. Steyer, J.-P. Delgenès and R. Escudié, *Rev. Environ. Sci. Bio/Technol.*, 2016, **15**, 499–547.
8. X. Xiong, K. Iris, D. C. Tsang, N. S. Bolan, Y. S. Ok, A. D. Igalavithana, M. Kirkham, K.-H. Kim and K. Vikrant, *Chem. Eng. J.*, 2019, **375**, 121983.
9. R. L. Skaggs, A. M. Coleman, T. E. Seiple and A. R. Milbrandt, *Renewable Sustainable Energy Rev.*, 2018, **82**, 2640–2651.
10. A. Nayak and B. Bhushan, *J. Environ. Manage.*, 2019, **233**, 352–370.
11. S. Grimberg, D. Hilderbrandt, M. Kinnunen and S. Rogers, *Bioresour. Technol.*, 2015, **178**, 226–229.
12. A. Sharma, *J. Hydrol.*, 2000, **239**, 249–258.
13. S. Chang, *Adv. Chem. Eng. Sci.*, 2014, **4**, 56–61.
14. J.-H. Jung, Y.-B. Sim, J. Ko, S. Y. Park, G.-B. Kim and S.-H. Kim, *Bioresour. Technol.*, 2022, **352**, 127094.
15. C. Deheri and S. K. Acharya, *Int. J. Hydrogen Energy*, 2022, **47**, 2862–2873.
16. X. Liu, Y. Yang, N. Wu, Y. Wei, H. Shan and H. Zhao, *Appl. Biochem. Biotechnol.*, 2022, 1–15.
17. H. Kadlimatti, B. R. Mohan and M. Saidutta, *Biomass Bioenergy*, 2019, **123**, 25–33.
18. S. Neha and N. Remya, *J. Environ. Manage.*, 2021, **297**, 113345.
19. R. Kaushik, G. K. Parshetti, Z. Liu and R. Balasubramanian, *Bioresour. Technol.*, 2014, **168**, 267–274.

20. K. Sakuragi, P. Li, M. Otaka and H. Makino, *Energies*, 2016, **9**, 106.
21. S. Neha and N. Remya, *Biomass Convers. Biorefin.*, 2022, 1–10.
22. T. Jarunglumlert, C. Prommuak, N. Putmai and P. Pavasant, *Int. J. Hydrogen Energy*, 2018, **43**, 634–648.
23. R. Jain, N. L. Panwar, S. K. Jain, T. Gupta, C. Agarwal and S. S. Meena, *Biomass Convers. Biorefin.*, 2022, 1–26.
24. J. Qin and H. Zeng, *Appl. Catal., B*, 2017, **209**, 161–173.
25. S. Shanmugam, A. Hari, A. Pandey, T. Mathimani, L. Felix and A. Pugazhendhi, *Fuel*, 2020, **270**, 117453.
26. J. Maroušek, *Fuel*, 2022, **328**, 125318.
27. Y. Lu, Y. Xu, S. Chen, B. Dong and X. Dai, *Environ. Pollut.*, 2020, **261**, 114085.
28. D. Feng, A. Xia, Q. Liao, A.-S. Nizami, C. Sun, Y. Huang, X. Zhu and X. Zhu, *Environ. Pollut.*, 2021, **272**, 116030.

CHAPTER 11

Techno-economic Analysis for Low Cost In-vessel Food Waste Composting at Universiti Malaysia Sabah

NUR AQEELA SYUHADAH AJI, MARIANI RAJIN, SARIAH SAALAH AND ABU ZAHRIM YASER*

Chemical Engineering Programme, Faculty of Engineering, Universiti Malaysia Sabah, Jalan UMS, 88400 Kota Kinabalu, Sabah, Malaysia.
*Email: zahrim@ums.edu.my

11.1 Introduction

Higher education institutions are one of the major generators of food waste. Tiew *et al.*[1] studied the waste composition at Universiti Kebangsaan Malaysia (UKM) and found that 54.8% of the waste generated was food waste, while 33% of total waste in higher education institutes consisted of food waste as reported by Torrijos *et al.*[2] An appropriate plan for managing campus waste should be implemented and all members of the campus community, including the top management, academic staff, management personnel, and students, should play their role in sustaining the green campus agenda.[3] Composting is an effective way to manage waste in universities as it reduces waste from landfilling, reduces the dependency on chemical fertilisers, and increases awareness among the campus community about environmental safeguarding. Besides, it can reduce the cost of waste management in universities.

Green Chemistry Series No. 78
Agri-food Waste Valorisation
Edited by Pankaj Chowdhary and Abhay Raj

Published by the Royal Society of Chemistry, www.rsc.org

The composting system can be classified based on the aeration method: active or passive. A passive aeration system can be static or dynamic. For passive aeration-static, the compost mixture is not turned, while the compost mixture for passive aeration-dynamic will be turned periodically. There are different opinions on the best system for composting. One of the disadvantages of static composting is that the temperature distribution is not uniform. Some factors need to be considered before choosing a suitable system. The common methods used for composting on campus are windrow and in-vessel systems. Windrow composting requires low cost but needs a large space for composting and is exposed to different weather and wild animals. Meanwhile, the in-vessel composting system is a fully enclosed technology method that uses less space and less odour and requires less time but is a bit expensive.

Implementing a composting system for waste management on campus has gained attention due to its being environmentally friendly and cost-efficient. Usually, all the organic waste generated on campus will be disposed of at a landfill. Some universities are already implementing composting systems at their campuses. The food waste generated at University Malaya (UM) was treated either by composting or anaerobic digestion.[4] The Faculty of Public Health/University of São Paulo (FSP/USP) started a pilot project for windrow composting in 2009, and more than 19 tons of organic waste were converted into compost from 2009 until the end of 2016.[5] Furthermore, the University of A Coruña (UDC) has had a composting program in place since 2011, with approximately 40 tons of food waste composted in 2019.[2] A rotary in-vessel with a capacity of 450 kg of food waste per day and 125 kg of wood chips per day was utilised to produce compost at Kean University and successfully generated around 13 377 kg of compost every year.[6] Meanwhile, the University of Nottingham Malaysia campus developed a composting system on campus that produced 550 kg of compost per month by composting food waste and leaves using the windrow method.[7]

One of the problems in implementing a food waste composting system is the varying characteristic of food waste. As stated by Yaser *et al.*,[3] food waste on campus had a C/N ratio of 14 to 27, a moisture content of 70% to 80%, a pH range of 4.1 to 7.4, an electrical conductivity of 1.7–8.9 $dS\,m^{-1}$, and nutrient content of N: 0.8–5.7%, P: 0.2–0.04%, and K: 0.6–0.8%. Due to the inconsistent characteristics of food waste, it is important to determine the food waste characteristics beforehand since it may affect the process mechanism. Yaser *et al.*[3] also mentioned that the type of waste produced, the amount of discharge based on activities, the design of the composting facilities, and the use of campus-scale composting as a living laboratory are all factors to consider before deciding on a technology.

The economic feasibility of a composting system on campus has been reported. Some studies found that their campus gained profit from the composting system, while some studies found a deficit from their composting system. Sanders *et al.*[8] evaluated economic data for a composting program at Texas State University. They found that the composting program led to cost savings in the transportation of waste. The economic performance of the composting system at Kean University was studied by Mu *et al.*[6] based on direct costs, savings, and income and indirect benefits, including social and environmental benefits. According to the study, the composting system could generate a profit of Malaysian Ringgit (MYR) 56 749 a year by selling vegetables grown with compost. Interestingly, a profit of MYR 101 246 could be generated if the educational and environmental benefits were included in the analysis.[6] Considering the economic performance will be different depending on the system and technology used, the objective of this study is to determine the economic feasibility of a composting system for food waste management at the Universiti Malaysia Sabah Kota Kinabalu campus.

11.2 Conceptual Process Design

The economic data collected for this study included the initial investment, the cost needed every year, and the total cash inflow per year. The ratio of benefit-to-cost analysis was also included to determine the economic viability of the composting system.[9] There were some assumptions made for the economic analysis. The total number of staff and students (resident and non-resident) of the UMS Kota Kinabalu campus was estimated to be around 2800 and 15 176, respectively. The food waste generation was calculated based on 0.12 kg per person per day for resident students and 0.06 kg per person per day for non-resident students and staff, considering they were at the campus only for half the day. Thus, approximately 309 532 kg of food waste is generated per year at the UMS KK campus. For dry leaves, it was estimated that 693 500 kg of dry leaves were generated per year.[10]

The calculation of the initial investment includes land clearing, site preparation, road construction, fencing, piping, wiring, drainage, machinery units, and other units that are needed for start-up. Solar, shredder, mixer, tong for rainwater collection, and weighing scale are some of the other equipment needed for starting up the composting system. Staff and laborer salaries, machinery maintenance and shredder fuel consumption, compost analysis (nutrients, phytotoxicity, heavy metals), electricity and water bill, and compost bagging were all included in the annual cost. The facility will be operated seven days a week with one staff and three laborers, with a salary of MYR 2800 per month and MYR 1500 per month, respectively. For cash inflow, the excess compost produced will be sold to the public for MYR 1.50 kg^{-1}. The net revenue value is obtained by subtracting the initial investment (normalised for 20 years) and cost per year from the cash inflow per year. The estimation calculation of the net revenue is given in eqn (11.1).

$$\text{Net revenue} = \text{Cash inflow per year} - (\text{Initial investment} + \text{Cost per year}) \tag{11.1}$$

From our previous study,[11] composting of food waste and dry leaves for 40 days using a passive aeration-static reactor achieved a maximum temperature of 47.8 °C. The pH value was 7.1 and the N, P, and K values of the final compost were within the recommended range. Pathogen analysis was also performed on the final compost since applying a pathogen-containing compost to agricultural soil will harm the environment and humans. In our study, no *Salmonella* spp. was detected in 25 g of a sample, and the concentration of *Escherichia coli* was 51 MPN per g in the final compost.

The compost produced from our study was used as growing media to evaluate the performance of the compost on the growth of dwarf crape jasmine (*Tabernaemontana divaricata*) for 150 days with different proportions of compost and soil. The application of compost positively affected the growth of dwarf crape jasmine, where the plant showed better growth in the treatment with compost (especially in 5–20%) than in the treatment with topsoil alone. Amongst the treatments, maximum plant growth was recorded in 5–20% of the compost medium, as illustrated in Figure 11.1. Plants with 5% compost had the greatest plant width increase (94.3%). Meanwhile, plants on growing media with 20% compost have the greatest plant height increase (58.5%), the highest dry weight value (7.2 ± 1.2 g), the highest number of leaves (184.3 ± 20.4), and the highest leaf area (4.7 ± 1.0 cm). Overall, from our previous study, the compost produced may be utilised to promote dwarf crape jasmine growth if the growing media contains a low compost dose.

Since the compost was successfully produced, it is necessary to study the feasibility of a composting system at Universiti Malaysia Sabah (UMS). The proposed location for the composting system is near Jalan Fakulti Perubatan, the UMS KK campus, as shown in Figure 11.2. Figure 11.3 shows the composter design for the composting system in UMS, which consists of two compartments, with a dimension of 1 m × 1 m × 1.5 m each, and the cost is MYR 4000 per composter. A passive aeration-static composting system has been applied in which no turning is required. Instead, the composter has been designed with holes at the back of the composter to facilitate natural air convection. A cycle of the composting process will take place in one compartment, and the retention time is 22 days. The feedstock consists of 75% (by weight) of food waste mixed with dry leaves. These were based on our previous study where compost was successfully produced using a passive aeration-static bioreactor with 75% food waste and 25% dry leaves.[11] The total weight of feedstock for each compost mixture is 400 kg. The food waste from the cafeteria on the campus and dry leaves around the campus will be sent to the composting system for free. Therefore, no transportation fee will be included. The mass reduction during the composting process was

Figure 11.1 Effect of compost proportion in media on the growth of dwarf crape jasmine after 150 days.

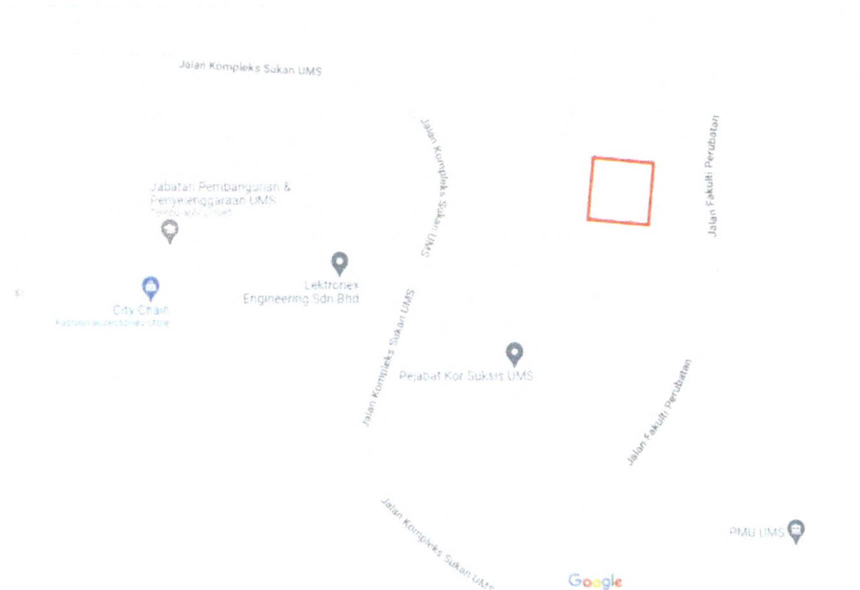

Figure 11.2 Location of composting plant in the Universiti Malaysia Sabah Kota Kinabalu campus. Map data © 2023 Google.

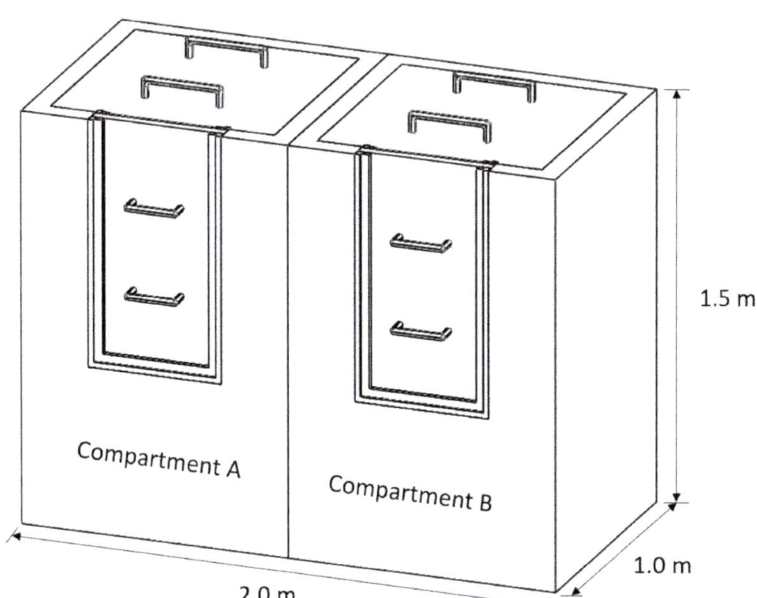

Figure 11.3 Composter for the composting system in the Universiti Malaysia Sabah Kota Kinabalu campus.

estimated to be 50% of the total weight of feedstock,[12] which means around 206 355 kg of compost will be produced each year.

Figure 11.4 shows the proposed layout for the composting system. The composting system consists of a segregation station to segregate the waste from uncompostable material; a dewatering station; a shredding station to shred the dry leaves; a mixing and weighing station; a bagging station for compost packaging assuming MYR 0.50 kg^{-1}; storage to store the compost produced; an office; a store; parking; a wetland; and a composter area. Based on the total waste produced, around 40 sets of composters (each set consisting of two compartments) are needed to process waste in UMS KK. The estimated total area of this composting system is around 400 m^2.

Table 11.1 shows the economic analysis for the composting system in UMS KK. A 400 m^2 building requires approximately MYR 66 355 for all stations, offices, and parking lots. Including contingencies of 10%, the total initial investment is MYR 424 330.50. The composting system at UMS KK will cost approximately RM 219 315.21 per year for salary, machinery maintenance, compost analysis, electricity and water bills, and bagging.

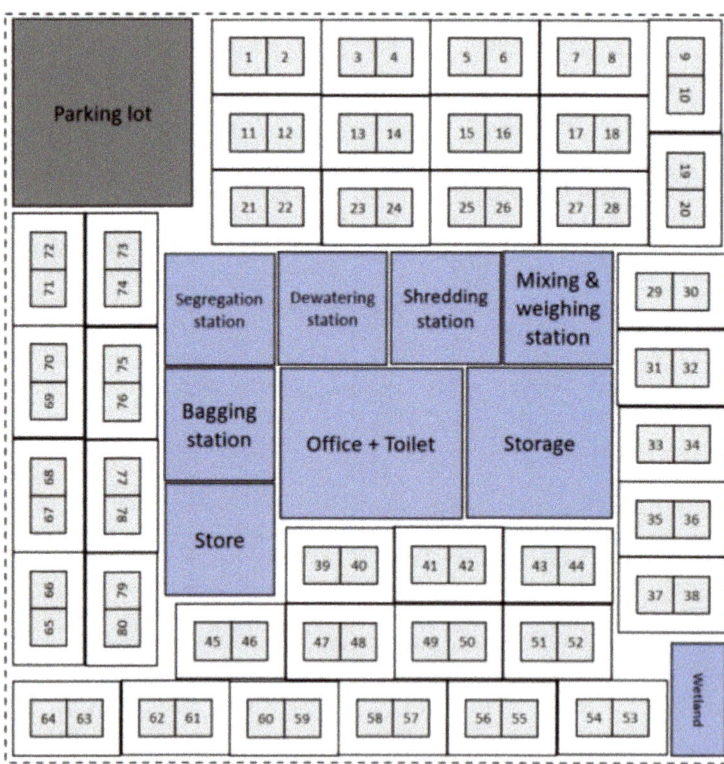

Figure 11.4 Plant layout for the composting system in the Universiti Malaysia Sabah Kota Kinabalu campus.

Table 11.1 Economic analysis of the composting system in UMS.

Items	Unit	Unit price (MYR)	Total cost (MYR)
Initial investment			
Land clearing		20 000.00	20 000.00
Site preparation		66 355.00	66 355.00
Road construction		10 000.00	10 000.00
Fencing		30 000.00	30 000.00
Piping		10 000.00	10 000.00
Wiring		10 000.00	10 000.00
Drainage		5000.00	5000.00
Effluent treatment (wetland)		5000.00	5000.00
Solar	2	15 000.00	30 000.00
Composter – 40 set	40	4000.00	160 000.00
Shredder	2	7000.00	14 000.00
Mixer	2	5000.00	10 000.00
Tong (2271 L)	3	2800.00	8400.00
Weighing scale	2	1000.00	2000.00
Auxiliary	1	5000.00	5000.00
Contingencies (10%)			38 575.50
		Total	**424 330.50**
Cost per year			
Staff (1 manager × 12 months)	12	2800.00	33 600.00
Labor (4 laborers × 12 months)	48	1500.00	72 000.00
Machinery maintenance and fuel consumption for shredder		20 000.00	20 000.00
Compost analysis		6000.00	6000.00
Electricity and water			2000.00
Bagging	0.5	196 354.92	98 177.46
Contingencies (10%)			23 177.75
		Total	**254 955.21**
Cash inflow per year			
Sell excess compost (RM1.50 kg^{-1})	196 354.9	1.50	**294 532.38**
Final costing and minimal return year			
Total investment (normalised for 20 years)		− sign: cost	−21 217
Total cost per year		+ sign: saving	−254 955
Total cash inflow per year			294 532
Total			**18 360.65**
Minimal years for return			**23.1**
Benefit-to-cost ratio			**1.07**

UMS KK is located on a 999 acre site and requires about 10 tons of chemical fertiliser per year.[7] The compost produced will be used for landscaping around the campus and the excess compost produced can be sold to the public at MYR 1.50 kg^{-1}. The total cost was calculated with the total investment normalised to 20 years.[13] The composting system installed on the UMS KK campus provides a minimum return of 23 years. A minimal return of 6 years for the implementation of a pilot-scale composting system was calculated by Bong *et al.*[13] The benefit-to-cost ratio is 1.07, which indicates that the composting system in UMS KK is economically viable. If the benefit-to-cost ratio value was greater than 1, the project offered more benefits compared to liabilities.[9]

The composting system at the UMS KK campus could gain a profit of MYR 18 360 per year based on the economic analysis. The estimation is considerably lower than that reported by Universiti Teknologi Malaysia, while higher as compared to International Islamic University Malaysia (IIUM). Kamyab *et al.*[14] estimated that composting 2700 tonnes of waste per year at Universiti Teknologi Malaysia (UTM) could generate a significant profit of MYR 1.6 million per year. It was estimated that around MYR 26 000 was spent on chemical fertilisers per year at the International Islamic University Malaysia (IIUM). Interestingly, by using compost for landscaping, IIUM saved approximately MYR 45 500 on chemical fertilisers from April 2018 to December 2019. Meanwhile, Keng *et al.*[7] evaluated the economic analysis for an aerated static pile composting system at the University of Nottingham, Malaysia and found a deficit of RM 5018 per year. The author stated that the main cause of the deficit might be due to Malaysia's low landfilling costs.

Based on these data, it is expected that the composting system at the UMS KK campus is economically viable. By using compost for landscaping around the UMS KK campus, the dependency on chemical fertilisers can be reduced.[15] Inorganic fertilisers or chemical fertilisers are expensive and affect the environment negatively if managed poorly.[16] On top of that, most of the fertilisers used in Malaysia come from outside the nation, and around RM 3.4 billion of fertilisers was imported to Malaysia in 2007.[17] A minimum of MYR 1.70 per kg of compost price is needed to increase the laborers to six people and gain a profit of around MYR 18 031 per year.

11.3 Conclusion

Composting is the best alternative for waste management on campus. In fact, some higher institutions have already implemented the composting system on their campuses. This study focused on the evaluation of the feasibility of a passive aeration-static composting system in UMS in terms of economic analysis. From our previous study, the compost produced using the passive aeration-static composting system was in the recommended range for pH, pathogens, and nutrients. Moreover, the compost can be used as a growing media for dwarf crape jasmine with a low dosage (5–20% by weight). The economic analysis in this study covered the initial investment, cost per year, and income per year. Based on the analysis, the composting system using passive aeration-static composters will be economically feasible on UMS campus KK and could possibly generate a profit of MYR 18 360 (USD 4160) per year. In addition, with a benefit-to-cost ratio value of more than one, it also suggests that the composting system at UMS is viable. Therefore, the implementation of composting systems in higher education institutions is a promising approach to organic waste management since it is an option that is both economically and environmentally feasible. Further detailed studies, including life cycle assessment, are recommended to better understand and prepare before implementing the composting system.

References

1. K.-G. Tiew, K. Watanabe, N. E. A. Basri and H. Basri, *Int. J. Adv. Sci., Eng. Inf. Technol.*, 2011, **1**, 675–678.
2. V. Torrijos, D. Calvo Dopico and M. Soto, *J. Clean.*, 2021, **315**, 128175.
3. A. Z. Yaser, J. Lamaming, E. Suali, M. Rajin, S. Saalah, Z. Kamin, N. N. Safie, N. A. S. Aji and N. Wid, *Int. J. Chem. Eng.*, 2022, **2022**, 6455889.
4. S. Yusoff, N. C. Guan, M. A. Azlan and N. S. B. K. Zaman, in *Natural Food Products and Waste Recovery*, ed. E. Carvajal-Millan, A. Z. Yaser and A. K. Haghi, Apple Academic Press, 2021, pp. 233–248.
5. J. R. Bringhenti, A. M. N. Korres, A. M. M. Moreira, S. M. Rocha and W. M. R. Günther, in *Towards Green Campus Operations: Energy, Climate and Sustainable Development Initiatives at Universities*, ed. W. Leal Filho, F. Frankenberger, P. Iglecias and R. C. K. Mülfarth, Springer International Publishing, Cham, 2018, pp. 159–173.
6. D. Mu, N. Horowitz, M. Casey and K. Jones, *Waste Manage.*, 2017, **59**, 476–486.
7. Z. X. Keng, S. Chong, C. G. Ng, N. I. Ridzuan, S. Hanson, G.-T. Pan, P. L. Lau, C. V. Supramaniam, A. Singh, C. F. Chin and H. L. Lam, *J. Clean.*, 2020, **261**, 121220.
8. J. Sanders, T. M. Waliczek and J.-M. Gandonou, *HortTechnology*, 2011, **21**, 639–646.
9. M. H. Abdul Rahman, T. Sadi, A. A. Ahmad, I. N. Masri, M. Mohammad Yusoff, H. Kamaruddin, N. A. Shakri, M. A. A. Hamid and R. Ab Malek, *Heliyon*, 2020, **6**, e04486.
10. Z. Shahudin, S. M. Zain, N. E. A. Basri, N. S. M. Zaini, N. F. M. Saad and H. Basri, *J. Teknologi*, 2013, **65**, 97–103.
11. N. A. S. Aji, A. Z. Yaser, J. Lamaming, M. A. M. Ugak, S. Saalah and M. Rajin, *J. Kejuruteraan*, 2021, **33**, 413–424.
12. S. Karnchanawong and N. Suriyanon, *Resour., Conserv. Recycl.*, 2011, **55**, 548–553.
13. C. P.-C. Bong, R. K. Y. Goh, J.-S. Lim, W. S. Ho, C.-T. Lee, H. Hashim, N. N. Abu Mansor, C. S. Ho, A. R. Ramli and F. Takeshi, *J. Environ. Manage.*, 2017, **203**, 679–687.
14. H. Kamyab, R. K. Y. Goh, J. H. Wong, J. S. Lim, T. Khademi, W. S. Ho, R. Ahmad, H. Hashim, C. S. Ho and C. T. Lee, *Chem. Eng. Trans.*, 2015, **45**, 577–582.
15. M. A. A. Samah, K. Yunus, M. R. M. Hussain and R. H. Basri, in *Natural Food Products and Waste Recovery*, ed. E. Carvajal-Millan, A. Z. Yaser and A. K. Haghi, Apple Academic Press, 2021, pp. 207–226.
16. A. Sharma and R. Chetani, *Int. J. Res. Appl. Sci. Eng. Technol.*, 2017, **5**, 677–680.
17. F.-W. Lai, M. K. Shad and H. A. Wahab, *Global Business Management Research*, 2017, **9**, 313–327.

CHAPTER 12

Techno-economic Analysis and Life Cycle Assessment of Value-added Products from Agri-food Waste

R. REBOLLEDO-LEIVA, M. T. MOREIRA AND S. GONZÁLEZ-GARCÍA*

CRETUS, Department of Chemical Engineering, School of Engineering, University of Santiago de Compostela, 15782 Santiago de Compostela, Spain
*Email: sara.gonzalez@usc.es

12.1 Introduction

Society is changing the resource consumption paradigm, moving from "take, make and dispose of" to "rethink, reuse and recovery" to face the climate change crisis and reduce the environmental consequences from extensive resource depletion. In this context, the circular economy (CE) appears with the aim of remodelling the life cycle of a product, wherein the decrease of environmental impacts is achieved, minimising resource consumption and waste generation.[1] Furthermore, this concept supports the promotion of industrial symbiosis, where industrial synergy is encouraged through the reuse of by-products within the industrial network, decreasing waste generation and raw material consumption of the system.[2] In this way, the sequential and consecutive use of resources (known as cascading principle) can increase their efficiency with less demand for fresh materials, promoting added value and job creation.

Green Chemistry Series No. 78
Agri-food Waste Valorisation
Edited by Pankaj Chowdhary and Abhay Raj

Published by the Royal Society of Chemistry, www.rsc.org

The bioeconomy framework encompasses the conversion of renewable biological resources (and waste streams) into value-added products, such as food, feed, bioenergy, and all kind of biobased materials and products.[3] In this sense, biomass obtained from agriculture, forestry, and marine environments is a carbon source with great availability worldwide and is suitable to obtain these products. Thus, it has been considered a sustainable renewable alternative to face growing greenhouse gas (GHG) emissions and the depletion of fossil fuels. Nevertheless, obstacles to efficient biomass use are its seasonality, geographic availability, and relatively low calorific value.[4] Therefore, a challenge for the bioeconomy is to keep the value of biomass during its cascading use, where energy purposes in the last place are recommendable.[3]

Biorefineries can contribute to maximising the use of biomass, including the wastes generated, from multiple conversion pathways, adopting a cascading approach that favours value-added and resource-efficient products over energy purposes. The biorefinery concept has been designed and utilized to process biomass feedstocks such as lignocelluloses, algae, and different types of wastes.[5] In this context, De Bhowmick *et al.*[6] proposed the use of lignocellulosic biorefinery as a platform to address the sustainable co-production of high-valued bio-products and biofuels, by recycling biomass and its wastes with the integration of conversion technologies. Furthermore, biorefineries using wastes (non-edible feedstocks and biogenic wastes) have also emerged as alternatives to produce bio-based products such as biopolymers, biofuels, and biochemicals.[7] The biomass selection for a biorefinery platform will be limited by the quality and quantity of the available supply. Hence, the first relevant task is estimating the available amount of biomass in the region under study and characterising it based on whatever purpose it may potentially serve, before proceeding with the valorisation strategies.[8] Furthermore, a high conversion yield and a minimum cost of co-product separation are the basic requirements for the success of biomass valorisations in biorefineries.[9] For the success of biorefineries, techno-economic analysis (TEA) is a well-known approach to assessing the technical and economic feasibility of a technology, combining energy and mass balances, with economic assessment through capital, operation, and input production costs.[10] In this manner, it is possible to determine the cost-effectiveness of the technology evaluated, compare technological options with the same purpose, or determine the minimum selling price (or maximum production costs) for competitive technology. In addition, special attention to the choice of transformation processes is crucial for avoiding creating a bio-product that could demand more energy or reagents than its fossil-based counterpart.[11] Generally, health and safety issues as well as emission control from the plant are considered during the design phase of a process. However, the prevention of potential pollution transfers from one life cycle phase to another is also relevant. In this regard, the life cycle assessment (LCA) can represent the most promising methodology for environmental assessment, because it allows considering the life cycle of the process or bioproduct, expanding the boundaries of the assessed system, by considering upstream and downstream processes and not just the operational stage.[12]

This chapter aims to present the progress of the joint use of TEA and LCA for the economic and environmental assessment of valorisation strategies of agri-food wastes into value-added products from a biorefinery perspective. Thus, this chapter will cover aspects such as agri-food waste properties, methodological aspects for the techno-economic and life cycle analysis, and a literature review about valorisation routes as well as the application of both perspectives in those studies.

12.2 Theoretical Backgrounds

12.2.1 Techno-economic Analysis

For a process design, it is crucial to assess the economic feasibility of a specific project or potential production alternatives. To do this, TEA evaluates both technical and economic aspects of an engineering project, comprising technical features such as market availability, lifetime, maintenance demand, skill requirement for operation, capacities range, transportation and installation, adaptability under specific conditions, and intrinsic risks.[13] Meanwhile, the economic approach deals with monetary value estimation, considering capital and operating costs, and revenues obtained from the evaluated project.[13,14]

12.2.1.1 Process Design and System Boundary

The process design begins with technological descriptions of the required conditions to operate the system and obtain the desired product. The technological descriptions include the capacity and location of the plant, the requirement of feedstocks and raw materials, and process descriptions including flow diagrams and information on products, by-products, and wastes. Then, boundaries specify the system to be evaluated as well as its inputs and outputs. Everything outside the system boundary can be described as the surrounding. Consequently, inputs flow from the surrounding to the system and outputs flow from the system to the surrounding, and all inputs and outputs must be balanced. In this sense, the system boundary includes all the required processes to transform feedstock into desired products.

12.2.1.2 Data Description for TEA

The precision of TEA will depend on the accuracy of the operating parameters and cost data used in the modelling process. Data on the operating parameters refer to the amounts of inputs, their conversion factors, operating conditions (*i.e.*, time, temperature, and pressure), products, by-products, and residual streams produced in each unit of operation. Based on the collected data, the balance of mass and energy should be done, considering also the estimation for the required utilities and the size and amount of equipment.

The operating data for the modelling process can be obtained from experimental studies, chemical engineering process modelling textbooks, scientific articles, pilot-scale tests, and commercial facilities. Concerning the cost data to perform the economic analysis, it should consider raw material purchase prices of feedstock, materials, and other consumables, engineering and construction charges, the purchase price of equipment, costs related to the installation and maintenance of equipment, utility cost, labour cost, taxes and insurance, and contingency costs. In this way, the economic analysis lets to determine the costs of the different involved factors, allowing set operational targets for process improvement at the early stage of design. Accordingly, this analysis leads the design process economically competitive by reducing the highest cost components through alternative production strategies and equipment, increasing product yield targets, and developing efforts to meet operational targets.

12.2.2 Life Cycle Assessment Methodology

Life cycle assessment is a well-known methodology for accounting and assessing the environmental impacts of services or products throughout their life cycle. Following the ISO 14040 series,[15] this methodology comprises four phases (see Figure 12.1): (i) goal and scope definition; (ii) life cycle inventory; (iii) life cycle impact assessment; and (iv) results interpretation. The results obtained from an LCA study allow the comparison of different products, systems, and scenarios; understanding the environmental impacts and identifying critical stages (*i.e.*, hotspots) of the life cycle, identifying potential trade-offs and opportunities for implementing improvement actions. Thus, LCA represents a fundamental tool for performing environmental analyses for decision-making processes.

12.2.2.1 Goal and Scope Definition

This stage is relevant because definitions made here will influence the whole procedure of the assessment. The purpose of this phase is to describe the

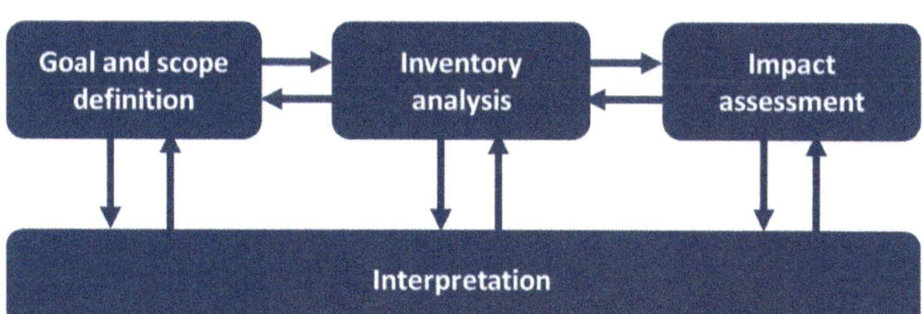

Figure 12.1 Life cycle assessment methodological framework. Adapted from ISO 14044:2006 – Environmental Management: Life Cycle Assessment; Requirements and Guidelines.

drivers and motivations as well as the decisions that are intended to support the study. Hence, the goal should clearly address aspects such as defining the reason for performing the study, product function and purpose, and the audience of interest (*e.g.*, organisations, costumers, producers, consumers, and policymakers, among others). The scope definition focuses on the limitations and assumptions considered, addressing clearly methodological aspects such as the functional unit, reference flow, system boundary, geographical location, limitations, allocation method, data sources, and assumptions.

Regarding these aspects, the functional unit is the reference unit in which environmental burdens of products and/or services are quantified, which always should be defined explicitly in any LCA study. This unit should represent the function provided by the evaluated product or service. Thus, the functional unit possibilities compare different products/services or systems to identify environmentally friendly options. Furthermore, a reference flow is a quantified amount of the product necessary to deliver the performance described by the functional unit. System boundary establishes all processes that belong to the product scheme considering foreground and background processes. Some approaches for boundaries are gate to gate, cradle to gate, cradle to consumer, and cradle to grave, among others. In the environmental evaluation, the boundaries definition is a relevant decision because how extended the system (*i.e.*, the number of processes involved) will determine the product profile, as well as allow for comparing different alternatives properly. For instance, a cradle-to-grave approach will consider all the processes, from the extraction of raw materials, passing through manufacturing, transportation, product use, and, ultimately, disposal. Meanwhile, a gate-to-gate approach only considers those activities from the factory entry gate to the exit gate. Allocation methods refer to how to allocate the environmental burdens when multiple products (main product, co-products, and by-products) are obtained.[16] Mass, economic, and energy criteria are commonly used for determining allocation factors.

12.2.2.2 Life Cycle Inventory (LCI) Analysis

This stage is an intensive and extensive process because of the data collection of processes within the boundaries. In this stage, all flows (*i.e.*, inputs and outputs) of the product system are identified, corresponding to raw materials and energy consumption, direct emissions, products, and waste (see Figure 12.2). Primary and secondary sources are traditionally used based on data availability. Primary data are taken through questionnaires, surveys, direct measurements, or experiments at process sites. Secondary data are obtained mainly through literature studies and commercial databases. Databases are useful for considering, for example, the environmental burdens of the background processes. Furthermore, sensitivity analysis should be performed to treat the potential uncertainties and assumptions made by practitioners to assure the credibility of the results.

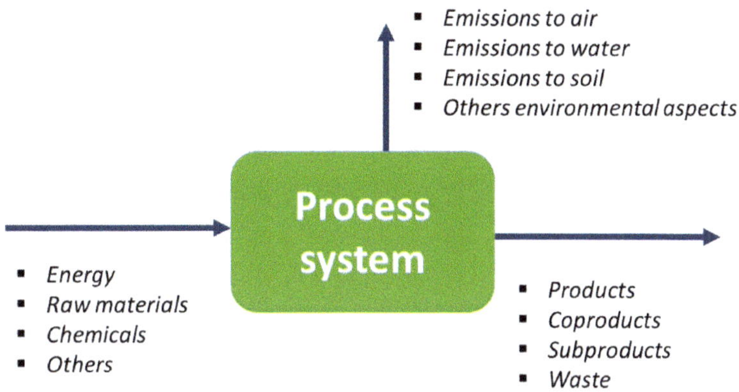

Figure 12.2 Identification of input and output flows for inventory analysis.

12.2.2.3 *Life Cycle Impact Assessment*

Impact methods transform information obtained from inventory balances into indicators that reflect potential impacts generated on the environment, natural resource availability, and human health. The ISO 14040 guidelines[15] distinguish the following mandatory steps: (i) selection of impact categories, category indicators, and characterisation models; (ii) assigning LCI results to impact categories based on their potential effects (classification), and (iii) calculation of category indicators by quantifying inventory flow contributions to impact categories (characterisation). After this stage, the environmental profile is obtained according to each impact category of interest. Different impact assessment methods such as CML 2001, IMPACT2002, ReCiPe 2016, and IPCC 2007/2013/2021 are used for reporting a set of impact indicators.

12.2.2.4 *Interpretation*

This stage interprets outcomes from the inventory and/or impact assessment stages and summarises the main findings that fulfil the goal and scope previously established. Figure 12.1 presents this stage as an iterative process with interaction with the remaining three stages, due to different issues that can be identified related to data availability and quality, sensitivity analysis, suitable definitions of the system boundaries or functional units, and completeness and consistency of the data.

12.3 Agri-food Waste

The kind of feedstock selected is a fundamental parameter during biomass conversion into a value-added product, because it determines the available amount of the desired component. Biomass contains mainly cellulose, hemicellulose, and lignin, and their contribution varies according to the feedstock type. Agricultural by-products come from different biomass sources, for example, the cultivation of wheat, corn, rice, and bagasse, among others, obtained after harvesting. Traditional biomass use can be animal feeding or as bedding for livestock, composting, or burning to reduce pests and weeds. The economic cost of biomass for potential use in a biorefinery plant can increase due to the seasonal availability of the biomass and for being an attractive alternative to traditional end-of-life. In this sense, potential trade-offs of the use of agricultural subproducts can be their low yields per hectare (compared to purpose-grown crops) and carefully regulated removal rate to maintain (or improve) soil carbon and organic matter stocks; there are other competing end uses which may be of higher priority locally (in particular for livestock).[17] For example, oil seed rape straw due to its high nitrogen content would preferably return to the soil to improve or maintain soil nitrogen stock.[18] Meanwhile, only a portion of cereal straw should be removed for biorefinery purposes to maintain soil carbon and organic matter stocks. This fraction may vary according to the soil type, climatic conditions (rainfall and temperature), the type of crop sequence, and crop yields.[17] The growing demand for energy and bioproducts motivates the promotion of a market for agricultural by-products, which should be carefully addressed by considering a suitable crop selection as feedstock based on their composition and the aim product (*i.e.*, selecting the subproduct that provides the highest quantity of the element required to generate the aim product) and avoiding crop diversity reduction, which is relevant to ensuring the provision of ecosystem services.[17]

Food waste (FW) is produced along the entire food supply chain, highlighting the consumption (56%) and the manufacturing stages (around 38%).[19] Its presence is predominant in the organic portion of municipal solid waste,[20] when it is diverted to landfilling with related economic and environmental implications. Nevertheless, composting, animal feed, and energy recovery (*e.g.*, anaerobic digestion)[21] are other feasible strategies. FW obtained from the food processing stage offers high compositional homogeneity, allowing their use to obtain added-value products through, for example, biorefinery platforms.[19] Some of the fundamental aspects to assess the viability of an FW biorefinery concept include[19] (i) estimating the availability of FW; (ii) determining whether the waste-based production process is less resource intensive than the manufacturing of the same products from a virgin material; and (iii) comparing different valorisation paths taking into account economic, environmental, and social perspectives. Furthermore, the market price of bioproducts is critical to determine the profitability of a biorefinery platform, which motivates performing a market analysis during the design stage.

12.3.1 Agri-food Waste Composition

Agri-food wastes are considered a valuable source because they are composed mainly of cellulose, hemicellulose, lignin, protein, and ash. Their composition ranges from about 35–50% for cellulose, 20–35% for hemicellulose, and 15–20% for lignin, and others, respectively.[22] A brief description of these elements is presented as follows:

(i) Cellulose: the main constituent of lignocellulosic biomass is a polysaccharide with a linear chain of D-glucose linked by β-(1,4)-glycosidic bonds. The strains are associated together to make cellulose fibrils, which are linked by intra- and inter-molecular hydrogen bonds.[23] Thus, cellulose is insoluble in water and most organic solvents.[24]
(ii) Hemicellulose: are heterogeneously branched biopolymers containing pentoses (β-D-xylose, α-L-arabinose), hexoses (β-D-mannose, β-D-glucose, α-D galactose), and/or organic acids (α-D-glucuronic, α-D-4-*O*-methylgalacturonicanda-D-galacturonic acids). Their amorphous, branched structure and lower molecular weight promote a relatively easily hydrolysation.[22,23] Operation conditions are relevant in the treatment of hemicelluloses; for example, temperature and retention time parameters must be monitored to avoid formation of inhibitors in the fermentation process (*e.g.*, furfurals and hydroxymethyl furfurals).[25]
(iii) Lignin: it provides rigidity and cohesion to the material cell wall, confers water impermeability to xylem vessels, and creates a physicochemical barrier against microbial damage.[26] Its structure does not contain sugar units, but phenylpropane units (such as syringyl, guaiacyl, and *p*-hydroxyphenol) linked in a complex matrix.[27]

12.3.2 Sources of Feedstocks

Biorefineries and their feedstocks related to agri-food waste can be categorised into first or second generation:

(i) First generation (1G): bioproducts (*e.g.*, biofuels) are produced from edible food crop parts such as starchy, sugary products, plant oils, *etc.* In the energy context, the edible parts can be converted to bioethanol, biodiesel, and biogas, among others. Nevertheless, the food and feed *versus* biofuels dilemma was a great concern in this area.
(ii) Second generation (2G): raw materials correspond to non-food crops (*e.g.*, energy crops), agricultural subproducts, and municipal waste. As for 1G, the competition of growing non-food crops (*e.g.*, for energy purposes) with food crops for the available land has limited the cultivation of non-food crops.

The biorefinery aim should encourage the feedstock selection from which the highest yield conversion could be possible to obtain. The different compositions of cellulose, hemicellulose, and lignin useful for bioproduct generation are presented in Table 12.1, for agricultural residues, and food industry wastes.

Table 12.1 Composition of different agri-food wastes.[28] Reproduced from ref. 28 with permission from Elsevier, Copyright 2020.

Source	Type	Cellulose (%)	Hemicellulose (%)	Lignin (%)
Agro-residues	Barley straw	36.0–43.0	24.0–33.0	6.3–13.1
	Barley hull	34	36	13.8–19.0
	Corn cobs	33.7–41.2	31.9–36.0	6.1–15.9
	Corn stalks	35.0–39.6	16.8–35.0	7.0–18.4
	Corn stover	30.6–43.9	19.1–28.9	16.7–21.8
	Extracted olive pomace	19	22	40
	Hemp stalk	52	25	17
	Jerusalem artichoke	20.9–25.9	4.5–5.4	5.1–5.7
	Maize	30.3	19.5	4.8
	Oat straw	31.0–35.0	20.0–26.0	10.0–15.0
	Palm oil frond	37.32	31.89	26.05
	Ray straw	36.0–47.0	19.0–24.5	9.9–24.0
	Rice husks	28.7–40.0	12.0–29.3	15.4–26.0
	Rice straw	29.2–38.1	23.0–31.1	17.0–26.4
	Soya stalks	34.5	24.8	19.8
	Sunflower	34.1–42.1	5.18–29.7	7.72–13.4
	Sorghum straw	32.0–35.0	24.0–27.0	15.0–21.0
	Sorghum bagasse	44.4	35.5	3.9
	Sugarcane bagasse	31.9–43.4	12.2–25.5	23.1–27.6
	Wheat straw	35.0–39.0	23.0–30.0	12.0–16.0
Food waste	Almond shell	27	30	36
	Spent grains from brewery	16.8–26.0	19.2–29.6	16.9–27.8
	Banana waste	13.2	14.8	14
	Coconut coir	44.2	22.1	32.8
	Coffee grounds	11.6–33.1	37.2–41.0	22.2–25.6
	Groundnut shell	37	18.7	28
	Hazelnut shell	30	23	38
	Pine nut shell	31	25	38
	Pistachio shell	15.2	38.2	29.4
	Sponge gourd fibers	66.6	17.4	15.5
	Walnut shell	23.3	20.4	53.5

12.4 Biorefinery Design and Performance Assessment

A biorefinery plant allows the conversion of biomass to bioproducts including bioenergy (fuels, heat, and power) and a diverse array of co-products (including materials and chemicals). A platform can be divided into three sections: pre-treatment, conversion, and recovery stages (see Figure 12.3). The conversion routes can be performed through a biochemical, design-chemical, or a combination of both approaches. Biochemical conversions use enzymes and microorganisms (yeast and bacteria), where biomass is first pre-treated and hydrolysed to mono-sugars (*i.e.*, glucose, xylose,

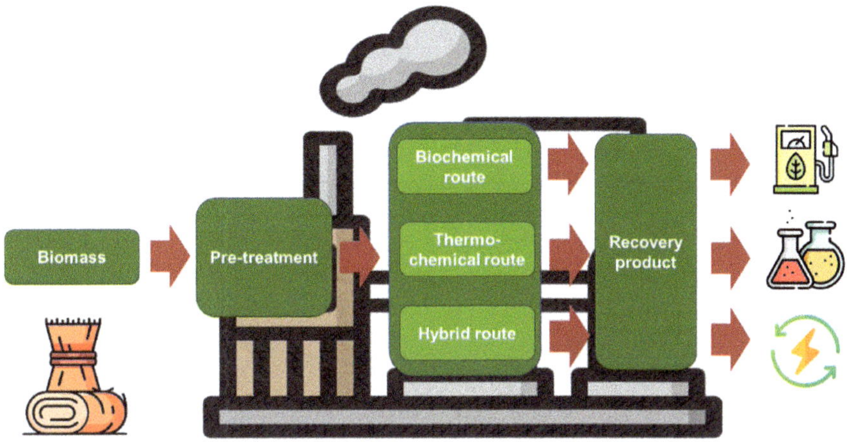

Figure 12.3 An integrated biorefinery on a life-cycle perspective, inspired by Azapagic.[29] Adapted from ref. 29 with permission from Elsevier, Copyright 2014.

arabinose, and galactose, among others). The mono-sugars are then fermented to obtain, based on the biocatalysts used, biofuels (*e.g.*, bioethanol, biobutanol, *etc.*) or chemicals (*e.g.*, lactic acid and succinic acid, among others). On the other hand, design-chemical conversion includes biomass combustion for heat and power, pyrolysis for bio-oil and biochar, hydrothermal liquefaction to bio-oils as a major product, and biomass gasification to syngas (mainly CO and H_2). Each conversion route will have to overcome technological issues before it can become a commercial reality, including feedstock availability, high conversion yield, and the successful scaling-up to large-scale plants.

12.4.1 Pre-treatment Processes

The first stage corresponds to the biomass pre-treatment to break down the lignocellulosic links and improves the polymeric sugar accessibility, releasing the cellulose and hemicellulose. In this way, it can be possible to transform polymeric sugars into monomeric sugars that can be further converted into value-added products. Pre-treatment can be considered a fundamental process in the cost-efficient conversion of lignocellulosic biomass;[28] thus, the selection of the pre-treatment process can significantly influence the overall cost and configuration design of the biorefinery.

Pre-treatment processes can be categorised into three route alternatives: physical, chemical, and biological.[26,30] The first one includes mechanical (*e.g.*, grinding, chipping, and milling), microwave and ultrasound irradiation, steam explosion, and liquid hot water treatments.[31] The second one requires acid or base chemicals for 20–30 min at high temperatures.[32] Some examples correspond to strong acids (HCl, H_2SO_4, and HNO_3), alkalis (NaOH, KOH, $Ca(OH)_2$, ammonia, and hydrazine), organic solvents (benzene, ether, cyclohexane, hexane, ethanol, and methanol), ozone (O_3), and ionic liquids (ILs).[31] Concerning the third one, it requires ligninolytic enzyme–producing bacteria or fungi to degrade the lignocellulosic complex resulting in the release of polymeric sugars in the solution.[33]

According to Saini *et al.*,[32] some advanced pre-treatment methods that attract researchers correspond to supercritical fluids, ionic liquids, and deep eutectic solvents. The supercritical fluid method employs compounds such as water, ammonia, and carbon dioxide (CO_2) as supercritical fluids. For example, pre-treatment with supercritical CO_2 includes an explosion step (*i.e.*, an abrupt reduction in pressure) to break down lignin, cellulose, and hemicellulose links.[34] A cost-effective pre-treatment and low concentrations of biomass degradation compounds are some advantages of this approach, while less enzymatic hydrolysis yields can be obtained compared to the steam explosion strategy.[32] The use of ionic liquids, also known as a green alternative to organic solvents, can dissolve carbohydrates and lignin from lignocellulosic biomass, simultaneously. Ionic liquids are organic salts composed of cations and anions (large organic cations and small inorganic anions) and have a melting point of 100 °C.[26] Furthermore, it is considered a low environmental impact strategy because of its low volatility and vapor pressure, high solvent power, and the chance of recovering the solvent employed.[31] However, the lack of knowledge regarding physicochemical techniques, high production cost, and toxicity issues represent some challenges to be addressed.[26] These motivated the last strategy, deep eutectic solvents, which are prepared by mixing a hydrogen acceptor and a hydrogen donor at 60–80 °C.[28] They improve the solubilisation of the lignocellulosic complex and selectively remove the lignin without affecting cellulose.[32] Some studies reported about 66%[35] and 90%[36] of lignin removal for poplar and corn stover pre-treatment, respectively. This solvent pre-treatment is judged as a green and cost-effective process due to its simplicity in the synthesis process, the low solvent requirement, and purification steps;[28] in addition, they are usually less toxic, easily biodegradable, and recyclable.[37]

12.4.2 Separation and Purification Processes

Separation and purification of biomass components and product streams, as well as their full integration with the overall process, is a relevant task. Here, a brief introduction of downstream processes is presented; for more detailed information related to separation and purification processes, please see the literature review performed by Huang *et al.*[38]

12.4.2.1 Equilibrium-based Separation Processes

(i) Absorption: it is used for the separation of desired gas components from a gas mixture into a liquid solvent phase. In biorefinery designs, absorption is commonly used for the removal of acid gases (such as H_2S and CO_2) from syngas prior to syngas synthesis.

(ii) Distillation: there are different distillation processes for liquid mixture separation: ordinary distillation, azeotropic distillation, and extractive distillation. In addition, molecular distillation is performed under high-vacuum conditions and is suitable for the fractionation and separation of chemicals from pyrolysis of bio-oils.[39] Steam distillation is utilized for isolating volatile organic compounds such as essential oils that are sensitive to high heat from plant materials. Moreover, it is used for direct separation of the desirable components from solid biomass feedstock (*i.e.*, not a liquid mixture). Steam distillation can separate light components of essential oils and bioactive compounds from biomass.[40]

(iii) Solvent extraction: a conventional separation process where one or more mixed solvents are used to extract the desirable component from the feed liquid phase to the solvent phase.

(iv) Supercritical fluid extraction: a supercritical fluid is applied to extract valuable solutes from a solid matrix or a liquid mixture under its supercritical condition. Carbon dioxide is the most used supercritical fluid in food, pharmaceutical, and chemical industries.[38]

12.4.2.2 Separation Based on Affinity

Adsorption, chromatography, and ion exchange are conventional sorption processes, where certain adsorbates are selectively moved from the fluid phase to the surface of insoluble, rigid particles suspended in liquid in a vessel or packed in a column. Both ion exchange and adsorption are used for the efficient removal of inhibitors from biomass hydrolysates.[38] Adsorption with different adsorbents can be applied for the separation and purification of biofuels and chemicals, for example, the dehydration of ethanol and biobutanol.[41]

12.4.2.3 Membrane Separation

These technologies have been widely researched for biofuel separation. From them, electrodialysis is commonly used for the separation of organic acids such as acetic and oxalic acid,[42] and citric acid[43] from their fermentation broths. Moreover, different membrane technologies have been widely studied, such as microfiltration (0.050–10 μm), ultrafiltration (1–100 nm), or nanofiltration (<2 nm), which can be selected depending on the size of the molecules to be separated. For example, nanofiltration is much better than ultrafiltration for separating hemicelluloses of small molecular weights from hydrolysates (*e.g.*, from the alkaline process).[44] Moreover, there are membrane technologies for gas separation and purification, which are employed, for example, to separate hydrogen from syngas or biogas, providing high purity of the H_2 product.[45]

12.4.2.4 Solid–Liquid Separation

(i) Conventional filtration: it is a well-known and commercially available solid–liquid separation technology. Biorefineries involve multiple solid–liquid separation procedures such as the separation of pre-hydrolysate and post-distillation slurry.
(ii) Solid–liquid extraction: a solvent or solvent mixture is used to extract valuable compounds from the solid feedstock. These extraction technologies include conventional solid–liquid, ultrasound-assisted, microwave-assisted, and pressurized subcritical liquid extraction.
(iii) Precipitation and crystallisation: they are two downstream operations wherein a solid phase biomolecule is formed from an aqueous solution. This solid-phase formation is produced by property change of the aqueous solution, *e.g.*, by shifting the temperature or pH, by adding polymers (either neutral or polyelectrolytes), salts, or organic solvents. The precipitation process occurs commonly at the beginning of a downstream process, and its main objective is to concentrate the biomolecule. Otherwise, crystallisation occurs usually at the end of the downstream process, aiming at achieving the final purification and/or formulation of the target bioproduct.[46]

12.4.3 LCA and TEA Through Simulation Tools

A process design of a biorefinery platform can be a complex task due to the arrangement of physical, chemical, biochemical, or thermochemical processes.[12] Thus, determining the best pathway from economic and environmental feasibility is essential for the potential implementation of the plant. In this sense, process simulation at the early phase of design appears as a suitable tool to obtain a comprehensive perspective of bio-based systems at an industrial scale. In this way, the main potential hotspots could be identified and improvement actions could be proposed to optimize them, seeking efficient raw materials and energy consumption. Process simulation software is a useful tool for biorefinery design; for instance, ProSimPlus®, Aspen Hysys®, Chemcad®, Unisim design®, or SuperPro designer® are available options. Besides process simulation, these tools also enable the performance of a techno-economic analysis. Thus, an emerging practice in the literature is the joint use of process simulation (with or without TEA) and LCA methodology for process design of environmentally feasible bio-based systems. The main objective is to simulate a process arrangement to obtain mass and energy balances and, consequently, the inventory data. In this sense, Azapagic *et al.*[47] proposed a methodological scheme called Process Design for Sustainability (PDfS). The first stage is "the project initiation", which seeks to identify relevant stakeholders and evaluate alternatives based on sustainability criteria. Then, "Preliminary Design" is accomplished to obtain mass and energy balances (for LCA) and the economic analysis to determine processing route feasibility. Finally, a detailed design (*e.g.*, equipment design, process control, and instrumentation) can be performed, with a full assessment of sustainability.

According to Julio *et al.*,[12] the combination of LCA and process simulation is an emerging and promising method to enhance the environmental and economic performance of industries, by proposing innovative and optimized solutions before platform construction. This approach has been increasingly used for supporting the operation of biorefining plants.

12.4.4 Multicriteria Analysis in Process Design and Optimization

Multi-objective optimization (or multicriteria or Pareto optimization) is a multiple-criteria decision-making approach to enhance simultaneously more than one objective function that can be in conflict, *e.g.*, increasing economic profit while reducing related environmental burdens. This approach is useful for the decision-making process related to the selection of the best processing route. The optimization modelling can improve multiple parameters, such as the process capacity, size of the unit process, flow rates, yields, and raw materials and energy consumption. Kniel *et al.*[48] presented one of the first studies that applied LCA for process design and optimization. They linked LCA and economic analysis to perform a multi-objective analysis for the optimization of a nitric acid plant. The authors asserted that LCA is one of the best approaches able to link unit processes, environmental burdens, and economic aspects. Other early studies[49,50] on the LCA and process application are associated with the LCA application in process selection, design, and optimization, as a tool for identifying clean technologies.

12.5 Search Procedure

To evaluate the progress of the joint use of TEA and LCA in the agri-food waste valorisation, a state-of-the-art analysis has been carried out. The procedure for searching and selecting the scientific articles on the topic is described as follows:

(i) Database search: a search was conducted in the Scopus database in June 2022, using the sets of keywords presented in Figure 12.4. The selected publications were those published from 2000 to 2021.
(ii) Article management: all manuscripts were handled through the Mendeley® software, where all articles duplicated, conference papers, reviews, and those published in a language other than English were removed.
(iii) Screening title, abstract, and keywords: all the titles, abstracts, and keywords of each article were read to determine their suitability for the research topic of this literature review. Articles related to energy

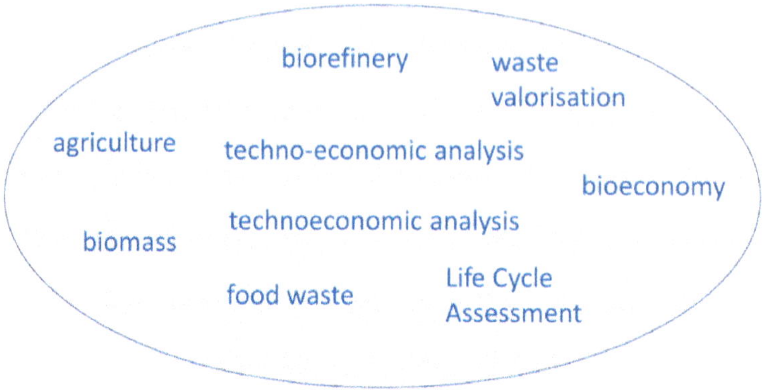

Figure 12.4 Keywords for search process in the Scopus database.

crops were not considered because they are cultivated for energy purposes, *i.e.*, they are not a subproduct of an agri-food system.

(iv) Full-text reading: after identifying and selecting articles, full texts were read. The decision to retain or exclude articles after a full reading was based on whether they are aligned with TEA and LCA application for agri-food waste valorisation or not.

(v) Research profile analysis: a descriptive analysis is performed for the sample of selected articles.

Following this procedure, a total of 39 articles were obtained, including original research and book chapters.

12.6 Descriptive Analysis

According to the sample of the articles identified, a great number of studies have been published recently, between 2020 and 2021 (see Figure 12.5). Moreover, there is a period, from 2014–2016, when no articles related to agri-food waste were found. On the other hand, corn and sugarcane biomass were the most used feedstocks for biorefinery systems with 29% each (see Figure 12.6). Regarding the journals where these articles were published, it is possible to indicate that there is a wide variety of sources, and they are, mainly, related to chemical engineering and energy areas (see Table 12.2).

12.7 Current Applications of TEA and LCA in the Valorisation of Agri-food Waste

Based on the sample selected, studies are classified according to their main aim performed: process design, economic and environmental assessment of bioproducts, multicriteria analysis, and simulation software proposed.

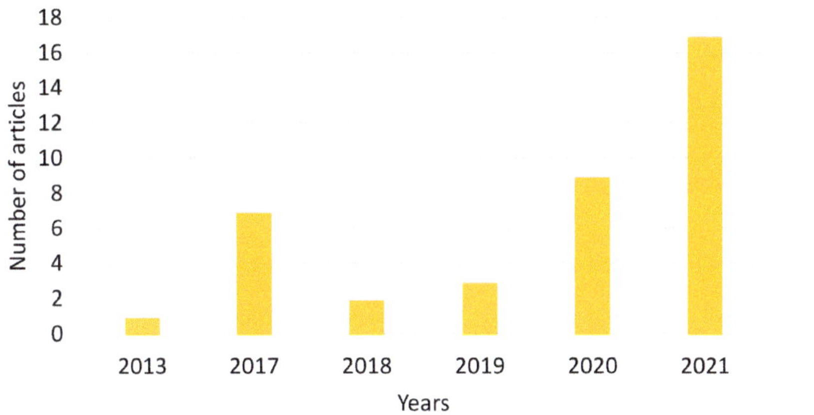

Figure 12.5 Number of articles on TEA and LCA of agri-food waste published per year.

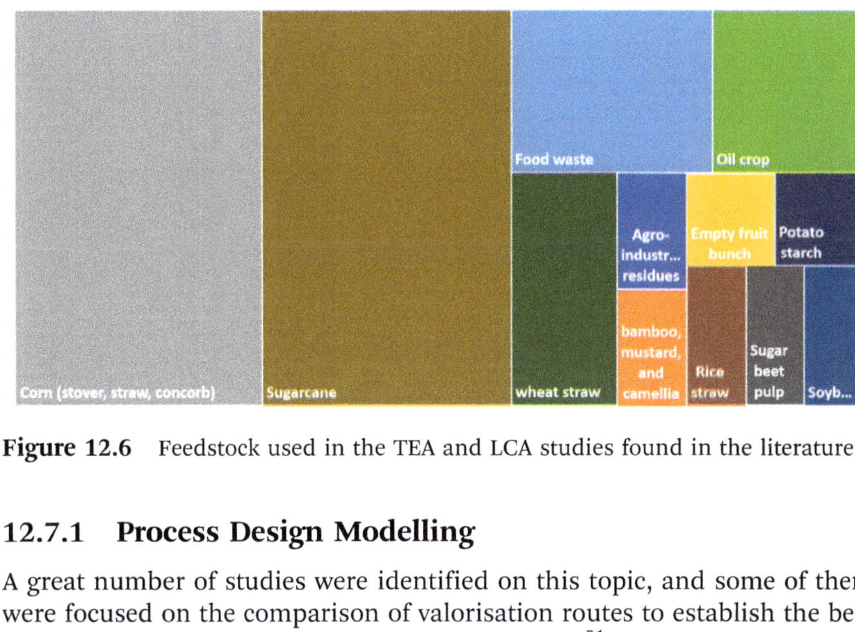

Figure 12.6 Feedstock used in the TEA and LCA studies found in the literature.

12.7.1 Process Design Modelling

A great number of studies were identified on this topic, and some of them were focused on the comparison of valorisation routes to establish the best processing alternative. For example, Souza *et al.*[51] compared three management strategies for heat-treated potato liquor (HTPL): (i) concentrating it *via* evaporation before using it as a fertilizer; (ii) fermentation to produce protein-rich biomass, and (iii) anaerobic digestion of the HTPL to produce biogas. The protein-rich biomass scenario obtained the lowest impact in five out of the six impact categories analysed. The exception was the global warming potential, which was higher than the biogas generation strategy. Mandegari *et al.*[52] studied eight alternative biorefineries coupled to a typical sugarcane mill to obtain the best biofuel/biochemical option, between producing ethanol, lactic acid, or methanol, and co-production of ethanol and

Table 12.2 Journals in which the studies have been published.

Journals	Frequency
ACS Sustainable Chemistry and Engineering	4
Energies	3
Applied Energy	2
Bioresource Technology	2
Energy	2
Energy Conversion and Management	2
Biomass and Bioenergy	1
Cell Reports Physical Science	1
Cellulose	1
Chemical Engineering Journal	1
Chemical Engineering Research and Design	1
Chemical Engineering Science	1
Computers and Chemical Engineering	1
Environmental Science and Technology	1
Fermentation	1
Food and Bioproducts Processing	1
Frontiers in Energy Research	1
GCB Bioenergy	1
Industrial and Engineering Chemistry Research	1
Industrial Crops and Products	1
Journal of Cleaner Production	1
Journal of Thermal Analysis and Calorimetry	1
Journal of the Japan Institute of Energy	1
Proceedings of the National Academy of Sciences	1
Processes	1
Renewable and Sustainable Energy Reviews	1
Waste Management	1

lactic acid. Moreover, the effect of the multi/co-production strategy with different energy supplementation methods was also analysed. Yang *et al.*[53] compared three ethylene manufacturing pathways, based on ethane-rich shale gas, corn stover, and corn grain, at five different scales. The shale-gas-based route obtained the lowest break-even ethylene prices, but the highest net GHG emissions with respect to corn stover and grain. Otherwise, the corn-stover-base pathway resulted in the lowest GHG emissions but the highest break-even ethylene prices.

Similarly, Pachón *et al.*[54] evaluated three alternatives to integrate a biorefinery platform into a common South African sugar mill coproducing biofuels, biochemicals, and electricity. Scenarios under study were (i) coproduction of sugar, ethanol fuel (from cellulose), and lactic acid (from hemicellulose); (ii) coproduction of sugar, ethanol fuel (from cellulose), and furfural (from hemicellulose); and (iii) integrated production of sugar, ethanol fuel (from cellulose and hemicellulose), and surplus electricity (from bagasse residues and green harvesting residues). For LCA analysis, the authors compared the use of relevant services provided by the products generated. For instance, for the first scenario, the reference system includes the distance travelled by a car operating under gasoline, sugar from sugarcane, and fossil-based lactic acid. The first scenario was the most profitable system, because of the suitable scale of lactic acid production. Moreover, it obtained the highest impact reduction in Global Warming Potential (GWP) and fossil depletion categories by substituting the reference system. Focusing on the anaerobic digestion process, Sahoo and Mani[55] compared three alternatives for producing compressed natural gas (CNG) from (i) integrated-state, (ii) stand-alone liquid-state, and (iii) solid-state technologies, using dairy-manure, food-waste, and miscanthus biomass feedstocks. LCA analysis was limited only to the GWP category, while the fossil energy ratio was considered for energy efficiency. Regarding fossil-based CNG, the systems evaluated obtained a reduction of 59–174% in GWP. Furthermore, stand-alone liquid-state or solid-state systems were economically competitive regarding the market price of fossil CNG. Thaore *et al.*[56] performed a bio-based production of pure glucaric acid from corn stover, considering two alternative routes for the oxidation of glucose to glucaric acid: (i) *via* heterogeneous catalytic oxidation with air and (ii) by homogeneous glucose oxidation using nitric acid. The authors identified that heterogeneously catalysed aerobic glucose oxidation is an economically and environmentally more sustainable route than nitric acid-mediated glucose oxidation, because of the higher glucaric acid yield in the first one, and the high GHG emissions related to nitric acid production in the latter. Finally, Bressanin *et al.*[57] compared a typical 1G biorefinery processing conventional sugarcane against the integration of a thermochemical technology with 1G sugarcane biorefinery, processing sugarcane, and eucalyptus biomass options. The results showed a GHG emission decrease (both for sugarcane and eucalyptus) of about 85–95% compared to the fossil counterpart, based on the product evaluated (hydrous ethanol, green diesel, renewable jet fuel, green gasoline, and electric energy).

Complete utilization of feedstock biomass was also considered in biorefinery designs. Liu *et al.*[58] evaluated the carbon and energy efficiencies of wheat straw utilization and a formic acid pre-treatment (Formiline biorefinery) compared with the conventional dilute acid pre-treatment to obtain ethanol, furfural, and lignin. They found that the Formiline biorefinery presents a higher carbon and energy utilization efficiency (58% and 43%) due to more complete utilization of biomass. Moreover, the authors expanded the Formiline biorefinery to include the conversion of lignin to phenol formaldehyde (LPF) resin and furfural to four chemicals, (*i.e.*, butadiene, surfactant lubricant, and jet fuel). The results illustrated that all expanded Formiline routes delivered GHG reduction, whilst surfactants and lubricant systems become more profitable than the Formiline baseline. Similarly, Pang *et al.*[59] evaluated the ethanol, xylose, and adhesive (EXA) biorefinery based on the xylose industry, enhancing traditional lignin valorisation. To do this, the authors integrated four conversion and separation stages to obtain these products with about 80% of total carbon utilization. TEA results showed that the revenue of EXA biorefinery increases significantly compared with the reference process, but it also obtained a higher total GHG emission.

New bio-based systems were also proposed. Sadhukhan and Martinez-Hernandez[60] proposed a system for the total valorisation of municipal solid waste (MSW), which contains a material recovery facility, a chemical section (comprising pulping, chemical conversion, separation, and purification of chemical products), an effluent treatment plant (for recycling water), a biochemical section (comprising anaerobic digestion for fertiliser and biogas production), and a combined heat and power (CHP) section. The authors declared environmental costs due to the use of fuels resourced from the MSW in the CHP system (*i.e.*, exhaust or flue gas from the combustion in the boiler). Moreover, they consider environmental benefits from the displacements of equivalent petroleum-derived products (levulinic acid, fertiliser, excess electricity for export). Zang *et al.*[61] analysed three different methanol production systems: methanol–ethanol coproduction, methanol–ammonia coproduction, and stand-alone methanol production systems, using CO_2 feedstock from ethanol plants, ammonia plants, and general market CO_2 supply. From these systems, the integrated methanol–ammonia system obtained the highest CO_2eq emissions per MJ; however, it was lower than conventional methanol produced from natural gas.

Based on the experimental results, Kadhum *et al.* 2017[62] explored the high solids loading effect for fed-batch enzymatic hydrolysis and fermentation. The solid loading was 19%, 30%, and 45% using corn stover and wheat straw as feedstocks. Their results showed that corn stover at 45% loading obtained the highest Return on Investment (ROI) at 8%, and the lowest global warming potential impacts (37.8 g CO_2eq MJ^{-1} ethanol produced). Baral *et al.*[63] assessed the implications of reducing the requirement of cellulase enzyme blend (alkali and Cellic CTec2) in the bioprocessing of sugarcane bagasse. They evaluated two approaches: (i) its partial replacement with an in-house enzyme cocktail by blending and (ii) performing extended hydrolysis with reduced loadings of the cellulase enzyme blend above its optimum temperature. Then, the authors analysed the economic and environmental impacts based on lactic acid as the targeted product. The authors found that with 40% reduced enzyme loadings, about 84% of glucan and 83% of xylan conversion yields were obtained from 20% solids in 48 h at 52.5 °C and pH 4.5. This optimization in consumption also lies in economic and environmental improvements with respect to the reference system.[64] The costs related to the pre-treatment process were reduced by 68%; thus, the cost of fermentable sugar was reduced by 41.3% (2.25 to 1.32 USD kg^{-1}). Moreover, the GWP category was reduced from 2.87 to 1.57 kg CO_2eq per kg of sugars produced from bagasse. Following enzyme influence on biorefinery performance, Nickel *et al.*[65] developed a multi-scale variability analysis, aiming at assessing the effects of input variations (enzymatic activity) on the performance indicators of a wheat straw-based ethanol biorefinery and comparing two alternative modes of operation, batch and multi-feed.

Other studies reviewed economic and environmental analysis from the literature. Pommeret *et al.*[66] summarised the techno-economic feasibility and the improved ecological impacts of food waste valorisation and biorefinery for a bio-based economy, analysing case-specific examples related to succinic acid production and energy generation, based on thermochemical processes conversion, *via* combustion/incineration, co-combustion or co-firing, gasification, pyrolysis, and hydrothermal carbonisation.

12.7.2 Chemical Products

Regarding biorefinery design for bioproducts, Putra *et al.*[67] evaluated the feasibility of oil palm empty fruit bunch (EFB) as a raw material for dissolving pulp and furfural co-production. The environmental evaluation followed a consequential LCA approach, arguing that EFB selection as a raw material decreases wood utilization (conventional raw material). Thus, the authors consider the amount of wood required for 1 kg of dissolving pulp (4.5 kg) as an avoided product. The results showed that co-producing 1 kg of dissolving pulp and 0.01 kg of furfural generated −1.218 kg CO_2 eq. Moreover, TEA indicated positive annual earnings from third-year production and earnings of 29.71 Japanese yen per kg of dissolving pulp production. Shaji *et al.*[68] performed a biorefinery design for producing succinic acid from sugarcane bagasse, considering acid pre-treatment, fermentation, followed by separation and purification processes. The TEA results showed economic infeasibility, expecting improvements in fermentation yields. However, GHG emissions were 1.39 kg of CO_2eq per kg succinic acid, with electricity as the hotspot, compared to the 1.9 kg of CO_2eq for the fossil-based route reported by Smidt *et al.*[69] Following with sugarcane feedstock bagasse, Munagala *et al.*[70] analysed the lactic acid production process. Results showed that 1 kg of lactic acid generated 4.62 kg CO_2eq, with a product cost of 2.9 USD kg^{-1}, and a payback time of six years. The pre-treatment and hydrolysis stages contributed 43% and 40% of the total cost, respectively. A key factor was sodium hydroxide (NaOH), which resulted in 75% of the variable operating cost of the pre-treatment stage and 39% of the total variable operating cost of the process. Zhou *et al.*[71] evaluated an ethylene glycol (EG) selective oxidation to glycolic acid (GA) process, which aims to realise the sustainable and cost-effective production of GA, where the bimetallic PtMn/MCM-41 catalysts with high efficiency were adopted to improve EG conversion and maximise the yields of GA. Finally, Salvachúa *et al.*[72] presented an integrated process to produce butyric acid (BA) from lignocellulosic sugars, evaluating different fermentation configurations and developing a product recovery process based on hybrid extraction-distillation. They showed that the minimum product selling price of bio-based BA is 55% of the selling price of petroleum-based BA.

12.7.3 Biofuels and Energy Production

Regarding energy generation systems, Amezcua-Allieri *et al.*[73] evaluated heat and electricity generation from sugarcane bagasse, compared to using fuel oil and electricity from the grid, both in the sugar production process. Results showed that the 100% of steam and electricity demand of sugar mill facilities is satisfied, as well as export electricity to the grid (about 30%). The cost per unit of generated bioenergy ranges from 0.026 to 0.019 USD $kW h^{-1}$ for small and large capacities, respectively. Impact reductions were identified in impact categories such as global warming, abiotic resource depletion, and acidification, among others. Nevertheless, growth in other categories were also obtained, for instance, human toxicity, terrestrial ecotoxicity, eutrophication, and photochemical ozone creation potential, due to the cultivation stage and pollutants in the combustion of bagasse. Li *et al.*[74] studied the techno-economic, energy consumption, and environmental performances of biomass-to-hydrogen (BTH) production with and without the methane tri-reforming (MTR) process from corn straw. Some of the results indicated that the energy efficiency of BTH with MTR is 17% higher than that without MTR. Furthermore, the GHG emissions of BTH with MTR were about 8% lower than those of BTH without MTR. Otherwise, Hamedani *et al.*[75] analysed the integration of an anaerobic digestion power plant, fed by zootechnical wastewater and agro-industrial residues, with a cogeneration system and a digestate-composting plant that produces soil fertilisers. Here, environmental analysis was also focused on GHG emissions, obtaining a value of −167.52 kg CO_2eq, associated with avoided mineral fertilisers consumption. Moreover, the payback time and net present values were eight years and 312 000 EU, considering externalities, respectively.

Concerning biofuel production, Pandey *et al.*[76] compared producing ethanol using loose (or baled) and pelleted forms of biomass. The environmental analysis was limited to the GWP category following a cradle-to-grave approach, based on the operative condition for the minimum ethanol selling price (MESP). According to the authors, all the scenarios analysed for pelleted biomass resulted in lower MESPs than any conditions used for loose biomass. Thus, the lowest MESP for loose and pelleted biomass was USD 4.41 and USD 3.83 per gal ethanol, respectively. On the other hand, pelleted biomass can decrease the GHG emissions of ethanol by about 50% compared to gasoline. On the other hand, Ansarinasab *et al.*[77] proposed to generate biofuel through fast pyrolysis and gasification, through a biorefinery that comprises hydrotreating, hydrocracking, and steam reforming sections. The design was evaluated through an exergy-based approach. Based on this analysis, it was demonstrated that employing compressors in the hydrotreating section and employing pumps in hydrocracking and steam reforming sections were those components that highly affect the cost of the system.

A bio-based jet fuel production was also analysed. Alam *et al.*[78] estimated the break-even price and life cycle carbon emissions of the sustainable aviation fuel derived from carinata (*Brassica carinata*) in the South-eastern United States. The authors estimated a 65% carbon savings compared with conventional aviation fuel. Nevertheless, without subsidies, carinata-based production was more expensive (0.85 to 1.28 $\$ L^{-1}$) than conventional aviation fuel (0.50 $\$ L^{-1}$). Similarly, Beal *et al.*[79] evaluated the feasibility of replacing petroleum jet fuel with domestic feedstocks in the United States, assessing the economic cost, GHG emissions, energy balance, as well as the land, water, and fertiliser impacts of nine alternative jet fuel pathways. Feedstocks correspond to terrestrial crops (soy, corn, wood, and grass), marine microalgae, and by-products (forestry residue, municipal solid waste, and waste oil) converted by extraction, fermentation, or thermo-chemical processes. Five routes from terrestrial crops and by-products represent over 50% reduction in lifecycle GHG emissions compared to petroleum jet fuel. Moreover, the largest arable land impacts are associated with terrestrial crops (*e.g.*, soy, corn stover, and grass). The economic results were influenced by capital costs, with the by-product scenarios obtaining the lowest capital investment. Resurreccion *et al.*[80] evaluated a cracking alternative (olefin metathesis, OMT) that converts plant oil to aviation biofuel using lipids from *Camelina sativa* L. (camelina). Thus, they assessed the camelina potential land area across the continental U.S. through geospatial analysis; evaluated the environmental impacts of the OMT technology and compared it with a hydro-processed renewable jet; and analysed the economics of an OMT-based biorefinery from a regional standpoint. Finally, Huq *et al.*[81] demonstrated the conversion of food waste-derived volatile fatty acids (VFAs) into sustainable aviation fuels (SAF). They estimated that by diverting food waste from landfills to avoid methane emissions, VFA–SAF could reduce about 165% of GHG emissions relative to fossil jet.

12.7.4 Multicriteria and Uncertainty Analysis

Concerning multicriteria analysis in biorefinery systems, Wang *et al.*[82] proposed multiobjective modelling for identifying an optimal design of hydrocarbon biorefinery *via* the gasification pathway under economic and environmental criteria. The Pareto-optimal solutions obtained showed that biofuels can be produced at \$3.17 per gasoline-equivalent gallon (GEG) and 20.92 kg CO_2eq GEG^{-1}. Kristianto and Zhu[83] designed a sustainable bioethanol supply chain based on rice straw, coupling LCA with techno-economic optimization to find optimum process planning. Similarly, Thompson *et al.*[84] proposed a multi-objective decision-making framework to compare techno-economic and environmental features of pyrolysis and hydrolysis technologies for furfural production from sugar beet pulp. Furthermore, Hafyan *et al.*[85] proposed an integrated biorefinery configuration of empty fruit bunch valorisation to obtain multiple biochemicals (xylitol, levulinic acid, succinic acid, vanillin, and guaiacol). The authors considered TEA and LCA analysis, as well as hazard identification ranking methods to estimate GWP, annual profit, fire explosion damage (FEDI), and toxicity damage (TDI) indexes. Then, trade-offs were identified through a multi-objective model for simultaneous profit maximisation and minimisation of GWP, FEDI, and TDI. For multiproduct generation, the optimal mass allocation of glucose was 90% to succinic acid and 10% to levulinic acid, whereas most of the lignin (70%) turned to guaiacol and the rest to vanillin. The yearly profit was 932 MUSD with GHG emissions of 284 t CO_2eq. Finally, Baral *et al.*[86] performed an uncertainty analysis of cost, life-cycle energy use, and GHG emissions related to a feedstock supply logistics system for butanol production based on corn stover.

12.7.5 Software Tools Proposed and Used

As it was mentioned previously, there are different software tools available for process simulation. From the sample of articles analysed, Aspen was the most used for biorefinery modelling. However, other alternatives have also been proposed, Cortes-Peña *et al.*[87] presented BioSTEAM (Biorefinery Simulation and Techno-Economic Analysis Modules), an open-source process simulator that enables biorefinery design, simulation, and TEA under uncertainty. After that, Shi and Guest[88] proposed an LCA framework of biorefineries modelled in this open-source software (BioSTEAM-LCA) to characterise their impacts based on the assumptions of design, technology, and parameter performance, among others. Then, Bhagwat *et al.*[89] used BioSTEAM to design, simulate, and perform TEA and LCA analysis of acrylic acid production *via* fermentation of sugars (glucose and xylose) to 3-hydroxypropionic acid (3-HP). Finally, Elias *et al.*[90] proposed the Retro-Techno-Economic-Environmental Analysis (RTEEA), a computational tool that identifies a range of feasible operations for industrial processes. Then, the authors applied RTEEA to establish windows of feasible operations for a biorefinery that produces first- and second-generation ethanol plus bioelectricity from sugarcane.

12.8 Concluding Remarks

Techno-economic analysis and life cycle assessment are useful methodologies for estimating the economic and environmental feasibility of products and/or process modelling at the early design stage. In this regard, determining viability aspects such as the production cost, revenue level, and environmental impacts are primary goals of TEA and LCA studies. These studies have been focused mainly on two aims: process and comparison design of valorisation routes, and the process design and comparison of bio-based energy (biofuels) and bioproducts (biochemicals) with their fossil-based counterpart. In this sense, process simulation has played a fundamental role in predicting operative performance and potential environmental profiles of bioproducts. A great number of articles were mainly centred on the global warming category in the environmental approach, motivated by the climate crisis emergency. However, one of the fundamental goals of the LCA methodology is the multicriteria assessment through all environmental categories available in the resource, ecosystem, and human health perspective. Thus, identifying potential trade-offs among impacts during process design is required for a comprehensive environmental assessment. In contrast, economic analysis was related to the conventional performance indicators, such as minimum sale price, payback time, and profit, among others. With lower attention, multicriteria optimization is another research line developed to identify the most sustainable design way for a biorefinery process. Furthermore, new software tools have also been proposed to extend available alternatives for this type of analysis. Incorporating the social dimension in future TEA and LCA studies will allow reaching a comprehensive sustainability assessment of bio-based systems. In this regard, social LCA can be a suitable methodology to apply as it also follows a life cycle perspective. Moving toward a carbon-neutral economy is needed to face the climate crisis, and the progress of new sustainable biorefinery platforms could lead a key role in this pathway.

Acknowledgements

This research has been supported by the Enhancing Diversity in Mediterranean Cereal Farming Systems (CerealMed) Project funded by PRIMA Programme and FEDER/Ministry of Science and Innovation–Spanish National Research Agency (PCI2020-111978), and the project Transition to sustainable agri-food sector bundling life cycle assessment and ecosystem services approaches (ALISE), funded by the Spanish National Research Agency (TED2021-130309B-I00). R. R. L., M. T. M., and S. G. G. belong to the Galician Competitive Research Groups (GRC)_ED431C-2021/37, co-funded by Xunta de Galicia and FEDER (EU).

References

1. E. MacArthur, *Ellen MacArthur Found.*, 2015, p. 20.
2. M. Chertow, *J. Ind. Ecol.*, 2007, **11**, 11–30.
3. M. Carus and L. Dammer, *Ind. Biotechnol.*, 2018, **14**, 83–91.
4. P. Basu, *Biomass Gasif. Des. Handb.*, 2010, pp. 27–63.
5. A. T. Ubando, C. B. Felix and W. H. Chen, *Bioresour. Technol.*, 2020, **299**, 122585.
6. G. De Bhowmick, A. K. Sarmah and R. Sen, *Bioresour. Technol.*, 2018, **247**, 1144–1154.
7. S. Venkata Mohan, *Ind. Wastewater Treat. Recycl. Reuse*, 2014, pp. 235–281.
8. D. Cardoen, P. Joshi, L. Diels, P. M. Sarma and D. Pant, *Resour., Conserv. Recycl.*, 2015, **102**, 39–48.
9. Z. H. Liu, R. K. Le, M. Kosa, B. Yang, J. Yuan and A. J. Ragauskas, *Renewable Sustainable Energy Rev.*, 2019, **105**, 349–362.
10. A. Zimmermann, J. Wunderlich, G. Buchner, L. Müller, K. Armstrong, S. Michailos, A. Marxen, H. Naims, F. Mason, G. Stokes and E. Williams, *Techno-Economic Assessment & Life-Cycle Assessment Guidelines for CO_2 Utilization*, 2018.
11. G. Clancy, M. Fröling and M. Svanström, *J. Clean. Prod.*, 2013, **39**, 372–385.

12. R. Julio, J. Albet, C. Vialle, C. Vaca-Garcia and C. Sablayrolles, *Biofuels, Bioprod. Biorefin.*, 2017, **11**, 373–395.
13. G. Towler and R. Sinnott, *Chemical engineering design: principles, practice and economics of plant and process design*, Butterworth-Heinemann, 2nd edn, 2013.
14. H. Silla, *Chemical process engineering: design and economics*, CRC Press, 2003.
15. IS0 14040, *Environmental Management. Principles and Framework*, Geneve, Switzerland, 2006.
16. S. Morais, A. A. Martins and T. M. Mata, *J. Energy Inst.*, 2010, **83**, 48–55.
17. K. Dufossé, W. Ben Aoun and B. Gabrielle, *Life-Cycle Assess. Biorefineries*, 2017, 77–96.
18. W. W. Wilhelm, J. M. F. Johnson, D. L. Karlen and D. T. Lightle, *Agron. J.*, 2007, **99**, 1665–1667.
19. J. Cristóbal, C. Caldeira, S. Corrado and S. Sala, *Bioresour. Technol.*, 2018, **259**, 244–252.
20. E. M. Barampouti, S. Mai, D. Malamis, K. Moustakas and M. Loizidou, *Renewable Sustainable Energy Rev.*, 2019, **110**, 298–314.
21. G. Capson-Tojo, M. Rouez, M. Crest, J. P. Steyer, J. P. Delgenès and R. Escudié, *Rev. Environ. Sci. Biotechnol.*, 2016, **15**, 499–547.
22. S. Haghighi Mood, A. Hossein Golfeshan, M. Tabatabaei, G. Salehi Jouzani, G. H. Najafi, M. Gholami and M. Ardjmand, *Renewable Sustainable Energy Rev.*, 2013, **27**, 77–93.
23. M. F. Li, Y. M. Fan, F. Xu, R. C. Sun and X. L. Zhang, *Ind. Crops Prod.*, 2010, **32**, 551–559.
24. R. P. Swatloski, S. K. Spear, J. D. Holbrey and R. D. Rogers, *J. Am. Chem. Soc.*, 2002, **124**, 4974–4975.
25. E. Palmqvist and B. Hahn-Hägerdal, *Bioresour. Technol.*, 2000, **74**, 17–24.
26. D. Kumari and R. Singh, *Renewable Sustainable Energy Rev.*, 2018, **90**, 877–891.
27. A. Demirbas, *J. Hazard. Mater.*, 2008, **157**, 220–229.
28. B. Kumar, N. Bhardwaj, K. Agrawal, V. Chaturvedi and P. Verma, *Fuel Process. Technol.*, 2020, **199**, 106244.
29. A. Azapagic, *Trends Biotechnol.*, 2014, **32**, 1–4.
30. S. Paul and A. Dutta, *Resour., Conserv. Recycl.*, 2018, **130**, 164–174.
31. V. K. Ponnusamy, D. D. Nguyen, J. Dharmaraja, S. Shobana, J. R. Banu, R. G. Saratale, S. W. Chang and G. Kumar, *Bioresour. Technol.*, 2019, **271**, 462–472.
32. R. Saini, C. S. Osorio-Gonzalez, K. Hegde, S. K. Brar, S. Magdouli, P. Vezina and A. Avalos-Ramirez, *Curr. Sustainable Energy Rep.*, 2020, **7**, 122–136.
33. H. M. Zabed, S. Akter, J. Yun, G. Zhang, F. N. Awad, X. Qi and J. N. Sahu, *Renewable Sustainable Energy Rev.*, 2019, **105**, 105–128.
34. L. V. Daza Serna, C. E. Orrego Alzate and C. A. Cardona Alzate, *Bioresour. Technol.*, 2016, **199**, 113–120.
35. G. C. Xu, J. C. Ding, R. Z. Han, J. J. Dong and Y. Ni, *Bioresour. Technol.*, 2016, **203**, 364–369.

36. Y. Liu, W. Chen, Q. Xia, B. Guo, Q. Wang, S. Liu, Y. Liu, J. Li and H. Yu, *ChemSusChem*, 2017, **10**, 1692–1700.
37. A. Satlewal, R. Agrawal, S. Bhagia, J. Sangoro and A. J. Ragauskas, *Biotechnol. Adv.*, 2018, **36**, 2032–2050.
38. H.-J. Huang, B. V. Ramarao and S. Ramaswamy, *Separation and purification technologies in biorefineries*, John Wiley & Sons, 2013.
39. Z. Guo, S. Wang, Y. Gu, G. Xu, X. Li and Z. Luo, *Sep. Purif. Technol.*, 2010, **76**, 52–57.
40. H.-J. Huang and S. Ramaswamy, in *Biorefinery Co-Products*, John Wiley & Sons, Ltd, 2012, pp. 37–53.
41. H. J. Huang, S. Ramaswamy, U. W. Tschirner and B. V. Ramarao, *Sep. Purif. Technol.*, 2008, **62**, 1–21.
42. Y. Wang, C. Huang and T. Xu, *J. Memb. Sci.*, 2011, **374**, 150–156.
43. Y. Wang, N. Zhang, C. Huang and T. Xu, *J. Memb. Sci.*, 2011, **385–386**, 226–233.
44. R. Schlesinger, G. Götzinger, H. Sixta, A. Friedl and M. Harasek, *Desalination*, 2006, **192**, 303–314.
45. J. M. Sánchez, M. M. Barreiro and M. Maroño, *Biomass Bioenergy*, 2011, **35**, S132–S144.
46. P. de, A. P. Filho, G. A. M. Hirata, É. O. Watanabe and E. A. Miranda, *Compr. Biotechnol.*, 2019, **2**, 725–738.
47. A. Azapagic, A. Millington and A. Collett, *Chem. Eng. Res. Des.*, 2006, **84**, 439–452.
48. G. E. Kniel, K. Delmarco and J. G. Petrie, *Environ. Prog.*, 1996, **15**, 221–228.
49. A. Azapagic, *Chem. Eng. J.*, 1999, **73**, 1–21.
50. A. Azapagic and R. Clift, *Comput. Chem. Eng.*, 1999, **23**, 1509–1526.
51. P. Souza, P. Brancoli, K. Bolton, A. Zamani and M. J. Taherzadeh, *Fermentation*, 2017, **3**(4), 56.
52. M. Mandegari, S. Farzad and J. F. Görgens, *Energy Convers. Manag.*, 2018, **165**, 76–91.
53. M. Yang, X. Tian and F. You, *Ind. Eng. Chem. Res.*, 2018, **57**, 5980–5998.
54. E. R. Pachón, P. Vaskan, J. F. Gorgens and E. Gnansounou, *Process design, techno-economic, and life-cycle assessments of selected sugarcane-based biorefineries: A case study in the South African context*, Elsevier, 2019.
55. K. Sahoo and S. Mani, *Renewable Sustainable Energy Rev.*, 2019, **115**, 109354.
56. V. B. Thaore, R. D. Armstrong, G. J. Hutchings, D. W. Knight, D. Chadwick and N. Shah, *Chem. Eng. Res. Des.*, 2020, **153**, 337–349.
57. J. M. Bressanin, B. C. Klein, M. F. Chagas, M. D. B. Watanabe, I. L. D. M. Sampaio, A. Bonomi, E. R. de Morais and O. Cavalett, *Energies*, 2020, **13**(17), 4576.
58. F. Liu, X. Dong, X. Zhao and L. Wang, *Energy Convers. Manag.*, 2021, **246**, 114653.
59. B. Pang, Z. Sun, L. Wang, W. J. Chen, Q. Sun, X. F. Cao, X. J. Shen, L. Xiao, J. L. Yan, P. J. Deuss, T. Q. Yuan and R. C. Sun, *Chem. Eng. J.*, 2021, **419**, 129565.

60. J. Sadhukhan and E. Martinez-Hernandez, *Bioresour. Technol.*, 2017, **243**, 135–146.
61. G. Zang, P. Sun, A. Elgowainy and M. Wang, *Environ. Sci. Technol.*, 2021, **55**, 5248–5257.
62. H. J. Kadhum, K. Rajendran and G. S. Murthy, *Bioresour. Technol.*, 2017, **244**, 108–116.
63. P. Baral, M. Munagala, Y. Shastri, V. Kumar and D. Agrawal, *Cellulose*, 2021, **28**, 6305–6322.
64. P. Baral, A. Pundir, V. Kumar, A. K. Kurmi and D. Agrawal, *Food Bioprod. Process.*, 2020, **124**, 72–81.
65. D. B. Nickel, R. Fornell, M. Janssen and C. J. Franzén, *Front. Energy Res.*, 2020, **8**, 55.
66. A. Pommeret, X. Yang, T. H. Kwan, E. A. Christoforou, P. A. Fokaides and C. S. K. Lin, *Techno-economic study and environmental assessment of food waste based biorefinery*, Springer International Publishing, 2017.
67. A. S. Putra, R. Noguchi, T. Ahamed and H. Ohi, *Nihon Enerugi Gakkaishi/J. Jpn. Inst. Energy*, 2020, **99**, 143–152.
68. A. Shaji, Y. Shastri, V. Kumar, V. V. Ranade and N. Hindle, *ACS Sustainable Chem. Eng.*, 2021, **9**, 12738–12746.
69. M. Smidt, J. den Hollander, H. Bosch, Y. Xiang, M. van der Graaf, A. Lambin and J.-P. Duda, in *Sustainability Assessment of Renewables-Based Products*, John Wiley & Sons, Ltd, 2015, pp. 307–321.
70. M. Munagala, Y. Shastri, K. Nalawade, K. Konde and S. Patil, *Waste Manag.*, 2021, **126**, 52–64.
71. X. Zhou, M. Zha, J. Cao, H. Yan, X. Feng, D. Chen and C. Yang, *ACS Sustainable Chem. Eng.*, 2021, **9**, 10948–10962.
72. D. Salvachua, P. O. Saboe, R. S. Nelson, C. Singer, I. McNamara, C. del Cerro, Y. C. Chou, A. Mohaghehi, D. J. Peterson, S. Haugen, N. S. Cleveland, H. R. Monroe, M. T. Guarnieri, E. C. D. Tan, G. T. Beckham, E. M. Karp and J. G. Linger, *Cell Rep. Phys. Sci.*, 2021, **2**(10), 100587.
73. M. A. Amezcua-Allieri, E. Martínez-Hernández, O. Anaya-Reza, M. Magdaleno-Molina, L. A. Melgarejo-Flores, M. E. Palmerín-Ruiz, J. A. Z. Eguía-Lis, A. Rosas-Molina, M. Enríquez-Poy and J. Aburto, *Food Bioprod. Process.*, 2019, **118**, 281–292.
74. G. Li, S. Wang, J. Zhao, H. Qi, Z. Ma, P. Cui, Z. Zhu, J. Gao and Y. Wang, *Energy*, 2020, **199**, 117488.
75. S. R. Hamedani, M. Villarini, A. Colantoni, M. Carlini, M. Cecchini, F. Santoro and A. Pantaleo, *Energies*, 2020, **13**(11), 2724.
76. R. Pandey, N. Nahar, S. W. Pryor and G. Pourhashem, *Energies*, 2021, **14**(9), 2528.
77. H. Ansarinasab, M. Mehrpooya and M. Sadeghzadeh, *J. Therm. Anal. Calorim.*, 2021, **145**, 1053–1073.
78. A. Alam, M. F. H. Masum and P. Dwivedi, *GCB Bioenergy*, 2021, **13**, 1800–1813.

79. C. M. Beal, A. D. Cuellar and T. J. Wagner, *Biomass Bioenergy*, 2021, **144**, 105881.
80. E. P. Resurreccion, J. Roostaei, M. J. Martin, R. L. Maglinao, Y. Zhang and S. Kumar, *Ind. Crops Prod.*, 2021, **170**, 113777.
81. N. A. Huq, G. R. Hafenstine, X. Huo, H. Nguyen, S. M. Tifft, D. R. Conklin, D. Stück, J. Stunkel, Z. Yang, J. S. Heyne, M. R. Wiatrowski, Y. Zhang, L. Tao, J. Zhu, C. S. McEnally, E. D. Christensen, C. Hays, K. M. van Allsburg, K. A. Unocic, H. M. Meyer III, Z. Abdullah and D. R. Vardon, *Proc. Natl. Acad. Sci.*, 2021, **118**(13), e2023008118.
82. B. Wang, B. H. Gebreslassie and F. You, *Comput. Chem. Eng.*, 2013, **52**, 55–76.
83. Y. Kristianto and L. Zhu, *Energy*, 2017, **141**, 2164–2176.
84. M. A. Thompson, A. Mohajeri and A. Mirkouei, *J. Cleaner Prod.*, 2021, **311**, 127695.
85. R. H. Hafyan, L. K. Bhullar, S. Mahadzir, M. R. Bilad, N. A. H. Nordin, M. D. H. Wirzal, Z. A. Putra, G. P. Rangaiah and B. Abdullah, *Processes*, 2020, **8**(7), 868.
86. N. R. Baral, C. Quiroz-Arita and T. H. Bradley, *Appl. Energy*, 2017, **208**, 1343–1356.
87. Y. Cortes-Peña, D. Kumar, V. Singh and J. S. Guest, *ACS Sustainable Chem. Eng.*, 2020, **8**, 3302–3310.
88. R. Shi and J. S. Guest, *ACS Sustainable Chem. Eng.*, 2020, **8**, 18903–18914.
89. S. S. Bhagwat, Y. Li, Y. R. Cortés-Peña, E. C. Brace, T. A. Martin, H. Zhao and J. S. Guest, *ACS Sustainable Chem. Eng.*, 2021, **9**, 16659–16669.
90. A. M. Elias, A. A. Longati, R. de Campos Giordano and F. F. Furlan, *Appl. Energy*, 2021, **282**, 116133.

Subject Index